ADVANCES IN MULTIVARIATE STATISTICAL ANALYSIS

THEORY AND DECISION LIBRARY

General Editors: W. Leinfellner and G. Eberlein

Series A: Philosophy and Methodology of the Social Sciences
Editors: W. Leinfellner (Technical University of Vienna)
G. Eberlein (Technical University of Munich)

Series B: Mathematical and Statistical Methods
Editor: H. Skala (University of Paderborn)

Series C: Game Theory, Mathematical Programming and Mathematical Economics
Editor: S. Tijs (University of Nijmegen)

Series D: System Theory, Knowledge Engineering and Problem Solving
Editor: W. Janko (University of Vienna)

SERIES B: **MATHEMATICAL AND STATISTICAL METHODS**

Editor: H. Skala (Paderborn)

Scope

The series focuses on the application of methods and ideas of logic, mathematics and statistics to the social sciences. In particular, formal treatment of social phenomena, the analysis of decision making, information theory and problems ofinference will be central themes of this part of the library. Besides theoretical results, empirical investigations and the testing of theoretical models of real world problems will be subjects of interest. In addition to emphasizing interdisciplinary communication, the series will seek to support the rapid dissemination of recent results.

ADVANCES IN MULTIVARIATE STATISTICAL ANALYSIS

Pillai Memorial Volume

Edited by

A. K. GUPTA

Department of Mathematics and Statistics,
Bowling Green State University, Ohio, U.S.A.

D. REIDEL PUBLISHING COMPANY

A MEMBER OF THE KLUWER ACADEMIC PUBLISHERS GROUP

DORDRECHT / BOSTON / LANCASTER / TOKYO

Library of Congress Cataloging in Publication Data

Advances in multivariate statistical analysis.

(Theory and decision library. Series B, Mathematical and Statistical methods)
"Pillai memorial volume."
Bibliography: p.
Includes index.
1. Multivariate analysis. 2. Sreedharan Pillai, K. C., 1920–1985.
I. Gupta, A. K. II. Series.
QA278.A28 1987 519.5′35 87–9888
ISBN 90–277–2531–4

Published by D. Reidel Publishing Company,
P.O. Box 17, 3300 AA Dordrecht, Holland.

Sold and distributed in the U.S.A. and Canada
by Kluwer Academic Publishers,
101 Philip Drive, Assinippi Park, Norwell, MA 02061, U.S.A.

In all other countries, sold and distributed
by Kluwer Academic Publishers Group,
P.O. Box 322, 3300 AH Dordrecht, Holland.

Printed in The Netherlands

To the memory of
Professor K. C. Sreedharan Pillai

Professor K. C. Sreedharan Pillai
(2.24.1920 - 6.5.1985)

TABLE OF CONTENTS

LIST OF CONTRIBUTORS

N. Balakrishnan, Department of Mathematics, McMaster University, Hamilton, Ontario, Canada

James O. Berger, Department of Statistics, Purdue University, West Lafayette, Indiana 47907

Shun-Yu Chen, Department of Statistics, Cheng-Kung University, Tainan, Taiwan

Louis J. Cote, Department of Statistics, Purdue University, West Lafayette, Indiana 47907

Hilmar Drygas, FB 17 (Math.), GhK, Postfach 101 380, D-3500 Kassel, F. R. Germany

Bernhard K. Flury, Department of Statistics, University of Berne, Sidlerstrasse 5, CH-3012 Berne, Switzerland

Y. Fujikoshi, Department of Mathematics, Hiroshima University, Hiroshima 730, Japan

N. C. Giri, Department of Mathematics and Statistics, Université de Montreal, Montreal, Canada

Leon Jay Gleser, Department of Statistics, Purdue University, West Lafayette, Indiana 47907

A. K. Gupta, Department of Mathematics and Statistics, Bowling Green State University, Bowling Green, Ohio 43403

Shanti S. Gupta, Department of Statistics, Purdue University, West Lafayette, Indiana 47907

C. G. Khatri, Department of Statistics, Gujrat University, Ahmedabad, Gujrat, India

K. Kocherlakota, Department of Statistics, University of Manitoba, Winnipeg, Manitoba, Canada

S. Kocherlakota, Department of Statistics, University of Manitoba, Winnipeg, Manitoba, Canada

Sadanori Konishi, The Institute of Statistical Mathematics, 4-6-7 Minami-Azabu, Minato-Ku, Tokyo 106, Japan

Yoshihiko Konno, Department of Mathematics, University of Tsukuba, Sakura mura, Ibaraki 305, Japan

P. R. Krishnaiah, Department of Mathematics and Statistics, University of Pittsburgh, Pittsburgh, Pennsylvania 15260

G. J. McLachlan, Department of Mathematics, University of Queensland, St. Lucia, Queensland 4067, Australia

Govind S. Mudholkar, Department of Statistics, University of Rochester, Rochester, New York 14627

Robb J. Muirhead, Department of Statistics, University of Michigan, Ann Arbor, Michigan 48109

D. K. Nagar, Department of Statistics, University of Rajasthan, Jaipur - 302015, India

B. N. Nagarsenker, Air Force Institute of Technology, Wright-Patterson Air Force Base, Ohio 45433

P. B. Nagarsenker, Air Force Institute of Technology, Wright-Patterson Air Force Base, Ohio 45433

Ingram Olkin, Department of Statistics, Stanford University, Stanford, California 94305

S. Panchapakesan, Department of Mathematics, Southern Illinois University, Carbondale, Illinois 62901

Prem Puri, Department of Statistics, Purdue University, West Lafayette, Indiana 47907

Andrew L. Rukhin, Department of Mathematics and Statistics, University of Massachusetts, Amherst, Massachusetts 01003

Stephen Samuels, Department of Statistics, Purdue University, West Lafayette, Indiana 47907

J. Schmidhammer, Department of Statistics, University of Tennessee, Knoxville, Tennessee 37996

Minoru Siotani, Department of Applied Mathematics, Science University of Tokyo, Kagurazaka, Shinjuku-ku, Japan

Milton Sobel, Department of Mathematics, University of California, Santa Barbara, California 93106

M. S. Srivastava, Department of Statistics, University of Toronto, Toronto, Ontario, Canada

Perla Subbaiah, Department of Mathematical Sciences, Oakland University, Rochester, Michigan 48063

Nariaki Sugiura, Department of Mathematics, University of Tsukuba, Sakura mura, Ibaraki 305, Japan

W. Y. Tan, Department of Mathematical Sciences, Memphis State University, Memphis, Tennessee 38152

PREFACE

The death of Professor K. C. Sreedharan Pillai on June 5, 1985 was a heavy loss to many statisticians all around the world. This volume is dedicated to his memory in recognition of his many contributions in multivariate statistical analysis. It brings together eminent statisticians working in multivariate analysis from around the world. The research and expository papers cover a cross-section of recent developments in the field. This volume is especially useful to researchers and to those who want to keep abreast of the latest directions in multivariate statistical analysis.

I am grateful to the authors from so many different countries and research institutions who contributed to this volume. I wish to express my appreciation to all those who have reviewed the papers. The list of people include Professors T. C. Chang, So-Hsiang Chou, Dipak K. Dey, Peter Hall, Yu-Sheng Hsu, J. D. Knoke, W. J. Krzanowski, Edsel Peña, Bimal K. Sinha, Dennis L. Young, Drs. K. Krishnamoorthy, D. K. Nagar, and Messrs. Alphonse Amey, Chi-Chin Chao and Samuel Ofori-Nyarko.

I wish to thank Professors Shanti S. Gupta and James O. Berger for their keen interest and encouragement. Thanks are also due to Cynthia Patterson for her help and Reidel Publishing Company for their cooperation in bringing this volume out.

Bowling Green, January 1987 A. K. Gupta

K. C. Sreedharan Pillai 1920-1985

K. C. Sreedharan Pillai, Professor of Statistics, died of a heart attack on June 5, 1985. He was in St. Elizabeth where he was taken after suffering a heart attack while playing golf. He is survived by his wife Kamalakshi, two sons, Mohanan and Anandan, a daughter, Mrs. Sudha Seethanathan, all residents of West Lafayette, and a brother, Raghavan, who resides in India. He was a devoted family man and maintained ties with his Indian relatives.

Sree was born on February 24, 1920 in Kerala, India. He attended the University of Travancore, later renamed Kerala University, receiving a B.Sc. in 1941 and an M.Sc. in 1945. He served as a lecturer in Kerala University from 1945 to 1951. In 1951 he came to the United States to study at Princeton University. After a year he transferred to the University of North Carolina where he received a Ph.D. in Statistics in 1954.

He began his career as an assistant statistician with the United Nations. He served in the United Nations from 1954 to 1962. Part of this service was as Senior Statistical Advisor and Visiting Professor at the University of the Philippines. In this capacity he founded the Statistical Center at the University of the Philippines. There he supervised many graduate students and continued to serve as an outside examiner of graduate dissertations for many years.

He came to Purdue in 1962 as a Professor of Statistics and Mathematics. In the 23 years he served Purdue, he directed the research of 15 Ph.D. students. He was also an active consultant on several projects both within and outside the University. He was a close friend of his students and maintained a correspondence with most of them, some of whom are in remote parts of the world.

Professor Pillai was a prolific researcher. His chief contributions to statistics were made in the field of multivariate statistical analysis. In particular, he obtained the probability distributions of statistics relating to several multivariate procedures. Perhaps his best known contribution is the widely used multivariate analysis of variance test which bears his name. His leadership in statistical research (nearly 80 published papers and 2 books) was recognized by his being named Fellow in both major statistical organizations in the United States, the American Statistical Association and the

Institute of Mathematical Statistics. He was also a life member of the Philippines Statistical Association and an elected member of the International Statistical Institute. After his death many letters were received from colleagues. One remark which exemplifies their sentiments was "It is a great loss--particularly for multivariate analysts. He had been a great inspiration to one and all. I am sure the Department of Statistics will have great difficulty filling his shoes."

Sree took an active interest in departmental affairs. He served on many of the committees that were created to revise graduate programs. He was a member of the graduate committee for many years and its chairman for two years. His remarks and advice on matters of curricula were often sought and always welcome.

His interests were not just confined to research, teaching and other academic pursuits. He was an avid golfer. His unique and unforgettable style charmed his playing companions and confused his opponents in the Purdue Staff League. His performances in the League matches were legendary.

Department of Statistics
Purdue University
West Lafayette
Indiana 47907

Louis J. Cote
Leon J. Gleser
Shanti Gupta
Prem Puri
Stephen Samuels

James O. Berger and Shun-Yu Chen

MINIMAXITY OF EMPIRICAL BAYES ESTIMATORS DERIVED FROM SUBJECTIVE HYPERPRIORS

1. INTRODUCTION

Let $X = (X_1, \ldots, X_p)^t$ have a p-variate normal distribution with unknown mean vector $\theta = (\theta_1, \ldots, \theta_p)^t$ and nonsingular known covariance matrix Σ. In estimating θ, a variety of shrinkage estimators have been proposed from decision-theoretic and Bayesian perspectives. It has been argued (cf. Berger (1980, 1982, 1985)) that minimax estimators developed in the decision-theoretic approach must usefully incorporate available prior information to offer significant advantages. Often the most attractive manner of doing this is to develop a Bayesian estimator which clearly incorporates such information, and then to establish the minimaxity of the estimator. In this paper we follow such a program for several hierarchical Bayesian situations with informative second stage prior distributions.

The usual hierarchical Bayes formulation for this problem (see Lindley and Smith (1972) or Berger (1985)) assumes that, given the p-vector μ and the $p \times p$ positive definite matrix A, the unknown θ has a $\mathcal{N}_p(\mu, A)$ (first stage) prior distribution. In addition, however, μ and A are considered unknown with a (second stage) prior distribution $\pi(\mu, A)$. One can then calculate that the corresponding (hierarchical) Bayes estimator of θ is

$$\delta^\pi(x) = x - \Sigma E^{\pi(\mu, A|x)}((\Sigma + A)^{-1}(x - \mu)), \tag{1.1}$$

where $\pi(\mu, A|x)$ is the posterior distribution of μ and A given x (see Berger (1985) for formulas). Verification of the minimaxity of such estimators has only been done for a few essentially degenerate choices of $\pi(\mu, A)$ (cf. Strawderman (1971) and Berger (1980)), partly because of the difficulty of mathematically working with $\pi(\mu, A|x)$. Somewhat more success has been achieved with the empirical Bayes approximation to the above estimator; one determines $\hat{\mu}$ and $\hat{A}$, the maximum likelihood estimates of μ and A with respect to the posterior distribution $\pi(\mu, A|x)$, and then considers the estimator in (1.1) with μ and

A. K. Gupta (ed.), Advances in Multivariate Statistical Analysis, 1–12.

A replaced by $\hat{\mu}$ and $\hat{A}$. (Extensive discussion of such approximations can be found in Lindley and Smith (1972) and Berger (1985).) In proving minimaxity for such empirical Bayes estimators, it is common to consider positive multiples of the shrinkage term, leading to a final form of the estimator of

$$\delta^{B*}(x) = x - t\, \Sigma\, (\Sigma + \hat{A})^{-1}(x - \hat{\mu}), \tag{1.2}$$

t being a positive scalar.

The most extensively studied special case of this formulation is that in which $\mu = \mu_0(1, \ldots, 1)^t$ and $A = \lambda^{-1} I_p$, μ_0 and λ unknown, being a model of the exchangeable scenario in which the θ_i are i.i.d. from an unknown distribution. If $\Sigma = \sigma^2 I_p$, the estimates of $\hat{\mu}$ and $\hat{A}$ are (letting $\underset{\sim}{1} = (1, \ldots, 1)^t$ and I_p be the $p \times p$ identity matrix)

$$\hat{\mu} = \bar{x}\underset{\sim}{1}, \quad \hat{A} = \max\{0, \frac{1}{p}\Sigma(x_i - \bar{x})^2 - \sigma^2\} I_p; \tag{1.3}$$

the estimator in (1.2) then becomes

$$\delta^{B*}(x) = \begin{cases} x - t(x - \bar{x}\underset{\sim}{1}) & \text{if } \sum_{i=1}^{p} (x_i - \bar{x})^2 \le p\sigma^2 \\ x - \frac{tp\sigma^2}{\Sigma(x_i - \bar{x})^2}(x - \bar{x}\underset{\sim}{1}) & \text{else.} \end{cases}$$

When $t = (p-3)/p$, this can be recognized as a truncated version of the usual James-Stein estimator which shrinks to a common mean, as analyzed in, say, Efron and Morris (1973).

When p is small it was argued in Berger (1982) that it might be preferable to use *subjective* estimates of μ_0 and λ (or μ and A in general) leading to the minimax robust generalized Bayes estimator in Berger (1980). The point is that the estimates in (1.3) will be very inaccurate for small p, and the overall risk performance can be substantially improved through use of subjective estimates of μ and A. (Note that all estimators here are minimax, so that the criterion of interest would be some measure of overall average risk improvement.)

The natural Bayesian solution to the above dilemma is to give μ_0 and λ (or μ and A in general) an *informative* (second stage) prior distribution which incorporates the available subjective information. This allows optimum estimation of μ_0 and λ, and hence optimum overall performance of the resulting estimator of θ. The difficulty is that verification of minimaxity of the resulting estimator of θ can be very difficult, even in the empirical Bayes case since the empirical

Bayes estimators of μ_0 and λ are only defined implicitly as solutions of likelihood equations.

In this paper we make substantial progress on a special case of the above problem. The specific scenario considered is the symmetric one where $\Sigma = \sigma^2 I_p, \sigma^2$ known, and $A = \lambda^{-1} I_p$. Thus we are assuming $X \sim \mathcal{N}_p(\theta, \sigma^2 I_p)$ and $\theta \sim \mathcal{N}_p(\mu, \lambda^{-1} I_p)$. The second stage prior distribution for (μ, λ) is as follows:

(i) Either μ is assumed to be known, or it is assumed to be of the form $\mu = B\gamma$, where B is a given matrix of rank q, and γ is unknown with noninformative prior $\pi(\gamma) \equiv 1$;

(ii) The distribution of λ is chosen to be Gamma $(\alpha, \beta/2)$ (independently of μ), where α and β are subjectively specified constants with $\beta \geq (\alpha - 1)\sigma^2$.

The gamma family of priors for λ is sufficiently general to allow reflection of most beliefs about λ. Note that λ^{-1} can be thought of as the common variance of the θ_i, and that

$$E[\lambda^{-1}] = \beta/[2(\alpha - 1)].$$

Recall that we assume $\beta \geq (\alpha - 1)\sigma^2$; thus only those priors for which $E[\lambda^{-1}] \geq \sigma^2/2$ are allowed. (This is a rather mild constraint, since the variance of the θ_i will typically be larger that σ^2, the sample variance.) Also,

$$\text{Var}(\lambda^{-1}) = \beta^2/[4(\alpha - 1)^2(\alpha - 2)].$$

Thus one could subjectively specify $E[\lambda^{-1}]$ (a "best guess" for the variability of the θ_i), and $\text{Var}(\lambda^{-1})$ (say, the square of the estimated accuracy of this "best guess"), and solve for the corresponding α and β. One could similarly allow for a more general subjective prior on μ, but we do not do so for two reasons. First, it seems to be somewhat less important than utilization of information about λ^{-1}. Mainly, however, we were unable to handle the ensuing complexity; proof of minimaxity of the resulting estimators is formidable.

2. RESULTS WHEN μ IS KNOWN

When μ is known, the joint density of X, θ, λ is

$$m(x, \theta, \lambda) \propto \lambda^{\frac{p}{2} + \alpha - 1} \exp\{-\frac{1}{2}[\frac{1}{\sigma^2}(x-\theta)'(x-\theta) + \lambda(\theta-\mu)'(\theta-\mu) + \beta\lambda]\}$$

and the marginal posterior density of λ is given X is

$$\pi(\lambda|x) \propto \frac{\lambda^{\alpha-1}}{(\sigma^2+\lambda^{-1})^{p/2}} \exp\{-\frac{1}{2}[(\lambda^{-1}+\sigma^2)^{-1}v+\beta\lambda]\} \tag{2.1}$$

where $v = ||x-\mu||^2 = (x-\mu)'(x-\mu)$.

From (2.1), it is easy to show that the MLE of $\lambda, \hat{\lambda}$, satisfies the equation $\ell(\hat{\lambda}) = 0$, where

$$\begin{aligned}\ell(\lambda) = (\beta\sigma^4)\lambda^3 + (2\beta\sigma^2 - 2(\alpha-1)\sigma^4)\lambda^2 \\ + (v - p\sigma^2 - 4(\alpha-1)\sigma^2 + \beta)\lambda - (p + 2(\alpha-1)).\end{aligned} \tag{2.2}$$

Note that the coefficient of λ^2 in $\ell(\lambda)$ is nonnegative (since $\beta \geq (\alpha-1)\sigma^2$), so that the equation $\ell(\lambda) = 0$ has a unique positive solution. The heirarchical estimator (1.2) can then be written

$$\delta^1(x) = x - \frac{t\hat{\lambda}\sigma^2}{(1+\sigma^2\hat{\lambda})}(x-\mu) \tag{2.3}$$

(recall we are assuming here that μ is known).

Although $\hat{\lambda}$ is quite complicated (being the solution to a cubic equation), it is possible to verify minimaxity of δ^1; that this can be done for such complicated estimators is one of the main messages of this paper.

Theorem 1. *Under sum of squares error loss, the estimator δ^1 is minimax for*

$$0 \leq t \leq \frac{2p}{t^*} - \frac{4}{p+2(\alpha-1)}, \tag{2.4}$$

where

$$t^* = \max\{p + 2(\alpha-1), \quad p + 4(\alpha-1) - \beta/\sigma^2\}. \tag{2.5}$$

Proof. The familiar Stein identity (see Stein (1981)) shows that, for any estimator

$$\delta(x) = x - \phi(x)$$

satisfying certain mild conditions (all of which are trivially satisfied by the estimators in this paper),

$$\begin{aligned} R(\theta,\delta) &= E_\theta|\theta - \delta(X)|^2 \\ &= p\sigma^2 + E_\theta[\mathcal{D}\phi(X)], \end{aligned} \tag{2.6}$$

where

$$\mathcal{D}\phi(x) = |\phi(x)|^2 - 2\sigma^2 \sum_{i=1}^{p} \frac{\partial}{\partial x_i}\phi_i(x). \tag{2.7}$$

For δ^1,

$$\mathcal{D}\phi(x) = -\sigma^4 \frac{t\hat\lambda}{(1+\sigma^2\hat\lambda)}\{2p + \frac{4v(d\hat\lambda/dv)}{(\hat\lambda+\sigma^2\hat\lambda^2)} - \frac{tv\hat\lambda}{(1+\sigma^2\hat\lambda)}\}.$$

Differentiating with respect to v in the equation $\ell(\hat\lambda) = 0$ and rearranging terms yield

$$\begin{aligned} \frac{d\hat\lambda}{dv} &= -\hat\lambda[3\beta\sigma^2\hat\lambda^2 + 4(\beta\sigma^2 - (\alpha-1)\sigma^4)\hat\lambda \\ &\qquad + (v - p\sigma^2 - 4(\alpha-1)\sigma^2 + \beta)]^{-1} \\ &= -\hat\lambda^2[2\beta\sigma^4\hat\lambda^3 + 2(\beta\sigma^2 - (\alpha-1)\sigma^4)\hat\lambda^2 + (p + 2(\alpha-1))]^{-1} \\ &\geq -\hat\lambda^2(p+2(\alpha-1))^{-1}. \end{aligned} \tag{2.8}$$

Also, from the equation $\ell(\hat\lambda) = 0$, it follows that

$$v\hat\lambda \leq (p\sigma^2 + 4(\alpha-1)\sigma^2 - \beta)\hat\lambda + (p + 2(\alpha-1)),$$

which implies that

$$\frac{v\hat\lambda}{1+\sigma^2\hat\lambda} \leq \frac{(p\sigma^2 + 4(\alpha-1)\sigma^2 - \beta)\hat\lambda + (p+2(\alpha-1))}{1+\sigma^2\hat\lambda} \leq t^*. \tag{2.9}$$

Hence

$$\mathcal{D}\phi(x) \leq -\sigma^4 \frac{t\hat\lambda}{(1+\sigma^2\hat\lambda)}\{2p + \frac{4v(-\hat\lambda^2)}{(\hat\lambda+\sigma^2\hat\lambda^2)[p+2(\alpha-1)]} - \frac{tv\hat\lambda}{(1+\sigma^2\hat\lambda)}\}$$

$$= -\frac{\sigma^4 t\hat{\lambda}}{(1+\sigma^2\hat{\lambda})}\{2p - \frac{v\hat{\lambda}}{(1+\sigma^2\hat{\lambda})}[\frac{4}{p+2(\alpha-1)} + t]\}$$

$$\leq -\frac{\sigma^4 t\hat{\lambda}}{(1+\sigma^2\hat{\lambda})}\{2p - t^*[\frac{4}{p+2(\alpha-1)} + t]\}. \tag{2.10}$$

From (2.6) it is clear that δ^1 is minimax (and has risk less than the minimax risk $p\sigma^2$), if $\mathcal{D}\phi(x) \leq 0$. Equation (2.10) assure that $\mathcal{D}\phi(x) \leq 0$ when (2.4) is satisfied (note that $\hat{\lambda} \geq 0$). This completes the proof. □

There is no clearly optimal choice of t in δ^1 for this problem. Certain asymptotic arguments suggest that the choice

$$t = \frac{p}{p+2(\alpha-1)}$$

is attractive for larger p or larger $\beta/2(\alpha-1)$, suggesting use of

$$\tilde{t} = \min\left\{\frac{2p}{t^*} - \frac{4}{p+2(\alpha-1)}, \quad \frac{p}{p+2(\alpha-1)}\right\}. \tag{2.11}$$

Note that $t^* = p + 2(\alpha-1)$ if $\beta \geq 2(\alpha-1)\sigma^2$ (which will occur when the "guess" for the variance of the θ_i exceeds σ^2), so that

$$\tilde{t} = \frac{\min\{2(p-2), p\}}{p+2(\alpha-1)} \qquad \text{if } \beta \geq 2(\alpha-1)\sigma^2. \tag{2.12}$$

3. RESULTS WHEN μ IS PARTIALLY UNKNOWN

When μ is known to be of the form $\mu = B\gamma$, where B is a given matrix of rank q and γ is unknown with noninformative prior $\pi(\gamma) = 1$, the joint (improper) density of X, λ, γ is

$$m(x,\lambda,\gamma) \propto (\sigma^2+\lambda^{-1})^{\frac{-p}{2}} \exp(-\frac{1}{2}(\sigma^2+\lambda^{-1})^{-1}(x-B\gamma)'(x-B\gamma))$$
$$\times\ \lambda^{\alpha-1}\exp(-\frac{\lambda\beta}{2})$$

$$\propto (\sigma^2 + \lambda^{-1})^{-\frac{p}{2}} \exp[-\frac{(x - B\gamma_x)'(x - B\gamma_x)}{2(\sigma^2 + \lambda^{-1})} - \frac{\lambda\beta}{2}]$$
$$\times \lambda^{\alpha-1} \exp[-\frac{1}{2}(\sigma^2 + \lambda^{-1})^{-1}(\gamma - \gamma_x)'B'B(\gamma - \gamma_x)],$$

where

$$\gamma_x = (B'B)^- B'x. \tag{3.1}$$

(Here D^- denotes the generalized inverse of D). It is clear that the MLE of γ is γ_x and the marginal posterior density of λ given x (with γ replaced by γ_x) is

$$\pi(\lambda|x) \propto \lambda^{\alpha-1}(\sigma^2 + \lambda^{-1})^{-\frac{p-q}{2}}$$
$$\times \exp[-\frac{1}{2}(\sigma^2 + \lambda^{-1})^{-1}(x - B\gamma_x)'(x - B\gamma_x) - \frac{\lambda\beta}{2}].$$

It follows as before that the MLE, $\hat{\lambda}$, is the solution of the equation $\ell_1(\lambda) = 0$, where

$$\ell_1(\lambda) = (\beta\sigma^4)\lambda^3 + (2\beta\sigma^2 - 2(\alpha - 1)\sigma^4)\lambda^2 \tag{3.2}$$
$$+ (x'Mx - (p - q)\sigma^2 - 4(\alpha - 1)\sigma^2 + \beta)\lambda - ((p - q) + 2(\alpha - 1))$$

and

$$M = I - B(B'B)^- B'. \tag{3.3}$$

The hierarchical estimator (1.2) can be written in this case as

$$\delta^2(x) = x - \frac{t\hat{\lambda}\sigma^2}{(1 + \sigma^2\hat{\lambda})}(x - B\gamma_x).$$

Theorem 2. *Under sum of squares error loss, δ^2 is minimax for*

$$0 \le t \le \frac{2(p - q)}{t^{**}} - \frac{4}{(p - q) + 2(\alpha - 1)}, \tag{3.4}$$

where

$$t^{**} = \max\{p - q + 2(\alpha - 1), \quad p - q + 4(\alpha - 1) - \beta/\sigma^2\}. \tag{3.5}$$

Proof. Analogous to that of Theorem 1. □

The suggested choice of t in δ^2 is

$$\tilde{\tilde{t}} = \min\left\{\frac{2(p-q)}{t^{**}} - \frac{4}{[p-q+2(\alpha-1)]}, \quad \frac{p-q}{[p-q+2(\alpha-1)]}\right\},$$

which for $\beta \geq 2(\alpha-1)$ becomes

$$\tilde{\tilde{t}} = \frac{\min\{2(p-q-2), p-q\}}{[p-q+2(\alpha-1)]}.$$

4. COMPARISONS AND CONCLUSIONS

Since all estimators considered in this paper are minimax, the main question of interest is to investigate overall average performance. Since we are mainly considering application in empirical Bayes or hierarchical Bayes scenarios, suppose that θ actually has a $\mathcal{N}_p(\mu, \tau^2 I)$ prior distribution. We consider $\sigma^2 = 1$ and the known μ case for simplicity; the case of unknown μ yields similar conclusions.

A convenient way to measure performance with respect to a prior π is through the relative savings loss discussed by Efron and Morris (1973); this is given by

$$RSL(\delta) = \frac{r(\pi,\delta) - r(\pi,\delta^\pi)}{r(\pi,\delta^0) - r(\pi,\delta^\pi)},$$

where $\delta^0(x) = x, \delta^\pi$ is the Bayes rule with respect to π, and

$$r(\pi,\delta) = E^\pi R(\theta,\delta).$$

RSL measures the additional overall risk incurred by using δ instead of the optimal δ^π, scaled by the total possible improvement over the standard estimator δ^0. Thus RSL near zero indicates optimal Bayesian performance with respect to π, while RSL near one indicates negligible overall improvement over δ^0.

The usual empirical Bayes estimator (for the known μ case) is

$$\delta^{J-S}(x) = x - \min\{1, \frac{(p-2)}{|x-\mu|^2}\}(x-\mu),$$

the James-Stein positive part estimator. This assumes no knowledge of τ^2, and performs reasonably well for any τ^2.

The new estimator $\delta^1(x)$ is similar to δ^{J-S}, except that it is designed to do particularly well for τ^2 near $E[\lambda^{-1}] = \beta/[2(\alpha-1)]$. Tables I and II indicate that this is indeed so, for larger p or $E[\lambda^{-1}]$.

(When p and $E[\lambda^{-1}]$ are small, the behavior is somewhat different.) For table I, $E[\lambda^{-1}] = 1$ while for table II, $E[\lambda^{-1}] = 3$. The choice of $\alpha = 3$ merely implies that the standard deviation of λ^{-1} equals the mean, corresponding to a situation of moderate uncertainty in the prior mean for λ^{-1}. In all cases, t was chosen using (2.12).

Table 1

RSL **for various** p**,** τ^2**, and** δ^{J-S} **and** δ^1**, when** $\alpha = 3.0$ **and** $\beta = 2(\alpha-1)$**.**

	p					
	3		6		10	
τ^2	$RSL(\delta^{J-S})$	$RSL(\delta^1)$	$RSL(\delta^{J-S})$	$RSL(\delta^1)$	$RSL(\delta^{J-S})$	$RSL(\delta^1)$
.5	.5393	.6331	.2427	.3080	.1480	.2040
1.0	.5472	.5579	.2642	.2156	.1699	.1274
1.5	.5547	.4994	.2804	.1618	.1824	.0940
2.0	.5612	.4541	.2920	.1335	.1892	.0845
2.5	.5669	.4188	.3003	.1217	.1931	.0875
3.0	.5718	.3914	.3064	.1202	.1954	.0965
3.5	.5760	.3700	.3110	.1253	.1968	.1080
4.0	.5798	.3535	.3145	.1344	.1978	.1201
4.5	.5832	.3408	.3173	.1459	.1984	.1318
5.0	.5862	.3311	.3195	.1586	.1988	.1428
5.5	.5889	.3238	.3213	.1719	.1991	.1529
6.0	.5913	.3185	.3227	.1854	.1993	.1619
6.5	.5935	.3148	.3239	.1986	.1994	.1701
7.0	.5956	.3124	.3250	.2115	.1996	.1773
7.5	.5975	.3111	.3258	.2238	.1996	.1838
8.0	.5992	.3106	.3265	.2357	.1997	.1896
8.5	.6009	.3109	.3272	.2470	.1998	.1947
9.0	.6024	.3118	.3277	.2577	.1998	.1993
9.5	.6038	.3132	.3282	.2678	.1998	.2034
10.0	.6051	.3150	.3286	.2775	.1999	.2070

Table 2

RSL for various p, τ^2, and δ^{J-S} and δ^1, when $\alpha = 3.0$ and $\beta = 6(\alpha - 1)$.

	p					
	3		6		10	
τ^2	$RSL(\delta^{J-S})$	$RSL(\delta^1)$	$RSL(\delta^{J-S})$	$RSL(\delta^1)$	$RSL(\delta^{J-S})$	$RSL(\delta^1)$
.5	.5393	.7739	.2427	.5124	.1480	.3922
1.0	.5472	.7199	.2642	.4200	.1699	.2964
1.5	.5547	.6736	.2804	.3499	.1824	.2306
2.0	.5612	.6338	.2920	.2965	.1892	.1857
2.5	.5669	.5994	.3003	.2561	.1931	.1554
3.0	.5718	.5695	.3064	.2254	.1954	.1353
3.5	.5760	.5433	.3110	.2022	.1968	.1222
4.0	.5798	.5204	.3145	.1849	.1978	.1142
4.5	.5832	.5002	.3173	.1722	.1984	.1097
5.0	.5862	.4824	.3195	.1630	.1988	.1077
5.5	.5889	.4666	.3213	.1566	.1991	.1074
6.0	.5913	.4525	.3227	.1525	.1993	.1084
6.5	.5935	.4400	.3239	.1501	.1994	.1103
7.0	.5956	.4288	.3250	.1492	.1996	.1127
7.5	.5975	.4188	.3258	.1493	.1996	.1155
8.0	.5992	.4099	.3265	.1504	.1997	.1185
8.5	.6009	.4019	.3272	.1522	.1998	.1218
9.0	.6024	.3947	.3277	.1545	.1998	.1250
9.5	.6038	.3883	.3282	.1573	.1998	.1283
10.0	.6051	.3825	.3286	.1604	.1999	.1316

Although the tables only deal with small and moderate τ^2, it is interesting to note that, as $\tau^2 \to \infty$, $RSL(\delta^1) \to RSL(\delta^*)$, where (assuming $\beta \geq 2(\alpha - 1)\sigma^2$ for convenience)

$$\delta^*(x) = x - \frac{\min\{2(p-2), p\}\sigma^2}{|x - \mu|^2}(x - \mu).$$

The RSL of δ^* is very similar to that of δ^{J-S}, especially when p is moderate or large. Thus, even if the prior information concerning the variance of the θ_i is completely wrong and τ^2 is huge, δ^1 will be comparable to δ^{J-S}.

We cannot give unqualified endorsement of δ^1 or δ^2 over the more familiar James-Stein type estimators, or over, say, the robust generalized Bayes estimators in Berger (1980), because there are too many variables to study all possibilities. Furthermore, from a practical perspective it may be questionable to demand complete minimaxity.

In any case, the results here are of theoretical interest because they
(i) deal for the first time in the "Stein estimation" literature with estimators which combine empirical Bayes type exchangeability structure with subjective inputs;
(ii)indicate that verification of minimaxity is possible even for highly complicated estimators which cannot even be easily written in closed form.

5. ACKNOWLEDGMENT

This research was supported by the National Science Foundation, Grant DMS-8401996.

James Berger
Statistics Department
Purdue University
West Lafayette, IN 47907

Shun-Yu Chen
Department of Statistics
Cheng-Kung University
Tainan, Taiwan

6. REFERENCES

Berger, J. (1980). 'A robust generalized Bayes estimator and confidence region for a multivariate normal mean.' *Ann. Statist.*, **8**, 716–761.

Berger, J. (1982). 'Selecting a minimax estimator of a multivariate normal mean.' *Ann. Statist.*, **10**, 81–92.

Berger, J. (1985). '*Statistical Decision Theory and Bayesian Analysis.* Springer-Verlag, New York.

Chen, S. Y. (1983). 'Restricted Risk Bayes Estimator.' Ph.D. Thesis. Department of Statistics, Purdue University.

Efron, B. and Morris, C. (1973). 'Stein's estimation rule and its competitors-an empirical Bayes approach.' *J. Amer. Statist. Assoc.*, **68**, 117–130.

James, W. and Stein, C. (1960). 'Estimation with quadratic loss.' *Proc. Fourth Berkeley Symp. Math. Stat. Prob.*, **1**, 361–379.

Lindley, D. V. and Smith, A. F. M. (1972). 'Bayes estimators for linear model (with discussion).' *J. Roy. Statist. Soc. B.*, **34**, 1–41.

Stein, C. (1981). 'Estimation of the mean of a multivariate normal distribution.' *Ann. Statist.*, **9**, 1135–1151.

Hilmar Drygas

QUASI-INNER PRODUCTS AND THEIR APPLICATIONS

1. INTRODUCTION, DEFINITION AND ELEMENTARY PROPERTIES.

Usually in statistics, in particular in estimation theory, a quadratic expression has to be minimized subject to some constraints. This can of course be done by calculus methods. However, if the arguments of the quadratic form are itself matrices, this method becomes very cumbersome and requires a lot of indices. Therefore another method, the use of quasi-inner products is more often used in statistical literature. An example is the forthcoming monograph by J. Kleffe and C.R. Rao on variance component estimation.

For this reason it seems to be appropriate that statisticians, in particular multivariate statisticians, become more acquainted with the properties of quasi-inner products. This paper is therefore devoted to the study of quasi-inner products and their applications. Quasi-inner products are a generalization of inner products but many of the properties of inner products are maintained. Moreover, applications in Statistics (Estimation) and Mathematical Programming show their usefulness.

1.1 Definition

Let V be a real or complex vector-space. A complex valued function $W(x,y)$ defined $V \times V$ is called a quasi-inner product if

(Q1) $W(x,y) - W(x,0)$ is linear function of y

(Q2) $W(x,y) = \overline{W(y,x)}$, where $\overline{\alpha}$ denotes the complex conjugate of the complex number α.

(Q3) $W(x,x)$ is a convex function of x.

Note that by (Q2), $W(x,x) = \Phi(x)$ is a real-valued function.

A. K. Gupta (ed.), Advances in Multivariate Statistical Analysis, 13–30.

1.2 Proposition

(a) $W(x,y)$ is a conjugate affine linear function of x in the following sense:

$$W(\alpha_1x_1+\alpha_2x_2,y) - W(0,y) = \bar{\alpha}_1(W(x_1,y)-W(0,y)) + \bar{\alpha}_2(W(x_2,y)-W(0,y))$$

(b) Let $W_o(x,y) = W(x,y) - W(x,0) - W(0,y) + W(0,0)$. Then $W_o(x,0) = W_o(0,y) = 0$ and $W_o(x,y)$ is a linear function of y. Moreover, $W_o(x,y) = \overline{W_o(y,x)}$.

The proof of (a) can be left to the reader. (b) follows from the fact that $W_o(x,y)$ is the difference of two linear functions. Later on, it will turn out, that $W_o(x,y)$ is the semi-inner product associated with the quasi-inner product $W(x,y)$.

1.3 Theorem

(a) (Parallelogram-identity)

$$W_o(x-y,x-y) = W(x,x) - 2 \operatorname{Re} W(x,y) + W(y,y)$$
$$= 2\{W(x,x)+W(y,y) - 2\, W(\tfrac{x+y}{2},\tfrac{x+y}{2})\}.$$

(b) (Polarisation equation)

$$W(x+\beta y,x+\beta y) = W(x,x) + |\beta|^2\, W_o(y,y) + 2 \operatorname{Re}\{\beta(W(x,y)-W(x,0))\}$$

(c) (Polarisation identity)

$$\operatorname{Re}\{W(x,y)-W(x,0)\} = \tfrac{1}{4}\{W(x+y,x+y) - W(x-y,x-y)\},$$

$$\operatorname{Re} W_o(x,y) = \tfrac{1}{4}\{W(x+y,x+y) + W(-y,-y) - W(y,y) - W(x-y,x-y)\}.$$

Proof

(a) Since W_o is linear and conjugate linear in all arguments it follows that

$$W_0(x-y,x-y) = W_0(x,x) - 2\,\mathrm{Re}\,W_0(x,y) + W_0(y,y) \quad (1.1)$$

On the other hand

$$W_0(x,x) - 2\,\mathrm{Re}\,W_0(x,y) + W_0(y,y)$$
$$= W(x,x) - 2\,\mathrm{Re}\,W(x,0)+W(0,0)-2\,\mathrm{Re}(W(x,y)$$
$$- W(0,y) - W(x,0) + W(0,0))+W(y,y) - 2\,\mathrm{Re}\,W(0,y)$$
$$+ W(0,0) = W(x,x)-2\,\mathrm{Re}\,(W(x,y)) + W(y,y). \quad (1.2)$$

We now remark that for a function f such that $f(x) - f(0)$ is linear in x, it follows that $f(\alpha x+\beta y) = \alpha f(x) + \beta f(y)$, if $\alpha + \beta = 1$. Applying this to the case $\alpha = \beta = \frac{1}{2}$ and $f(y) = W(x,y)$ we get finally

$$W(\tfrac{x+y}{2}, \tfrac{x+y}{2}) = \tfrac{1}{4} W(x,x) + \tfrac{1}{2}\,\mathrm{Re}\,W(x,y) + \tfrac{1}{4}W(y,y) \quad (1.3)$$

and therefore

$$2\{W(x,x)+W(y,y)-2W(\tfrac{x+y}{2}, \tfrac{x+y}{2})\}$$
$$= W(x,x)- 2\,\mathrm{Re}\,W(x,y) + W(y,y) = W_0(x-y,x-y) \quad (1.4)$$

(b)
$$W(x+\beta y,x+\beta y) = W(x,x+\beta y) + \bar{\beta}W(y,x+\beta y)-\bar{\beta}W(0,x+\beta y)$$
$$= W(x,x)+\beta W(x,y)-\beta W(x,0+\bar{\beta}\,W(y,x) + |\beta|^2W(y,y)$$
$$- |\beta|^2W(y,0) - \bar{\beta}W(0,x) - |\beta|^2\,W(0,y) + |\beta|^2W(0,0)$$
$$= W(x,x) + |\beta|^2W_0(y,y) + 2\,\mathrm{Re}\,(\beta(W(x,y)-W(x,0))).$$

(c) This is an immediate consequence of the polarisation equation (b).

1.4 Definition

(a) Let V be a real or complex vector-space. A complex-valued function, defined on $V \times V$, is called a non-negative quasi-inner product if it fullfills the axioms (Q1) and (Q2) of affinity and hermitecity in Definition 1.1 and, moreover, $W(x,x) \geq 0$ for all $x \in V$.

(b) A non-negative quasi-inner product is called a semi-inner product if $W(x,0) = 0\ \forall\ x \in V$. If, moreover, $W(x,x) > 0$ for all $x \in V$, $x \neq 0$, then W is called an inner product. (It may be noted that then $W(x,y)$ is an inner product in the usual sense).

1.5 Theorem

(a) Let $W(x,y)$ be a function defined on $V \times V$, where V is a real or complex vector-space. If W meets the two axioms (Q1) and (Q2) in Definition 1.1 then it is a quasi-inner product if and only if W_o is non-negative, i.e., W_o is a semi-inner product.

(b) A non-negative quasi-inner product is always a quasi-inner product.

Proof

(a)
$$W(\alpha x+\beta y,\alpha x+\beta y) = \alpha^2 W(x,x) + \alpha\beta(W(x,y)+W(y,x)) + \beta^2 W(y,y),$$

if $0 \leq \alpha$, $\beta \leq 1$, $\alpha + \beta = 1$. Thus

$$W(\alpha x+\beta y,\alpha x+\beta y) = \alpha^2 W(x,x) + 2\alpha\beta \operatorname{Re} W(x,y) + \beta^2 W(y,y) \leq \alpha W(x,x) + \beta W(y,y) \quad (1.5)$$

if and only if

$$\alpha(1-\alpha)[W(x,y)-2 \operatorname{Re} W(x,y) + W(y,y)] \geq 0.$$

If this holds for all $\alpha \in [0,1]$ then necessarily $W(x,x) - -2 \operatorname{Re} W(x,y) + W(y,y) \geq 0$. By the parallelogram-identity the latter expression is identical to $W_o(x-y,x-y) \geq 0$. Thus $W(x,x)$ is convex iff W_o is non-negative, i.e., a semi-inner product.

(b) If W meets the axioms of affinity and hermitecity (Q1 and Q2), then we get from the polarisation equation if W is non-negative:

$$0 \leq W(x+\beta y,x+\beta y) = W(x,x) + |\beta|^2 W_o(y,y) + 2 \operatorname{Re}(\beta(W(x,y)-W(x,0))). \quad (1.6)$$

If we choose β as a real number and divide both sides of the equation (1.6) by $\beta^2 > 0$, we get

$$0 \leq \frac{1}{\beta^2} W(x,y) + W_o(y,y) + 2\beta^{-1}\mathrm{Re}(W(x,y)-W(x,0)). \tag{1.7}$$

By passing to the limit $\beta \to \infty$ we get $W_o(y,y) \geq 0$. W_o is therefore a semi-inner product. Consequently $W(x,x)$ is convex and therefore $W(x,y)$ is a quasi-inner product.

1.6 Theorems

(a) (Generalized Cauchy-Schwarz inequality)

$$|W(x,y)-W(x,0)|^2 \leq W(x,x) \cdot W_o(y,y) \tag{1.8}$$

if W is a non-negative quasi-inner product.

(b) (Generalized triangle inequality)

Let $W(x,x) = \Phi(x)$. Then

$$(\Phi(x+y))^{1/2} \leq ((\Phi(x))^{1/2} + (2(\Phi(y)-\Phi(0)) - 4(\Phi(\tfrac{y}{2}) - \Phi(0)))^{1/2}$$

Proof

(a) From the polarisation equation we get for real β $0 \leq W(x,x) + \beta^2 W_o(y,y) + 2\beta\,\mathrm{Re}(W(x,y)-W(x,0))$. This inequality can hold iff the discriminant of the quadratic expression on the right side is non-negative. This is equivalent to $|\mathrm{Re}(W(x,y)-W(x,0))|^2 \leq W(x,x) \cdot W_o(y,y)$. Now let $W(x,y)-W(x,0) = e^{i\varphi}|W(x,y)-W(x,0)|$ and let $a = e^{-i\varphi}$. Then $|W(x,y)-W(x,0)|^2 = (\mathrm{Re}(a(W(x,y)-W(x,0))))^2 = (\mathrm{Re}(W(x,ay) - W(x,0)))^2 \leq (W(x,x))W_o(ay,ay) = W(x,x)W_o(y,y)$, since $|a| = 1$.

(b) From the Cauchy-Schwarz inequality follows in the usual way that

$$(\Phi(x+y))^{1/2} \leq ((\Phi(x))^{1/2} + (W_o(y,y))^{1/2}).$$

The relation $W_o(y,y) = 2(\Phi(y)-\Phi(0)) - 4(\Phi(\frac{y}{2}) - \Phi(0))$ follows from the parallelogram-identity, QED.

It may be noted that $\psi_1(x) = 2(\Phi(x)-(0))-4(\Phi(\frac{x}{2})-\Phi(0))\geq 0$ follows from the convexity of Φ.

2. CONSTRUCTION OF A QUASI-INNER PRODUCT FROM THE CORRESPONDING QUADRATIC FORM.

From the parallelogram-identity

$$W(x,x) - 2\,\mathrm{Re}\,W(x,y) + W(y,y) = 2\{W(x,x) + W(y,y) - W(\tfrac{x+y}{2},\tfrac{x+y}{2})\} \tag{2.1}$$

it follows that

$$\mathrm{Re}\,W(x,y) = \tfrac{1}{2}\{4\Phi(\tfrac{x+y}{2}) - \Phi(x) - \Phi(y)\}, \tag{2.2}$$

where $\Phi(x) = W(x,x)$. Since

$$\mathrm{Re}(W(x,iy)-W(x,0)) = \mathrm{Re}(i(W(x,y)-W(x,0)) = -\mathrm{Im}((W(x,y)-W(x,0)) \tag{2.3}$$

and

$$\mathrm{Im}(W(x,0)) = \mathrm{Im}(W(x,0)-W(0,0)) = \mathrm{Re}(W(ix,0)-W(0,0)) = \mathrm{Re}(-i(W(x,0)-W(0,0))) \tag{2.4}$$

it follows that

$$\begin{aligned}\mathrm{Im}(W(x,y)) &= -\mathrm{Re}(W(x,iy)-W(x,0)) + \mathrm{Re}(W(ix,0))-W(0,0)\\ &= -\mathrm{Re}\,W(x,iy) + \mathrm{Re}\,W(x,0) + \mathrm{Re}(W(ix,0) - W(0,0))\\ &= -\tfrac{1}{2}\{4\Phi(\tfrac{x+iy}{2} - 4\Phi(\tfrac{x}{2}) - 4\Phi(\tfrac{ix}{2}) + \Phi(ix) - \Phi(iy)+4\Phi(0)\}\\ &= -\tfrac{1}{2}\{4(\psi(\tfrac{x+iy}{2}) - \psi(\tfrac{x}{2}) - \psi(\tfrac{ix}{2})) + \psi(ix) - \psi(iy)\},\end{aligned} \tag{2.5}$$

where $\psi(x) = \Phi(x) - \Phi(0)$.

Thus we can construct $W(x,y)$ from $\Phi(x)$ by the formula

$$W(x,y) = \tfrac{1}{2}\{4\Phi(\tfrac{x+y}{2}) - \Phi(x) - \Phi(y) - i[4(\psi(\tfrac{x+iy}{2}) - \psi(\tfrac{ix}{2}) - \psi(\tfrac{x}{2})) + \psi(ix) - \psi(iy)]\} \tag{2.6}$$

The question now arises under which conditions on

ϕ (or ψ, respectively) a quasi-inner product is defined by (2.6) in such a way that $W(x,x) = \Phi(x)$. The following theorem generalizes a theorem by P.Jordan und J.v.Neumann (1935).

2.1 Theorem

Let $\Phi(x)$, $x \in V$ be a real valued function which meets one of the following condtions:

(a) $\Phi(x)$ is continuous in the sense that $\lim_{n\to\infty} \Phi(\lambda_n x+y) = \Phi(\lambda x+y)$ if $\lim_{n\to\infty} \lambda_n = \lambda$; $x,y \in V$.

(b) $\Phi(x) \geq 0 \ \forall x \in V$ and $\Phi(x)$ obeys the generalized triangle inequality

$$(\Phi(x+y))^{1/2} \leq (\Phi(x))^{1/2} + (2\psi(y)-4\psi(\tfrac{y}{2}))^{1/2}.$$

Then by (2.6) a function $W(x,y)$ obeying the axioms (Q1) and (Q2) of Definition 1.1 and coinciding with $\Phi(x)$ if $x = y$ can be obtained if and only if the following functional equations are valid:

(i) $\psi(x) + \psi(y) = 2(\psi(\frac{x+y}{2}) - \psi(\frac{x-y}{2})) + \psi(x-y)$

(ii) $0 = \sum_{i=1}^{k} \mu_i \psi(\lambda_i x)$ if μ_i are real numbers and

$$\sum_{i=1}^{k} \mu_i |\lambda_i|^2 = \sum_{i=1}^{k} \lambda_i \mu_i = 0.$$

Proof

(a) From the parallelogram-identity we get

$$2\Phi(x)+2\Phi(y) = 4\Phi(\tfrac{x+y}{2}) + W_o(x-y,x-y)$$

$$= 4\Phi(\tfrac{x+y}{2}) +\Phi(x-y)+\Phi(0)- 2 \operatorname{Re} W(x-y,0)$$

$$= 4\Phi(\tfrac{x+y}{2})+\Phi(x-y) -\{4\Phi(\tfrac{x-y}{2})-\Phi(x-y)-\Phi(0)\}+\Phi(0)$$

$$= 4(\Phi(\tfrac{x+y}{2}) -\Phi(\tfrac{x+y}{2})) + 2\Phi(x-y) + 2\Phi(0). \qquad (2.7)$$

This is evidently equivalent to the functional equation (i).

(b) From the polarisation equation it follows that

$$\psi(\lambda_i x) = \phi(0+\lambda_i x) - \Phi(0) = |\lambda_i|^2 W_o(x,x)$$
$$+ 2\,\mathrm{Re}(\lambda_i(W(0,x)-W(0,0)) = |\lambda_i|^2 W_o(x,x)$$
$$+ 2\,\mathrm{Re}(\lambda_i)\mathrm{Re}(W(0,x)-W(0,0))$$
$$- 2\,\mathrm{Im}(\lambda_i)\mathrm{Im}(W(0,x)-W(0,0)). \quad (2.8)$$

From this evidently the functional equation (ii) is obtained if μ_i are reals.

(c) Now let Φ (or ψ, respectively) obey the functional equations (i) and (ii). We define W(x,y) by formula (2.6). Then

$$W(x,x) = \Phi(x) - 4i[\psi(\tfrac{(i+1)x}{2}) - \psi(\tfrac{ix}{2}) - \psi(\tfrac{x}{2})]. \quad (2.9)$$

We apply the functional equation (ii) to

$$\lambda_1 = (1+i)/2,\ \lambda_2 = i/2,\ \lambda_3 = 1/2,$$
$$\mu_1 = 1,\ \mu_2 = \mu_3 = -1.$$

Then it is readily verified that $\sum_{i=1}^{3}|\lambda_i|^2\mu_i = \sum_{i=1}^{3}\lambda_i\mu_i = 0.$

Thus the second term on the right hand side of (2.9) vanishes and $W(x,x) = \Phi(x)$.

Next we want to demonstrate that $W(x,y) = \overline{W(y,x)}$. Since evidently Re W(x,y) = Re W(y,x) it is enough to show that Im(W(x,y)+W(y,x)) = 0. From formula (2.6) we get:

$$-\tfrac{1}{2}[\mathrm{Im}\,W(x,y)+\mathrm{Im}\,W(y,x)] = -\tfrac{1}{2}\,\mathrm{Im}(W(x,y)+W(y,x))$$
$$= \psi(\tfrac{x+iy}{2}) + \psi(\tfrac{y+ix}{2}) - \psi(\tfrac{ix}{2}) - \psi(\tfrac{x}{2}) - \psi(\tfrac{iy}{2}) - \psi(\tfrac{y}{2}). \quad (2.10)$$

From the functional equation (i) we get

$$\psi(\tfrac{x}{2}) + \psi(\tfrac{iy}{2}) = 2\psi(\tfrac{x+iy}{2}) - 2\psi(\tfrac{x-iy}{4}) + \psi(\tfrac{x-iy}{2}). \quad (2.11)$$

$$\psi(\tfrac{y}{2}) + \psi(\tfrac{ix}{2}) = 2\psi(\tfrac{y+ix}{2}) - 2\psi(\tfrac{y-ix}{2}) + \psi(\tfrac{y-ix}{2}). \quad (2.12)$$

In view of $-i(x+iy) = y - ix$, $-i(y+ix) = x - iy$ we get

$$-\tfrac{1}{2}\{\mathrm{Im}\,W(x,y) + \mathrm{Im}\,W(y,x)\} =$$

$$= \psi(\frac{x+iy}{2}) - 2\psi(\frac{x+iy}{4}) + 2\psi(\frac{-i(x+iy)}{4}) - \psi(\frac{-i(x+iy)}{2})$$
$$+ \psi(\frac{y+ix}{2}) - 2\psi(\frac{y+ix}{4}) + 2\psi(\frac{-i(y+ix)}{4}) - \psi(\frac{-i(y+ix)}{2}) \tag{2.13}$$

We apply functional equation (ii) to

$$\lambda_1 = \frac{1}{2}, \lambda_2 = \frac{1}{4}, \lambda_3 = -\frac{i}{4}, \lambda_4 = \frac{-i}{2},$$
$$\mu_1 = 1, \mu_2 = -2, \mu_3 = 2, \mu_4 = -1.$$

Then it is readily verified that $\sum_{i=1}^{4} |\lambda_i|^2 \mu_i = \sum_{i=1}^{4} \lambda_i \mu_i = 0.$

Consequently the sum in the first and second line of the right hand side of (2.13) vanishes. It is thus proved that $W(x,y) = \overline{W(y,x)}$.

Finally we get from formula (2.6) that

$$W(x,y) - W(x,0) = \frac{1}{2}\{4\Phi(\frac{x+y}{2}) - 4\Phi(\frac{x}{2}) - (\Phi(y)-\Phi(0))$$
$$-i[4\psi(\frac{x+iy}{2}) - 4\psi(\frac{x}{2}) - \psi(iy)]\}$$
$$= \frac{1}{2}[W_1(x,y)-iW_1(x,iy)], \tag{2.14}$$

where

$$W_1(x,y) = 4(\psi(\frac{x+y}{2}) - \psi(\frac{x}{2})) - \psi(y). \tag{2.15}$$

Therefore, in order to prove linearity of $W(x,y)-W(x,0)$ in y it is enough to show linearity of $W_1(x,y)$ with respect to y. We get

$$W_1(x,y_j) = 4\psi(\frac{x+y_j}{2}) - 4\psi(\frac{x}{2}) - \psi(y_j), \ j = 1,2 \tag{2.16}$$

and

$$W_1(x,y_1+y_2) - \sum_{j=1}^{2} W_1(x,y_j)$$
$$= 4(\psi(\frac{x+y_1+y_2}{2}) + \psi(\frac{x}{2})) - 4(\psi(\frac{x+y_1}{2}) + \psi(\frac{x+y_2}{2}))$$
$$- (\psi(y_1+y_2) - \psi(y_1) - \psi(y_2)). \tag{2.17}$$

By applying functional equation (i) to the first two summands, the second two summands and $\psi(y_1)+\psi(y_2)$ we get

$$W_1(x,y_1+y_2) - \sum_{j=1}^{2} W_1(x,y_j) = -8\psi(\frac{y_1+y_2}{4}) + 4\psi(\frac{y_1+y_2}{2})$$

$$- 4\psi(\frac{y_1-y_2}{2}) - (\psi(y_1+y_2) - 2\psi(\frac{y_1+y_2}{2}) + 2\psi(\frac{y_1+y_2}{2})$$

$$- \psi(y_1-y_2)) = -\psi(y_1+y_2) + 6\psi(\frac{y_1+y_2}{2}) - 8\psi(\frac{y_1+y_2}{4})$$

$$+ \psi(y_1-y_2) - 6\psi(\frac{y_1-y_2}{2}) + 8\psi(\frac{y_1-y_2}{4}). \qquad (2.18)$$

We now apply functional equation (ii) to $k = 3$, $\lambda_1 = 1$, $\lambda_2 = \frac{1}{2}$, $\lambda_3 = \frac{1}{4}$, $\mu_1 = 1$, $\mu_2 = -6$, $\mu_3 = 8$. Then it is readily verified that $\sum_{i=1}^{3} \mu_i |\lambda_i|^2 = \sum_{i=1}^{3} \mu_i \lambda_i = 0$. Thus the two lines of the right hand side of (2.18) must vanish, which proves $W_1(x,y_1,y_2) = \sum_{j=1}^{2} W_1(x,y_j)$.

Evidently $W(x,y) - W(x,0)$ vanishes if $y = 0$ and therefore $W(x,-y) - W(x,0) = -(W(x,y) - W(x,0))$. Moreover, $W(x,iy) - W(x,0) = \frac{1}{2}[W_1(x,iy)-iW_1(x,-y)] = \frac{1}{2}[W_1(x,iy) + iW_1(x,y)] = \frac{i}{2}[W_1(x,y)-iW_1(x,iy)] = i(W(x,y)-W(x,0))$. To prove linearity of W_1 with respect to y it is therefore now enough to show $W_1(x,ay) = aW_1(x,y)$ if $a \in \mathbb{R}^+$

If Φ is continuous in the sense of (a), then the linearity follows at first for all natural numbers, then for all rational numbers and finally for all reals $a \in \mathbb{R}^+$ just as in Greub (1976, Kap. VIII, §1, p. 150-152).

Under the conditions of(b) we first remark that $\psi_1(x) =: 2\psi(x) - 4\psi(\frac{x}{2})$ is homogeneous of degree two, i.e., $\psi_1(\lambda x) = |\lambda|^2\psi_1(x)$. To see this we apply functional equation (ii) to $\lambda_1 = \lambda$, $\lambda_2 = \frac{\lambda}{2}$, $\lambda_3 = 1$, $\lambda_4 = \frac{1}{4}$, $\mu_1 = 2$, $\mu_2 = -4$, $\mu_3 = -2|\lambda|^2$, $\mu_4 = 4|\lambda|^2$. Then it is readily verified that $\sum_{i=1}^{4} |\lambda_i|^2\mu_i = \sum_{i=1}^{4} \lambda_i\mu_i = 0$ which gives the assertion .

Now let $a \in \mathbb{R}^+$ and $n(k)$, $m(k)$ be sequences of real numbers such that $\lim_{k\to\infty} 2^{-n(k)}m(k) = a$. From $W_1(x,y_1,y_2) = W_1(x,y_1) + W_1(x,y_2)$ it follows ($y_1 = y_2 = \frac{y}{2}$), that $W_1(x,\frac{y}{2}) = \frac{1}{2}W_1(x,y)$ and by induction $W_1(x,my) = mW_1(x,y)$,

$W_1(x,a_k y) = a_k W_1(x,y)$, where $a_k = 2^{-n(k)}m(k)$. Thus $\lim_{k\to\infty} W_1(x,a_k y) = aW_1(x,y)$. The proof will be finished if we show that

$$\lim_{k\to\infty} |W_1(x,a_k y) - W_1(x,ay)| = 0. \tag{2.19}$$

This is done by the triangle inequality (employing the homogeneity of $\psi_1(y)$) and the inequality

$$|\Phi^{1/2}(x) - \Phi^{1/2}(y)| \leq \psi_1{}^{1/2}(x-y), \tag{2.20}$$

which follows in the usual way from the homogeneity of $\psi_1(y)$. The following estimate can be obtained in a way similar to Weidmann (1976, pp. 17-19).

$$\begin{aligned} |W_1(x,a_k y) - W_1(x,ay)| &\leq |a_k - a|(\psi_1(y))^{1/2} \cdot \\ &\cdot (\Phi^{1/2}(x) + |a_k|(\psi_1(y))^{1/2}) + |a_k - a|(\psi_1(y))^{1/2} \cdot \\ &\cdot (\Phi^{1/2}(0) + |a_k|(\psi_2(y))^{1/2}) \to 0 \end{aligned} \tag{2.21}$$

as $k \to \infty$. This finishes the proof of the theorem. Q.E.D.

A formula for reproducing $W(x,y)$ from $\Phi(x)$ can also be obtained from the polarisation identity (Theorem 1.3(c)). This yields the formula

$$\begin{aligned} W(x,y) = \tfrac{1}{4}[&\psi(x+y)+\psi(x)-\psi(-x)-\psi(x-y)-i(\psi(x+iy)+ \\ &+\psi(-ix)-\psi(ix)-\psi(x-iy)]. \end{aligned} \tag{2.22}$$

The addition theorem is easier obtained from this formula. However, $W(y,x) = \overline{W(x,y)}$ is more difficult to prove and requires an extension of the functional equation (ii).

3. APPLICATIONS IN STATISTICS.

Quasi-inner products can be successfully applied in linear estimation theory, in particular in unbiased estimation. Examples can be found in the author's work (1969, 1970, 1972, 1975). Also other authors, e.g. Kleffe and Rao, Klonecki and Zontek have adopted this concept in the theory of linear models. The fundamental theorem is the projection theorem proved in Drygas (1972, p. 375/376).

3.1 Theorem

Let $W(x,y)$ be a quasi-inner product on V and $C \subseteq V$ a convex set, then for $c_o \in C$ the relation

$$\min_{c \in C} W(c,c) = W(c_o,c_o)$$

holds if and only if

$$\mathrm{Re}(W(c_o,c_1-c_o) - W(c_o,0)) \geq 0 \;\forall\; c_1 \in C. \tag{3.1}$$

If C is a linear manifold then (3.1) is equivalent to

$$\mathrm{Re}(W(c_o,f) - W(c_o,0)) = 0 \;\forall\; f \in F = C - C. \tag{3.1a}$$

We consider a typical application of theorem 3.1: Let $L \subseteq \mathbb{R}^n$ be a linear subspace and consider the linear model

$$y = l + \sigma\varepsilon,\; E(\varepsilon) = 0,\; E(\varepsilon\varepsilon') = Q,\; l \in L. \tag{3.2}$$

$y'Ay$ is called an invariant estimator of σ^2 iff $Al = 0$ for all $l \in L$. Under the assumption of quasi-normality and invariance we get (see Drygas (1970, pp. 98-103)):

$$\begin{aligned} E((y'Ay-\sigma^2)^2) &= \sigma^4 E((\varepsilon'A\varepsilon-1)^2) \\ &= \sigma^4\{(\mathrm{tr}AQ-1)^2 + 2\mathrm{tr}(AQAQ)\}. \end{aligned} \tag{3.3}$$

The minimization of (3.3) with respect to A gives the best invariant quadratic estimaor of σ^2. This can easily be done by using the quasi-inner product

$$W(A,B) = (\mathrm{tr}AQ-1)(\mathrm{tr}BQ-1) + 2\mathrm{tr}(AQBQ). \tag{3.4}$$

Clearly, $W(A,B)$ is a non-negative quasi-inner product and $E((y'Ay-\sigma^2)^2) = \sigma^4 W(A,A)$, while

$$\begin{aligned} W(A,B) - W(A,0) &= (\mathrm{tr}AQ-1)\mathrm{tr}BQ + 2\mathrm{tr}(QAQB) \\ &= \mathrm{tr}([(\mathrm{tr}AQ-1)Q + 2QAQ]B) \end{aligned} \tag{3.5}$$

Here $C = \{A:AL = 0\}$, implying the A is optimal iff $(\mathrm{tr}A-1)Q + 2QAQ$ is orthogonal to all B meeting $BL = 0$. Now it is easy to verify that

$$A = (f+2)^{-1}\,(I-G)'Q^-(I-G) \tag{3.6}$$

is a solution to equation (3.5) if Q^- is a symmetric g-inverse of Q, G is a BLUE (Best Linear Unbiased Estimator) of Ey in the linear model $Ey \in L$, Cov $y = \sigma^2 Q$ and $f = \dim(\mathrm{im}(Q)) - \dim(\mathrm{im}(Q) \cap L) = \dim(\mathrm{im}(Q)+L) - \dim(L)$. Indeed $\mathrm{tr}(AQ) = (f+2)^{-1}f$, $(\mathrm{tr}AQ-1) = -2(f+2)^{-1}$, $QAQ = (f+2)^{-1}(I-G)Q$. Thus

$$(\mathrm{tr}(AQ)-1)Q + 2QAQ = -2(f+2)^{-1}fQ + 2QAQ = -2(f+2)^{-1}GQ \tag{3.7}$$

and $W(A,B) - W(B,0) = -2(f+2)^{-1}\mathrm{tr}(GQB)$. But $GQB = 0$ if $BL = 0$.

In Drygas (1970, p. 99-103) the use of quasi-inner product was avoided by showing that

$$W(A,B) = W_o(A-A_o, A-A_o), \tag{3.8}$$

where $A_o = (\dim(\mathrm{im}\ Q))^{-1}Q^-$ and $W_1(A,B)$ is the semi-inner product

$$W_o(A,B) = \mathrm{tr}(AQ)\mathrm{tr}(BQ) + 2\mathrm{tr}(AQBQ). \tag{3.9}$$

A representation of the form (3.8) (possibly plus a constant) is always possible for non-negative quasi-inner products. But you need not find such a representation in order to be ready for applying the projection theorem.

Sometimes quasi-inner products could be used as an alternative to convex analysis. An example of this kind is lemma 3.8 in La Motte (1982), p. 249, which is nothing else than the projection-theorem.

4. APPLICATIONS TO MATHEMATICAL PROGRAMMING.

In this section we want to apply the projection theorem to find the minimizers of $\Phi(x) = W(x,x)$, where $W(x,y) = \frac{1}{2}x'By + \frac{1}{2}p'x + \frac{1}{2}p'y + d$ is a quasi-inner product on $\mathbb{R}^n$. Clearly B is a symmetric n.n.d. (non negative definite) matrix. We want to show that the Kuhn-Tucker theorem for quadratic programming problems and the separation theorem can be obtained from the projection theorem for quasi-inner products.

4.1 Theorem

Let $W(x,y)=\frac{1}{2}x'By+\frac{1}{2}p'x+\frac{1}{2}p'y+d$. If B is positive definite, then $\Phi(x)=W(x,x)$ has a minimum on every closed convex set C.

Proof

Let $\sigma=\inf_{x\in C}\Phi(x)>-\infty$ (since $p\in \mathrm{im}(B)$) and let $x_n\in C$ be such that $\lim_{n\to\infty}\Phi(x_n)=\sigma$. Then $(x_n+x_m)/2\in C$ by convexity and therefore $\Phi(\frac{x_n+x_m}{2})\geq\sigma$. Let n,m be so large that $\Phi(x_n),\Phi(x_m)\leq\sigma+\frac{\varepsilon^2}{4}$. Then by the parallelogram-identity

$$\begin{aligned} W_o(x_n-x_m,x_n-x_m) &= \frac{1}{2}(x_n-x_m)'B(x_n-x_m) \\ &= 2\{\Phi(x_n)+\Phi(x_m)-2\Phi(\frac{x_n+x_m}{2})\}\leq\varepsilon^2. \end{aligned} \tag{4.1}$$

Since B is positive definite, $B^{1/2}$ is so, too. (4.1) shows that $\{B^{1/2}x_n\}$ is a Cauchy-sequence, having the limit y. Then it follows that $\lim_{n\to\infty}x_n=B^{-1/2}y=x_o\in C$, since C is closed. Finally $\lim_{n\to\infty}\Phi(x_n)=\Phi(x_o)=\sigma$, Q.E.D.

4.2 Remark

The preceding theorem is incorrect if B is not positive definite. A simple example is $B=0$, $p=(1,0)'$ and

$$C=\{(x,y):x\geq 0,\ y\geq 0,\ xy\geq 1\}. \tag{4.2}$$

4.3 Theorem (Separation theorem for convex sets).

Let $C\neq\emptyset$ be a closed and convex subset of $\mathbb{R}^n$ and $a\notin C$. Then there exists an element $b\in\mathbb{R}^n$, $b\neq 0$ and $c\in\mathbb{R}$ such that for $\phi(x)=b'x$

(1) $\phi(x)>c\ \forall\ x\in C$

(2) $\phi(a)<c$

Proof

Let $W(x,y) = (a-x)'(a-y)$. $W(x,y)$ is a non-negative quasi-inner product and $W_o(x,x) = x'x$ is positive-definite. By theorem 4.1 $\Phi(x)$ attains a minimum at some point $x_o \in C$. By the projection theorem the minimum is characterized by

$$\begin{aligned} &W(x_o, x-x_o) - W(x_o, 0) \\ &= (x_o-a)'(x-x_o) \geq 0 \ \forall \ x \in C. \end{aligned} \tag{4.3}$$

Let $b = (x_o-a) \neq 0$ and $c = (x_o-a)'a + \frac{1}{2}\|x_o-a\|^2$. Then from (4.3) it follows that

$$(x_o - a)'x = b'x > c = (x_o-a)'a + \frac{1}{2}\|x_o-a\|^2 \tag{4.4}$$

$$(x_o-a)'a = b'a < c = (x_o-a)'a + \frac{1}{2}\|x_o-a\|^2. \tag{4.5}$$

4.4 Farkas' theorem

Necessary and sufficient for $a \varepsilon \{x : x'u \geq 0 \ \forall \ u : A'u \geq 0\}$ is that $a = Ax$ for some $x \geq 0$. ($a \geq b$ for vectors, a,b iff the relation holds for all components of the two vectors).

The proof of necessity of the above assertion is based on the Separation Theorem 4.3 and can be found in the text-books on Mathematical Programming, e.g., Stoer-Witzgall (1970). The sufficiency simply follows from $(Ax)'u = x'A'u \geq 0$ if $x, A'u \geq 0$.

4.5 Theorem (Main theorem of quadratic programming)

Let the optimization problem

$$\text{Min } \Phi(x) = p'x + \frac{1}{2} x'Bx + d, \tag{4.6}$$

B n.n.d. and symmetric, subject to the constraints

$$Ax \leq b, \ b \in \mathbb{R}^m, \ A \in \mathbb{R}^{m \times n} \tag{4.7}$$

be given. Then $x^{(o)}$ yields an optimum if and only if

$$Ax^{(o)} = b-y^{(o)}, \ y^{(o)} \geq 0 \text{ (Feasibility)} \tag{4.8}$$

and

$$\exists u^{(o)} \geq 0 \text{ such that}$$
$$\begin{array}{ll}(a) & Bx^{(o)} + p = -A'u^{(o)} \\ (b) & (y^{(o)})'u^{(o)} = 0\end{array} \quad \text{(Optimality)} \tag{4.9}$$

Proof

$W(x,y) = \frac{1}{2}x'By + \frac{1}{2}p'x + \frac{1}{2}p'y + d$ is an quasi-inner product if B is n.n.d. and symmetric. Since $\Phi(x) = W(x,x)$ and $C = \{x : Ax \leq b\}$ is a convex set the projection theorem 3.1 applies. This yields that $x^{(o)}$ is optimal iff $x^{(o)} \in C$ and

$$\begin{aligned} &W(x^{(o)},x-x^{(o)}) - W(x^{(o)},0) \\ &= \tfrac{1}{2}(Bx^{(o)}+p)'(x-x^{(o)}) \geq 0 \;\forall\; x : Ax \leq b. \end{aligned} \tag{4.10}$$

$x^{(o)} \in C$ is, of course, equivalent to (4.8). Now we show that condition (4.9) is sufficient for optimality. Indeed

$$\begin{aligned} &(Bx^{(o)}+p)'(x-x^{(o)}) = (-A'u^{(o)})'(x-x^{(o)}) \\ &= (u^{(o)})'Ax^{(o)} - (u^{(o)})'(Ax) = (u^{(o)})'(b-y^{(o)}) \\ &\;\; - (u^{(o)})'Ax = (u^{(o)})'(b-Ax) \geq 0, \end{aligned} \tag{4.11}$$

since $(u^{(o)})'y^{(o)} = 0$, $u^{(o)}, b - Ax \geq 0$ for all $x \in C$.

To prove necessity, let us split up $y^{(o)}$ according to the positive and the vanishing components. Thus let

$$A = \left(\frac{A_1}{A_2}\right), \; b = \left(\frac{b_1}{b_2}\right), \; b-Ax^{(o)} = \left(\frac{y_1}{y_2}\right), \tag{4.12}$$

where $A_1 \in \mathbb{R}^{r\times n}$, $A_2 \in \mathbb{R}^{m-r\times n}$, $b_1 \in \mathbb{R}^r$, $b_2 \in \mathbb{R}^{m-r}$.

Let $A_1x^{(o)} = b_1$, $A_2x^{(o)} \ll b_2$ (i.e., there is inequality for all components) or

$$\begin{aligned} &(y_2)_i = (y^{(o)})_{i+r} = (b_2)_i - (A_2x^{(o)})_i > 0, \\ &i = 1,2,\ldots,m-r. \end{aligned} \tag{4.13}$$

Now let z be any vector such that $-A_1z \geq 0$ and

$$\Gamma = \Gamma(z) = \{i\in\{1,2,\ldots,m-r\} : (A_2z)_i > 0\} \tag{4.14}$$

and let

$$0 < \lambda \leq \min_{i \in \Gamma} \frac{(y_2)_i}{(A_2 z)_i} \tag{4.15}$$

(If $A_2 z \leq 0$ choose λ arbitrary but $\lambda > 0$). Now our assertion is that $x = x^{(o)} + \lambda z \in C$. Clearly

$$A_1 x = A_1 x^{(o)} + \lambda A_1 z \leq A_1 x^{(o)} = b_1. \tag{4.16}$$

But $A_2 x = A_2 x^{(o)} + \lambda A_2 z \leq b_2$ holds as well. This is clear if $(A_2 z)_i \leq 0$ and follows for $(A_2 z)_i > 0$ by the very choice of λ in (4.15). Thus

$$(Bx^{(o)}+p)'(x-x^{(o)}) = (Bx^{(o)}+p)'\lambda z \geq 0 \tag{4.17}$$

and in view of $\lambda > 0$ also $(Bx^{(o)}+p)'z \geq 0$ for all z such that $-A_1 z \geq 0$. It follows from Farkas' theorem that there exists $u_1 \geq 0$ such that

$$Bx^{(o)} + p = -A_1' u_1. \tag{4.18}$$

Now let $u^{(o)} = ((u^{(1)})', 0')'$. Then evidently $u^{(o)} \geq 0$, $(u^{(o)})' y^{(o)} = 0$ (since $y_1 = 0$) and $Bx^{(o)} + p = -A_1' u_1 - A_2' 0$ $= -(A_1 : A_2)' u^{(o)}$. This finishes the proof of the theorem, Q.E.D.

The theorem can also be applied to the special case B = 0, then yielding the duality theorem for linear programs.

FB 17 (Math.), GhK
Postfach 101 380
D-3500 Kassel
F.R. Germany

REFERENCES

Drygas, H. 'Gauss-Markov estimation and Best Linear Minimum Bias Estimation' Report Nr. 91, Studiengruppe für Systemforschung, Heidelberg (1969).

Drygas, H. 'The coordinate-free approach to Gauss-Markov estimation', Springer, Berlin (1970).

Drygas, H. 'The estimation of residual variance in regression Analysis', Math. Operationsforschung **3**, (1972), p. 373-388.

Drygas, H. 'Estimation and prediction for linear models in general Spaces', Math. Operationsforschung und Statistik **6** (1975), p. 301-324

Greub, W. 'Lineare Algebra', Springer, Berlin (1976).

P. Jordan and J.v. Neumann 'On inner products in linear metric spaces', Annals of Mathematics **36**, (1935), p. 719-723.

Klonecki, W. and Zontek, S. 'The structure of admissible linear estimators'. To appear in Journal of multivariate analysis.

La Motte, L.R. 'Admissibility in Linear Estimation', Annals of Statistics, **10** (1982), p. 245-255.

Rao, C.R. and Kleffe, J. 'Estimation in variance component models', Forthcoming.

Stoer-Witzgall 'Convexity and Optimization in Finite Dimensions I', Springer, Berlin (1970).

Weidmann, J. 'Lineare Operatoren in Hilberträumen', Teubner, Stuttgart (1976).

Bernhard K. Flury

A HIERARCHY OF RELATIONSHIPS BETWEEN COVARIANCE MATRICES

1. INTRODUCTION

In multivariate methods involving several populations, such as discriminant analysis or MANOVA, equality of all covariance matrices is a frequent assumption. If a test for equality of the covariance matrices suggests that this assumption does not hold, the usual reaction is to estimate the covariance matrices individually in each group. For k populations and p variables this means that the number of parameters estimated increases by (k-1)p(p-1)/2, which is quadratic in p. In many practical applications (as in the example given in section 4), this is not satisfactory, for two reasons: First, the k covariance matrices, although not being identical, may exhibit some common structure. Second, in parametric model fitting, the "principle of parsimony" (Dempster, 1972, p. 157) suggests that parameters should be introduced sparingly and only when the data indicate that they are needed.

Dempster (1972) applied the principle of parsimony to the estimation of a single covariance matrix by setting selected elements of its inverse equal to zero, a technique called "covariance selection". More popular methods of saving parameters in the case of a single population are usually summarized under the label "patterned covariance matrices", and the interested reader is referred to the recent review by Szatrowski (1985) and references therein.

Modelling several covariance matrices simultaneously under constraints on the parameter space has received rather little attention in the statistical literature, except for the model of proportionality. Proportional covariance matrices in the two-sample case and for normal populations have been studied by Federer (1951), Khatri (1967), Pillai et al. (1969), Kim (1971), Rao (1983), and Guttman et al. (1985). More recently, the case of $k \geq 2$ groups has been investigated by Flury (1986a) and Eriksen (1986). Dargahi-Noubary (1981) and Owen (1984) used proportionality of co-

A. K. Gupta (ed.), Advances in Multivariate Statistical Analysis, 31–43.

variance matrices in the context of nonlinear discrimination.

A more general type of relationship between several covariance matrices is given by the common principal component (CPC) model, which is based on the assumption that the characteristic vectors (but not necessarily the characteristic roots) of k covariance matrices are identical (Flury, 1984). A further generalization is the partial CPC model (Flury, 1987), in which only q out of p characteristic vectors are common to all groups.

The following section is to define these models mathematically and to establish a hierarchy among them. In section 3, a decomposition of the log-likelihood ratio statistic for equality of k covariance matrices is proposed. This decomposition is useful to select an appropriate model for given data, as will be illustrated in section 4.

2. A HIERARCHY OF MODELS

We are now going to define five levels of similarity between k covariance matrices $\Psi_1, \ldots, \Psi_k$ of dimension $p \times p$. It will always be assumed in the sequel that all Ψ_i are positive definite and symmetric.

Level 1: Equality of all Ψ_i.

The number of parameters to be estimated is $p(p+1)/2$.

Level 2: Proportionality of all Ψ_i, that is,

$$\Psi_i = \rho_i \Psi_1 \quad (i = 2, \ldots, k)$$

for some positive constants $\rho_2, \ldots, \rho_k$. (2.1)

The number of parameters is $p(p+1)/2 + k - 1$.

Level 3: The common principal component (CPC) model

$$\Psi_i = \beta \Lambda_i \beta' \quad (i = 1, \ldots, k) \tag{2.2}$$

where β is an orthogonal $p \times p$-matrix, and

$$\Lambda_i = \mathrm{diag}(\lambda_{i1}, \ldots, \lambda_{ip}). \tag{2.3}$$

The number of parameters on level 3 is $p(p-1)/2$ (for the

orthogonal matrix β) plus kp (for the diagonal matrices Λ_i). The CPC model can also be written as

$$\Psi_i = \lambda_{i1}\beta_1\beta_1' + \lambda_{i2}\beta_2\beta_2' + \ldots + \lambda_{ip}\beta_p\beta_p' \quad (i=1,\ldots,k), \tag{2.4}$$

where the β_j are the columns of β.

The representation (2.4) of the Ψ_i suggests a further modification, which has mainly been motivated by practical examples. Frequently in applications of principal component analysis the investigator is mostly interested in the first components and discards the last ones, provided their variances are relatively small. Similarly, one may wish to estimate only a few (say q) common components, the remaining p-q ones being possibly different from group to group. An appropriate model could then be defined as follows.

<u>Level 4:</u> The partial CPC model.

For a fixed integer $q < p-1$, let

$$\Psi_i = \lambda_{i1}\beta_1\beta_1' + \ldots + \lambda_{iq}\beta_q\beta_q' + \lambda_{i,q+1}\beta_{q+1}^{(i)}\beta_{q+1}^{(i)'} + \ldots + \lambda_{ip}\beta_p^{(i)}\beta_p^{(i)'}, \quad (i=1,\ldots,k), \tag{2.5}$$

where β_1 to β_q are the common characteristic vectors of all Ψ_i, and $\beta_{q+1}^{(i)}$ to $\beta_p^{(i)}$ are specific to each group.

It is tacitly assumed in (2.5) that the characteristic vectors are ordered such that the common ones are labeled 1 to q. If we define the orthogonal matrices $\beta^{(i)}$ as

$$\beta^{(i)} := (\beta_1, \beta_2, \ldots, \beta_q, \beta_{q+1}^{(i)}, \ldots, \beta_p^{(i)}), \tag{2.6}$$

then (2.5) can also be written as

$$\Psi_i = \beta^{(i)}\Lambda_i\beta^{(i)'}, \quad \Lambda_i = \mathrm{diag}(\lambda_{i1}, \ldots, \lambda_{ip}), \tag{2.7}$$

but (2.5) exhibits the basic idea underlying the partial CPC model somewhat more clearly.

Two further remarks are in order concerning this model. First, by the orthogonality of all $\beta^{(i)}$, the partial CPC model with $q = p-1$ components implies the ordinary CPC model of level 3. Second, there is not just one partial CPC model, but a whole family, some of which are nested hierarchically. Within the convention that the common components be labeled 1 through q, it is clear that the model with q common components implies the model with q-1 common components.

The number of parameters in the partial CPC model is as follows: kp parameters for the diagonal matrices Λ_i, $p(p-1)/2 - (p-q)(p-q-1)/2$ parameters for the common characteristic vectors β_1 to β_q, and $k(p-q)(p-q-1)/2$ parameters for the specific vectors $\beta_{q+1}^{(i)}$ to $\beta_p^{(i)}$. The total number of parameters is thus $p(p-1)/2 + kp + (k-1)(p-q)(p-q-1)/2$.

As stated above, if we set $q = p-1$ or $q = p$, then the partial CPC model coincides with the ordinary CPC model. The other extreme, namely $q = 0$, leads to

Level 5: $\Psi_1, \ldots, \Psi_k$ are arbitrary (positive definite) covariance matrices.

The number of parameters is $kp(p+1)/2$ in this case.

There is a modification of the partial CPC model, called the common space model (Flury, 1987), which ranks in the hierarchy between levels 4 and 5. This model can be expressed as follows: Let

$$\Psi_i = \beta^{(i)} \Lambda_i \beta^{(i)'} \qquad (2.8)$$

denote the spectral decomposition of Ψ_i, where $\Lambda_i = \mathrm{diag}(\lambda_{i1}, \ldots, \lambda_{ip})$, and the orthogonal matrices $\beta^{(i)}$ are partitioned as

$$\beta^{(i)} = (\alpha_1^{(i)}, \alpha_2^{(i)}) . \qquad (2.9)$$

Here, $\alpha_1^{(i)}$ has dimension $p \times q$, and $\alpha_2^{(i)}$ has dimension $p \times (p-q)$, that is, the columns of $\alpha_1^{(i)}$ and $\alpha_2^{(i)}$ contain q and p-q characteristic vectors, respectively. The hypothesis of common space spanned by q principal components in each group is then defined by the constraints

$$\alpha_1^{(i)'} \alpha_2^{(m)} = 0 \quad \text{for all } i, m \leq k . \tag{2.10}$$

Another way of putting this hypothesis is:

> There exist orthogonal matrices $H_1^{(i)}$ of dimension $q \times q$, and orthogonal matrices $H_2^{(i)}$ of dimension $(p-q) \times (p-q)$, $i = 2, \ldots, k$, such that $\alpha_1^{(i)} = \alpha_1^{(1)} H_1^{(i)}$ and $\alpha_2^{(i)} = \alpha_2^{(1)} H_2^{(i)}$. (2.11)

This expresses the fact that in the common space model the first q (or last p-q) characteristic vectors in group i span the same subspace as those in group 1.

Common space analysis is an alternative to a method proposed by Krzanowski (1979, 1982). Krzanowski's method consists essentially of a descriptive comparison of the subspaces spanned by the first q principal components in each group, the principal component transformation being computed individually in each sample. For a detailed comparison of the two methods, see the discussion in sections 3 and 4 of Flury (1987).

3. A DECOMPOSITION OF THE LOG-LIKELIHOOD RATIO STATISTIC FOR EQUALITY OF k COVARIANCE MATRICES

In this section we are going to propose a decomposition of the log-likelihood ratio statistic for equality of several covariance matrices into components due to non-proportionality, inequality of the principal components across groups, etc. This decomposition parallels the "analysis of deviance" in generalized linear models; see, for instance, McCullagh and Nelder (1983, Section 2.3). Another close analog is the fitting of regression equations in several groups, as has been noticed by Krzanowski (1984, p. 166): One may compare, for instance, regression equations fitted separately in each group to equations obtained under constraints of parallelism or coincidence of regression lines across groups. Yet another analog is model selection in the analysis of multivariate contingency tables by the log-linear model; see, for instance, Fienberg (1977, ch. 5).

Let now $S_1,\ldots,S_k$ denote stochastically independent sample covariance matrices from normal samples of size $N_i = n_i + 1 \geq p$, such that $n_i S_i$ is distributed as central Wishart on n_i degrees of freedom and parameter matrix Ψ_i. It is well known that the log-likelihood ratio statistic for equality of all Ψ_i versus the alternative of unrelatedness is

$$X^2_{total} = \sum_{i=1}^{k} n_i \log \frac{\det S}{\det S_i} , \tag{3.1}$$

where $S = (n_1 S_1 + \ldots + n_k S_k)/(n_1 + \ldots + n_k)$ is the pooled sample covariance matrix. Collecting the relevant results on the maximum of the likelihood function under the various models (Flury, 1984, 1986a, 1987), it is seen that the log-likelihood ratio criterion for a model A (null hypothesis) versus a hierarchically lower model B (alternative hypothesis) always has the form

$$X^2(A\,|\,B) = \sum_{i=1}^{k} n_i \log \frac{\det \hat{\Psi}_i}{\det \tilde{\Psi}_i}$$

$$= \sum_{i=1}^{k} n_i \sum_{j=1}^{p} (\log \hat{\lambda}_{ij} - \log \tilde{\lambda}_{ij}) , \tag{3.2}$$

where the MLE's marked by "^" refer to model A, those marked "~" refer to model B, and λ_{i1} to λ_{ip} are the characteristic roots of Ψ_i. The asymptotic null distribution of all statistics (3.2) is chi square on a number of degrees of freedom that corresponds to the difference in number of parameters between the two models. Formula (3.1) is a special case of (3.2).

Going step by step through the hierarchy of models, we can now use (3.2) to decompose X^2_{total} as follows:

$$
\begin{aligned}
X^2_{total} = {} & X^2 \text{ (inequality of proportionality constants} \mid \text{proportionality)} \\
& + X^2 \text{ (deviation from proportionality} \mid \text{CPC)} \\
& + X^2 \text{ (non-equality of the last p-q components} \mid \text{CPC(q))} \\
& + X^2 \text{ (non-equality of the first q components} \mid \text{CS(q))} \\
& + X^2 \text{ (non-equality of subspaces spanned by q components) .}
\end{aligned} \qquad (3.3)
$$

Here, CPC(q) refers to the partial CPC model with q common components, and CS(q) refers to the common space model with subspaces of dimension q and p-q, respectively. Of course, the integer q must be the same in both CPC(q) and CS(q), because otherwise the two models would not necessarily be in hierarchical order. Instead of using the common space model, one may also establish a decomposition into several hierarchically ordered partial CPC models.

The decomposition (3.3) is also summarized in table I.

Table I

A decomposition of the log-likelihood ratio statistic for equality of k covariance matrices of dimension p x p. The log-likelihood ratio statistics associated with each line have the form (3.2).

model A (higher)	model B (lower)	degrees of freedom
equality	proportionality	k-1
proportionality	CPC	(p-1)(k-1)
CPC	CPC(q) (1 ≦ q ≦ p-2)	$\frac{1}{2}$ (k-1)(p-q)(p-q-1)
CPC(q)	CS(q)	$\frac{1}{2}$ (k-1)q(q-1)
CS(q)	arbitrary covariance matrices	(k-1)q(p-q)

Two cautionary remarks are in order regarding this decomposition. First, formal hypotheses testing may not be appropriate if we let the data determine what model to fit. In the terminology of Selvin and Stuart (1966), this process of model fitting is a "fishing trip", and since it is the big fish that get caught in the net it is not reasonable to test them for average size. The various "partial chi squares" should therefore be used rather descriptively, and we will compare their relative magnitude by dividing them by the associated number of degrees of freedom. Second, the "partial chi squares" may not be independent, even under the null hypothesis of equality of all Ψ_i. Take, for instance, three nested models A, B and C such that A implies B and B implies C. We know that

$$X^2(A \mid C) = X^2(A \mid B) + X^2(B \mid C) \tag{3.4}$$

and that the degrees of freedom add as well. Suppose that model A is correct. Then it is tempting to assume that the two partial chi squares on the right-hand side of (3.4) should be independent. Unfortunately no such statement has been established yet. In fact, it is not very difficult to construct examples of stochastically dependent chi square variables where the sum behaves as if they were independent - see Flury (1986b) for details.

4. APPLICATION

Airoldi and Hoffmann (1984) took various skull measurements on two species of voles (small rodents): Microtus californicus and Microtus ochrogaster. The animals were further grouped by sex. We consider here the three variables (1) LENGTH, (2) WIDTH, and (3) HEIGHT of the skulls. Since there is considerable age variation in the data, we followed the usual practice of transforming the data logarithmically, which can be justified by the relationship of log-measurements to models of growth - see, e.g., Morrison (1976, p. 295). The raw data were kindly provided by J.P. Airoldi of the University of Berne.

Table II displays the four sample covariance matrices and associated degrees of freedom. Table III gives the decomposition of X^2_{total} as proposed in formula (3.3) and table

I. Since the dimension is $p = 3$, there is no common space model to be considered. (For $p < 4$, the common space model always coincides with a partial CPC model). The CPC(1) model in table III is such that the single common component is the one associated with the largest characteristic root in each group.

Table II

Sample covariance matrices in four groups of voles, multiplied by 10^4. Variables are as described in the text.

M. californicus, male ($n_1 = 172$)

$$\underset{\sim}{S}_1 = \begin{bmatrix} 112.01 & 106.64 & 52.97 \\ 106.64 & 108.13 & 54.75 \\ 52.97 & 54.75 & 33.86 \end{bmatrix}$$

M. californicus, female ($n_2 = 140$)

$$\underset{\sim}{S}_2 = \begin{bmatrix} 86.08 & 81.66 & 40.24 \\ 81.66 & 85.54 & 42.08 \\ 40.24 & 42.08 & 26.66 \end{bmatrix}$$

M. ochrogaster, male ($n_3 = 87$)

$$\underset{\sim}{S}_3 = \begin{bmatrix} 65.40 & 60.23 & 24.69 \\ 60.23 & 62.27 & 23.47 \\ 24.69 & 23.47 & 16.33 \end{bmatrix}$$

M. ochrogaster, female ($n_4 = 75$)

$$\underset{\sim}{S}_4 = \begin{bmatrix} 88.66 & 79.11 & 41.32 \\ 79.11 & 80.57 & 38.61 \\ 41.32 & 38.81 & 23.97 \end{bmatrix}$$

Table III

Decomposition of X^2_{total} in the vole example

Models: higher	Models: lower	X^2	df	X^2/df
equality	proportionality	.71	3	.24
proportionality	CPC	37.75	6	6.29
CPC	CPC(1)	8.51	3	2.84
CPC (1)	arbitrary cov. mat.	6.98	6	1.16
equality	arbitrary cov. mat.	53.95	18	

From table III we see that equality of all covariance matrices can be clearly rejected: X^2_{total} = 53.95 on 18 degrees of freedom . It appears also that the CPC(1) model fits well, while the fit of the ordinary CPC model is questionable. The large value of X^2_{total} is mostly due to the part for non-proportionality, given the CPC model. On the other hand, the part due to inequality, given proportionality, is small. This indicates that once we consider the covariance matrices as proportional, we would have to accept equality as well. Of couse one has to be careful in interpreting the "partial chi squares": If the hierarchically lower model (in this case, proportionality) is wrong, then the distribution of the log-likelihood ratio statistic itself may no longer be chi square, since the model is misspecified.

From table III we may decide to fit a partial CPC model. Table IV displays the maximum likelihood estimates of the CPC(1) model. The four orthogonal matrices $\hat{\beta}^{(i)}$ have, by their construction, an identical first column. (Note: MLE's were computed to four exact decimal digits, but only three are displayed in the table).

Table IV

Partial CPC's in four groups of voles

---

a) MLE's of characteristic vectors and roots

M. californicus, male

$$\hat{\beta}^{(1)} = \begin{pmatrix} .674 & -.519 & -.526 \\ .660 & .103 & .744 \\ .332 & .849 & -.412 \end{pmatrix}$$

$$\hat{\lambda}_{1j} = 244.26 \quad 7.22 \quad 2.52$$

M. californicus, female

$$\hat{\beta}^{(2)} = \begin{pmatrix} .674 & -.551 & -.492 \\ .660 & .150 & .736 \\ .332 & .821 & -.465 \end{pmatrix}$$

$$\hat{\lambda}_{2j} = 188.40 \quad 6.49 \quad 3.40$$

M. ochrogaster, male

$$\hat{\beta}^{(3)} = \begin{pmatrix} .674 & -.156 & -.722 \\ .660 & -.311 & .683 \\ .332 & .937 & .107 \end{pmatrix}$$

$$\hat{\lambda}_{3j} = 133.55 \quad 6.91 \quad 3.54$$

M. ochrogaster, female

$$\hat{\beta}^{(4)} = \begin{pmatrix} .674 & -.370 & -.640 \\ .660 & -.087 & .746 \\ .332 & .925 & -.186 \end{pmatrix}$$

$$\hat{\lambda}_{4j} = 183.87 \quad 3.82 \quad 5.51$$

b) MLE's of population covariance matrices

M. californicus, male

$$\hat{\Psi}_1 = \begin{bmatrix} 113.57 & 107.32 & 51.93 \\ 107.32 & 107.97 & 53.32 \\ 51.93 & 53.32 & 32.46 \end{bmatrix}$$

M. californicus, female

$$\hat{\Psi}_2 = \begin{bmatrix} 88.35 & 82.07 & 39.93 \\ 82.07 & 84.12 & 40.87 \\ 39.93 & 40.87 & 25.81 \end{bmatrix}$$

M. ochrogaster, male

$$\hat{\Psi}_3 = \begin{bmatrix} 62.67 & 58.01 & 28.55 \\ 58.01 & 60.55 & 27.47 \\ 28.55 & 27.47 & 20.79 \end{bmatrix}$$

M. ochrogaster, female

$$\hat{\Psi}_4 = \begin{bmatrix} 86.28 & 79.31 & 40.42 \\ 79.31 & 83.26 & 39.17 \\ 40.42 & 39.17 & 23.66 \end{bmatrix}$$

The estimates $\hat{\Psi}_i$ under the partial CPC model agree closely with the matrices S_i of table II, but they have one exactly identical characteristic vector.

DEDICATION

I wish to dedicate this work to the late K.C.S. Pillai, whom I admired as one of the greatest multivariate statisticians, and whose advice enabled me to do research into problems of relationship between covariance matrices.

Department of Statistics
University of Berne
Sidlerstrasse 5
CH-3012 Berne
Switzerland

REFERENCES

AIROLDI, J.P., and HOFFMANN, R.S. (1984): 'Age variation in voles (Microtus californicus, Microtus ochrogaster) and its significance for systematic studies'. Occasional Papers of the Museum of Natural History. The University of Kansas, Lawrence, No. 111, pp. 1-45.

DARGAHI-NOUBARY, G.R. (1981): 'An application of discrimination when covariance matrices are proportional .

Australian Journal of Statistics, 23, 38-44.

DEMPSTER, A.P. (1972): 'Covariance selection'. Biometrics, 28, 157-175.

ERIKSEN, P.S. (1986): 'Proportionality of covariance matrices'. The Annals of Statistics, 14, to appear.

FEDERER, W.T. (1951): 'Testing proportionality of covariance matrices'. Annals of Mathematical Statistics, 22, 102-106.

FIENBERG, S.E. (1977): The Analysis of Cross-Classified Categorical Data. The MIT Press, Cambridge, MA.

FLURY, B. (1984): 'Common principal components in k groups'. Journal of the American Statistical Association, 79, 892-898.

FLURY, B. (1986a): 'Proportionality of k covariance matrices'. Statistics and Probability Letters, 4, 29-33.

FLURY, B. (1986b): 'On sums of random variables and independence'. The American Statistician, 40, 214-215.

FLURY, B. (1987): 'Two generalizations of the common principal component model'. Biometrika, 74, to appear.

GUTTMAN, I., KIM, D.Y., and OLKIN, I. (1985): 'Statistical inference for constants of proportionality'. In: Multivariate Analysis VI, ed. P.R. Krishnaiah, North Holland, New York, pp. 257-280.

KHATRI, C.G. (1967): 'Some distribution problems connected with the characteristic roots of $S_1S_2^{-1}$'. Annals of Mathematical Statistics, 38, 944-948.

KIM, D.Y. (1971): 'Statistical inference for constants of proportionality between covariance matrices'. Technical Report No. 59, Stanford University, Department of Statistics.

KRZANOWSKI, W.J. (1979): 'Between-groups comparison of principal components'. Journal of the American Statistical Association, 74, 703-707. (Correction note: 1981, 76, 1022).

KRZANOWSKI, W.J. (1982): 'Between-group comparison of principal components - some sampling results'. Journal of Statistical Computation and Simulation, 15, 141-154.

KRZANOWSKI, W.J. (1984): 'Principal component analysis in the presence of group structure'. Applied Statistics, 33, 164-168.

McCULLAGH, P., and NELDER, J.A. (1983): Generalized Linear Models. Chapman and Hall, London.

MORRISON, D.F. (1976, 2nd ed.): Multivariate Statistical Methods. McGraw-Hill, New York.

OWEN, A. (1984): 'A neighbourhood-based classifier for LANDSAT data'. The Canadian Journal of Statistics, 12, 191-200.

PILLAI, K.C.S., AL-ANI, S., and JOURIS, G.M. (1969): 'On the distribution of the ratios of the roots of a covariance matrix and Wilks' criterion for tests of three hypotheses'. Annals of Mathematical Statistics, 40, 2033-2040.

RAO, C.R. (1983): 'Likelihood ratio tests for relationships between covariance matrices'. In: Studies in Econommetrics, Time Series and Multivariate Statistics, eds. S. Karlin, T. Amemiya and L.A. Goodman, Academic Press, New York, pp. 529-543.

SELVIN, H.C., and STUART, A. (1966): 'Data-dredging procedures in survey analysis'. The American Statistician, 20, no. 3, 20-23.

SZATROWSKI, T.H. (1985): 'Patterned covariances'. In: Encyclopedia of Statistical Sciences, vol. 6, eds. S. Kotz and N.L. Johnson, Wiley, New York, pp. 638-641.

Y. Fujikoshi, P.R. Krishnaiah and J. Schmidhammer

EFFECT OF ADDITIONAL VARIABLES IN PRINCIPAL COMPONENT ANALYSIS, DISCRIMINANT ANALYSIS AND CANONICAL CORRELATION ANALYSIS

1. INTRODUCTION

In a number of situations, it is of interest to find out whether the addition of a new set of variables gives additional information for inference. For example, in the area of principal component analysis, it is of interest to find out whether the new variables contribute to explanation of the variation among experimental units. In the area of multi-group discriminant analysis, it is important to find out whether the addition of new variables contributes to the discrimination between the groups. Similarly, in the area of canonical correlation analysis, it is of interest to find out as to whether the addition of variables to one or both sets of variables contributes to the degree of association between the two sets of variables.

Rao (1966) considered the effect of additional variables on the efficiency of estimates and the power of the test under multivariate regression model. Recently, Wijsman (1984) derived asymptotic distribution of the increase in the largest sample canonical correlation when some variables are added. In Section 3 of this paper, we first derive asymptotic distributions of changes in functions of the eigenvalues of the sample covariance matrix. Asymptotic distributions of changes in functions of the eigenvalues of the multivariate analysis of variance (MANOVA) matrix when some variables are added are derived in Section 4. In Section 5, we derive asymptotic distributions of changes in certain functions of the sample canonical correlations when new variables are added to one or both sets of original variables. The above results are derived under the assumption that the underlying distribution is multivariate normal. Further results are given in Section 6.

A. K. Gupta (ed.), Advances in Multivariate Statistical Analysis, 45–61.

2. PRELIMINARIES

In this section, we state the three lemmas which are needed in the sequel.

LEMMA 2.1 [Cramer (1946, p.366)]. Let $\underset{\sim}{x}_n$ be a p-component random vector and $\underset{\sim}{\mu}$: p×1 be a fixed vector. Assume $\sqrt{n}(\underset{\sim}{x}_n - \underset{\sim}{\mu})$ converges to $N(\underset{\sim}{0},\Sigma)$ in law. Let $f(\underset{\sim}{x})$ be continuously differentiable in a neighborhood of $\underset{\sim}{\mu}$ and let $\underset{\sim}{\xi} = \partial f(\underset{\sim}{x})/\partial \underset{\sim}{x}|_{\underset{\sim}{x}=\underset{\sim}{\mu}}$. Then the limiting distribution is the same as the one of $\sqrt{n}\underset{\sim}{\xi}'(\underset{\sim}{x}_n - \underset{\sim}{\mu})$, which is $N(0,\underset{\sim}{\xi}'\Sigma\underset{\sim}{\xi})$.

LEMMA 2.2 Let nS be distributed as a Wishart distribution $W_p(\Sigma,n)$ and B be distributed as a noncentral Wishart distribution $W_p(\Sigma,q:\Omega)$. Assume $\Omega = O(n) = n\Theta$. Then the limiting distributions of $V = \sqrt{n}(S - \Sigma)$ and $U = \sqrt{n}(\frac{1}{n}B - \Theta)$ are multivariate normal. Further the limiting distributions of tr CV and tr CU are $N(0,\sigma_1^2)$ and $N(0,\sigma_2^2)$ respectively, where C is a symmetric matrix of order p×p, $\sigma_1^2 = 2\,\mathrm{tr}(C\Sigma)^2$ and $\sigma_2^2 = 4\,\mathrm{tr}\,C^2\Theta$.

This lemma is well known and is proved by considering the characteristic functions of V and U.

LEMMA 2.3. Let $S_{1,n}$ and $S_{2,n}$ be sequences of symmetric matrices of order p×p such that the limiting distributions of $V_{1,n} = \sqrt{n}(S_{1,n} - \Lambda)$ and $V_{2,n} = \sqrt{n}(S_{2,n} - I_p)$ are multivariate normal, where $\Lambda = \mathrm{diag}(\lambda_1, \ldots, \lambda_p)$, $\lambda_1 \geq \ldots \geq \lambda_p$ and I_p is the identity matrix of order p×p. Let $\ell_1 \geq \ldots \geq \ell_p$ and $d_1 \geq \ldots \geq d_p$ be the eigenvalues of $S_{1,n}$ and $S_{1,n}S_{2,n}^{-1}$, respectively. Suppose that the α-th largest eigenvalue λ_α of Λ is simple. Then the limiting distributions of $\sqrt{n}(\ell_\alpha - \lambda_\alpha)$ and $\sqrt{n}(d_\alpha - \lambda_\alpha)$ are the same as the ones of $(V_{1,n})_{\alpha\alpha}$ and $(V_{1,n})_{\alpha\alpha}$ respectively, where $(A)_{\alpha\beta}$ denotes the (α,β)-th element of matrix A.

This lemma has been essentially proved in the papers of Hsu (1941a,b) and Anderson (1963) who treated the general case of multiple roots.

3. PRINCIPAL COMPONENT ANALYSIS

In the area of principal component analysis, it is of interest to find out a small number of principal components which would adequately explain the variation among experimental units. In the population, the variance of i-th important principal component is the i-th largest eigenvalue of the population covariance matrix. If these eigenvalues are small, then the corresponding principal components are unimportant. In a number of situations, it is of interest to find out as to whether the addition of some variables will increase the variances of the first few important principal components. Similarly, it is of interest to find out whether there is significant increase in the ratio of the i-th largest eigenvalue to the trace of the covariance matrix if some variables are added. So, we will derive asymptotic distributions of increases in certain functions of the eigenvalues of the sample covariance matrix when a new set of variables is added.

Let $\underset{\sim}{x}$: p×1 be distributed as $N(0,\underset{\sim}{\Sigma})$. We partition $\underset{\sim}{x} = (\underset{\sim}{x}_1', \underset{\sim}{x}_2')$, $\underset{\sim}{x}_1$: $p_1 \times 1$ and

$$\Sigma = \begin{bmatrix} \Sigma_{11} & \Sigma_{12} \\ \Sigma_{21} & \Sigma_{22} \end{bmatrix}, \quad \Sigma_{11}: p_1 \times p_1. \tag{3.1}$$

Suppose the vector $\underset{\sim}{x}_1$ is augmented to $\underset{\sim}{x}$. Let λ_α and $\tilde{\lambda}_\alpha$ be the α-th largest roots of Σ_{11} and Σ, respectively. Then

$$\tilde{\lambda}_\alpha \geq \lambda_\alpha, \quad (\alpha = 1,\ldots,p_1)$$

which follows from the Poincaré separation theorem (see,

e.g., Rao (1973, p.64)).

We are interested in the increases $\delta_\alpha = \tilde{\lambda}_\alpha - \lambda_\alpha$. Let S be the sample covariance matrix based on a sample of size $N = n + 1$. We partition S as in (3.1)

$$S = \begin{bmatrix} S_{11} & S_{12} \\ S_{21} & S_{22} \end{bmatrix}. \tag{3.2}$$

The sample quantities corresponding to δ_α are

$$d_\alpha = \tilde{\ell}_\alpha - \ell_\alpha, \quad (\alpha = 1,\ldots,p_1)$$

where ℓ_α and $\tilde{\ell}_\alpha$ are the α-th largest roots of S_{11} and S, respectively. We consider the distribution of

$$J = \sqrt{n}\{f(d_1,\ldots,d_{p_1}) - f(\delta_1,\ldots,\delta_{p_1})\}. \tag{3.3}$$

We assume A1: $f(\underset{\sim}{d})$ is continuously differentiable in a neighborhood of $\underset{\sim}{d} = \underset{\sim}{\delta}$ where $\underset{\sim}{d} = (d_1,\ldots,d_{p_1})'$ and $\underset{\sim}{\delta} = (\delta_1,\ldots,\delta_{p_1})'$. Let

$$\underset{\sim}{c} = (c_1,\ldots,c_{p_1})' = \frac{\partial}{\partial \underset{\sim}{d}} f(\underset{\sim}{d})\Big|_{\underset{\sim}{d}=\underset{\sim}{\delta}}. \tag{3.4}$$

Let H_{11}: $p_1 \times p_1$ be an orthogonal matrix such that

$$H_{11}\Sigma_{11}H_{11}' = \Lambda_{11} = \mathrm{diag}(\lambda_1,\ldots,\lambda_{p_1}).$$

Since ℓ_α and $\tilde{\ell}_\alpha$ are invariant under the transformation

$$\underset{\sim}{x} \to \begin{bmatrix} H_{11} & 0 \\ 0 & 1 \end{bmatrix} \underset{\sim}{x}$$

we may assume

$$\Sigma = \Lambda$$
$$= \begin{bmatrix} \Lambda_{11} & \Lambda_{12} \\ \Lambda_{21} & \Lambda_{22} \end{bmatrix} \tag{3.5}$$

where $\Lambda_{12} = H_{11}\Sigma_{12}$, $\Lambda_{21} = \Sigma_{21}H'_{11}$ and $\Lambda_{22} = \Sigma_{22}$. Let

$$S = \Lambda + \frac{1}{\sqrt{n}} V \tag{3.6}$$

and

$$\tilde{S} = \Gamma' S \Gamma$$
$$= \tilde{\Lambda} + \frac{1}{\sqrt{n}} \Gamma' V \Gamma \tag{3.7}$$

where Γ is an orthogonal matrix such that $\Gamma'\Lambda\Gamma = \tilde{\Lambda} = \mathrm{diag}(\tilde{\lambda}_1,\ldots,\tilde{\lambda}_p)$. Since ℓ_α and $\tilde{\ell}_\alpha$ are the α-th largest roots of $S_{11} = \Lambda_{11} + (1/\sqrt{n})V_{11}$ and $\tilde{S} = \tilde{\Lambda} + (1/\sqrt{n})\Gamma'V\Gamma$, we obtain by Lemma 2.3 that the asymptotic distribution of $\sqrt{n}(d_\alpha - \delta_\alpha)$ is the same as that of

$$g_\alpha = (\Gamma'V\Gamma)_{\alpha\alpha} - (V)_{\alpha\alpha} \tag{3.8}$$

if λ_α and $\tilde{\lambda}_\alpha$ are simple. Using Lemma 2.1 we obtain that the asymptotic distribution of J is the same as that of

$$\sum_{\alpha=1}^{p_1} c_\alpha g_\alpha = \mathrm{tr}\, AV \tag{3.9}$$

where

$$A = \Gamma D_c \Gamma' - D_c,\ D_c = \text{diag}(c_1, \ldots, c_{p_1}, 0, \ldots, 0). \qquad (3.10)$$

This implies the following.

THEOREM 3.1. Let nS be distributed as a Wishart distribution $W_p(\Sigma, n)$. Let $d_\alpha = \tilde{\ell}_\alpha - \ell_\alpha$ and $\delta_\alpha = \tilde{\lambda}_\alpha - \lambda_\alpha$, where ℓ_α, $\tilde{\ell}_\alpha$, λ_α and $\tilde{\lambda}_\alpha$ are the α-th largest roots of S_{11}, S, Σ_{11} and Σ. Assume a function $f(d_1, \ldots, d_{p_1})$ satisfies the assumption A1 and all the roots λ_α and $\tilde{\lambda}_\alpha$ $(\alpha = 1, \ldots, p_1)$ are simple. Then

$$\sqrt{n}\{f(d_1, \ldots, d_{p_1}) - f(\delta_1, \ldots, \delta_{p_1})\} \xrightarrow{D} N(0, \sigma^2)$$

as $n \to \infty$, where

$$\sigma^2 = 2\ \text{tr}(A\Lambda)^2$$

$$= 2 \sum_{\alpha=1}^{p_1} c_\alpha^2(\lambda_\alpha^2 + \tilde{\lambda}_\alpha^2) - 4 \sum_{\alpha,\beta=1}^{p_1} c_\alpha c_\beta \tilde{\lambda}_\alpha^2 \gamma_{\beta\alpha}^2$$

and $\Gamma = (\gamma_{\alpha\beta})$.

COROLLARY 3.1.1. When λ_α and $\tilde{\lambda}_\alpha$ are simple,

$$\sqrt{n}\{(\tilde{\ell}_\alpha - \ell_\alpha) - (\tilde{\lambda}_\alpha - \lambda_\alpha)\} \xrightarrow{D} N(0, \sigma^2)$$

as $n \to \infty$, where $\sigma^2 = 2(\lambda^2 + \tilde{\lambda}^2) - 4\tilde{\lambda}_\alpha^2 \gamma_{\alpha\alpha}^2$.

4. EFFECT OF ADDITIONAL VARIABLES IN DISCRIMINANT ANALYSIS

In the area of discriminant analysis, it is of interest to find out as to whether the addition of a new set of variables will make a significant contribution on the discriminant functions. This problem can be investigated by examining the increases due to the additional variables in certain functions of the eigenvalues of the MANOVA matrix. So, we will study the asymptotic distributions of the above increases in the sample.

Let W and B be independently distributed as a Wishart distribution $W_p(\Sigma,n)$ and a noncentral Wishart distribution $W_p(\Sigma,p:\Xi)$. We partition

$$W = \begin{bmatrix} W_{11} & W_{12} \\ W_{21} & W_{22} \end{bmatrix}, \qquad B = \begin{bmatrix} B_{11} & B_{12} \\ B_{21} & B_{22} \end{bmatrix},$$
$$\Sigma = \begin{bmatrix} \Sigma_{11} & \Sigma_{12} \\ \Sigma_{21} & \Sigma_{22} \end{bmatrix}, \qquad \Xi = \begin{bmatrix} \Xi_{11} & \Xi_{12} \\ \Xi_{21} & \Xi_{22} \end{bmatrix}, \tag{4.1}$$

with W_{11}: $p_1 \times p_1$, B_{11}: $p_1 \times p_1$, Σ_{11}: $p_1 \times p_1$ and Ξ_{11}: $p_1 \times p_1$. Let ℓ_α, $\tilde{\ell}_\alpha$, ω_α and $\tilde{\omega}_\alpha$ be the α-th largest roots of $B_{11}W_{11}^{-1}$, BW^{-1}, $\Xi_{11}\Sigma_{11}^{-1}$, BW^{-1}, $\Xi_{11}\Sigma_{11}^{-1}$ and $\Xi\Sigma^{-1}$, respectively. Then

$$d_\alpha = \tilde{\ell}_\alpha - \ell_\alpha \geq 0,$$
$$n\delta_\alpha = \tilde{\omega}_\alpha - \omega_\alpha \geq 0, \quad (\alpha = 1,\ldots,p_1) \tag{4.2}$$

which follows from the Poincaré separation theorem in the case of two matrices (also see Gabriel (1968)).

Let

$$T = HL$$
$$= \begin{bmatrix} H_{11} & 0 \\ 0 & I \end{bmatrix} \begin{bmatrix} L_{11} & 0 \\ L_{21} & L_{22} \end{bmatrix}, \tag{4.3}$$

where $L\Sigma L' = I$ and H_{11} is an orthogonal matrix such that $H_{11}L_{11}\Xi_{11}L_{11}'H_{11}' = \Omega_{11} = \mathrm{diag}(\omega_1,\ldots,\omega_{p_1})$. Since ℓ_α and $\tilde{\ell}_\alpha$ are invariant under the transformation $B \to TBT'$ and $W \to TWT'$, we may assume

$$\Sigma = I,$$

$$\Xi = \Omega = \begin{bmatrix} \Omega_{11} & \Omega_{12} \\ \Omega_{21} & \Omega_{22} \end{bmatrix} \tag{4.4}$$

with $\Omega_{11} = \mathrm{diag}(\omega_1, \ldots, \omega_{p_1})$, $\Omega_{12} = H_{11}L_{11}(\Xi_{11}L_{21}' + \Xi_{12}L_{22}')$, $\Omega_{21} = \Omega_{12}'$ and $\Omega_{22} = (L_{21}\Xi_{11} + L_{22}\Xi_{21})L_{21}' + (L_{21}\Xi_{12} + L_{22}\Xi_{22})L_{22}'$. We assume

$$\text{A2: } \Omega = O(n)$$

$$= n\Theta = n\begin{bmatrix} \Theta_{11} & \Theta_{12} \\ \Theta_{21} & \Theta_{22} \end{bmatrix} \tag{4.5}$$

where $\Theta_{11} = \mathrm{diag}(\theta_1,\ldots,\theta_{p_1})$. Let Γ be an orthogonal matrix such that

$$\Gamma'\Theta\Gamma = \tilde{\Theta} = \mathrm{diag}(\tilde{\theta}_1,\ldots,\tilde{\theta}_p) \tag{4.6}$$

with $\tilde{\theta}_1 \geq \ldots \geq \tilde{\theta}_p$. Then $n\tilde{\theta}_\alpha = \tilde{\omega}_\alpha$. Let

$$\frac{1}{n}B = \Theta + \frac{1}{\sqrt{n}}U,$$

$$\frac{1}{n}W = I + \frac{1}{\sqrt{n}}W. \tag{4.7}$$

Then it is easily seen that ℓ_α and $\tilde{\ell}_\alpha$ are the α-th largest roots of

$$|\Theta_{11} + \frac{1}{\sqrt{n}}U_{11} - \ell(I - \frac{1}{\sqrt{n}}V_{11})| = 0$$

and

$$|\tilde{\Theta} + \frac{1}{\sqrt{n}} \Gamma'U\Gamma - \tilde{\ell}(I + \frac{1}{\sqrt{n}} \Gamma'V\Gamma)| = 0$$

respectively, where U_{11}: $p_1 \times p_1$ and V_{11}: $p_1 \times p_2$ are the submatrices of U and V partitioned as in (4.1).

From Lemma 2.3, it is seen that the asymptotic distribution of $\sqrt{n}(d_\alpha - \delta_\alpha)$ is the same as that of

$$g_\alpha = (\Gamma'U\Gamma)_{\alpha\alpha} - \theta_\alpha(\Gamma'V\Gamma)_{\alpha\alpha} - (V)_{\alpha\alpha} + \theta_\alpha(V)_{\alpha\alpha} \qquad (4.8)$$

if ω_α and $\tilde{\omega}_\alpha$ are simple. Using Lemma 1, we obtain that the asymptomatic distribution of $\sqrt{n}\{f(d_1,\ldots,d_{p_1}) - f(\delta_1,\ldots,\delta_{p_1})\}$ is the same as that of

$$\sum_{\alpha=1}^{p_1} c_\alpha g_\alpha = \text{tr } A^{(1)}V + \text{tr } A^{(2)}U \qquad (4.9)$$

where $A^{(1)} = D_{c\theta} - D_{c\tilde{\theta}}\Gamma'$, $A^{(2)} = \Gamma D_c\Gamma' - D_c$, $D_c = \text{diag}(c_1,\ldots,c_{p_1}, 0,\ldots,0)$, $D_{c\theta} = \text{diag}(c_1\theta_1,\ldots,c_{p_1}, 0,\ldots,0)$, etc. This implies the following.

THEOREM 4.1. Let W and B be independently distributed as a Wishart distribution $W_p(I,n)$ and a noncentral Wishart distribution $W_p(\Sigma,q:\Xi)$, respectively. Let $d_\alpha = \tilde{\ell}_\alpha - \ell_\alpha$ and $n\delta_\alpha = \tilde{\omega}_\alpha - \omega_\alpha$, where ℓ_α, $\tilde{\ell}_\alpha$, ω_α and $\tilde{\omega}_\alpha$ are the α-th largest roots of $B_{11}W_{11}^{-1}$, BW^{-1}, $\Xi_{11}\Sigma_{11}^{-1}$ and $\Xi\Sigma^{-1}$, respectively. Assume that the assumptions A1 and A2 are satisfied, and ω_α and $\tilde{\omega}_\alpha$ $(\alpha = 1,\ldots,p_1)$ are simple. Then

$$\sqrt{n}\{f(d_1,\ldots,d_{p_1}) - f(\delta_1,\ldots,\delta_{p_1})\} \xrightarrow{D} N(0,\sigma^2)$$

as $n \to \infty$, where

$$\sigma^2 = 2\,\mathrm{tr}\{D_{c\theta} - \Gamma D_{c\tilde{\theta}}\Gamma'\}^2 + 4\,\mathrm{tr}\{\Gamma D_c \Gamma' - D_c\}^2\Theta$$

$$= 2\sum_{\alpha=1}^{p_1} c_\alpha^2\{\theta_\alpha(\theta_\alpha + 2) + \tilde{\theta}_\alpha(\tilde{\theta}_\alpha + 2)$$

$$- 4\sum_{\alpha,\beta=1}^{p_1} c_\alpha c_\beta \tilde{\theta}_\beta \gamma_{\alpha\beta}^2(\tilde{\theta}_\alpha + 2)$$

and $\Gamma = (\gamma_{\alpha\beta})$.

COROLLARY 4.1.1. When ω_α and $\tilde{\omega}_\alpha$ are simple

$$\sqrt{n}\,[(\tilde{\ell}_\alpha - \ell_\alpha) - (\tilde{\theta}_\alpha - \theta_\alpha)] \xrightarrow{D} N(0,\sigma^2)$$

as $n \to \infty$, where $\sigma^2 = 2\theta_\alpha(\theta_\alpha + 2) + 2\tilde{\theta}_\alpha(\tilde{\theta}_\alpha + 2) - 4\tilde{\theta}_\alpha(\theta_\alpha + 2)\gamma_{\alpha\alpha}^2$.

5. EFFECT OF ADDITIONAL VARIABLES ON CANONICAL VARIABLES

In this section, we study asymptotic distributions of certain statistics useful in studying the effect of additional variables on the canonical correlations.

Consider two sets of variables $\underset{\sim}{x}_1$: $p_1 \times 1$ and $\underset{\sim}{y}_1$: $q_1 \times 1$. We assume $p_1 \leq q_1$. Let $\rho_1 \geq \ldots \geq \rho_{p_1} \geq 0$ be the canonical correlations between $\underset{\sim}{x}_1$ and $\underset{\sim}{y}_1$. We shall augment the variates $\underset{\sim}{x}_1$ and $\underset{\sim}{y}_1$ to $\underset{\sim}{x}$: $p \times 1$ and $\underset{\sim}{y}$: $q \times 1$ by adding extra variates $\underset{\sim}{x}_2$: $p_2 \times 1$ and $\underset{\sim}{y}_2$: $q_2 \times 1$, respectively. We assume that $(\underset{\sim}{x}', \underset{\sim}{y}')$ is distributed as $N_{p+q}[\mu, \Sigma]$. We partition Σ as

$$\Sigma = \begin{bmatrix} \Sigma_{xx} & \Sigma_{xy} \\ \Sigma_{yx} & \Sigma_{yy} \end{bmatrix}, \quad \Sigma_{xx}: p\times p. \tag{5.1}$$

Let $\tilde{\rho}_\alpha$ be the α-th largest canonical correlation between $\underset{\sim}{x}$ and $\underset{\sim}{y}$. Then

$$\delta_\alpha = \tilde{\rho}_\alpha - \rho_\alpha \geq 0, \quad (\alpha = 1,\ldots,p_1), \tag{5.2}$$

which has been shown in Fujikoshi (1982). Let

$$S = \begin{bmatrix} S_{xx} & S_{xy} \\ S_{yx} & S_{yy} \end{bmatrix}$$

be the sample covariance matrix based on a sample of size $N = n + 1$. Let r_α and $\tilde{r}_\alpha$ be the α-th largest canonical correlations between $\underset{\sim}{x}_1$ and $\underset{\sim}{y}_1$ and between $\underset{\sim}{x}$ and $\underset{\sim}{y}$. We consider the asymptotic distribution of

$$\sqrt{n}\{f(d_1,\ldots,d_{p_1}) - f(\delta_1,\ldots,\delta_{p_1})\}$$

where $d_\alpha = \tilde{r}_\alpha - r_\alpha$. Let L_1 and L_2 be the lower triangular matrices such that

$$L_1\Sigma_{xx}L_1' = I_p, \quad L_2\Sigma_{yy}L_2' = I_q. \tag{5.3}$$

We partition

$$L_1 = \begin{bmatrix} L_{1\cdot 11} & 0 \\ L_{1\cdot 21} & L_{1\cdot 22} \end{bmatrix}, \quad L_2 = \begin{bmatrix} L_{2\cdot 11} & 0 \\ L_{2\cdot 21} & L_{2\cdot 22} \end{bmatrix}. \tag{5.4}$$

Let $H_{1\cdot 11}$: $p_1\times p_1$ and $H_{2\cdot 11}$: $p_2\times p_2$ be the orthogonal matrices chosen so that

$$H_{1\cdot 11}L_{1\cdot 11}\Sigma_{xy\cdot 11}L'_{2\cdot 11}H'_{2\cdot 11} = P_{11} = \begin{bmatrix} \rho_1 & & 0 \\ & \ddots & \\ & & \rho_{p_1} \end{bmatrix}, \tag{5.5}$$

where

$$\Sigma_{xy} = \begin{bmatrix} \Sigma_{xy\cdot 11} & \Sigma_{xy\cdot 12} \\ \Sigma_{xy\cdot 21} & \Sigma_{xy\cdot 22} \end{bmatrix}, \quad \Sigma_{xy\cdot 11}: p_1 \times q_1. \tag{5.6}$$

Then r_α and r_α are invariant under the transformation

$$\underset{\sim}{x} \to \begin{bmatrix} H_{1\cdot 11}L_{1\cdot 11} & 0 \\ L_{1\cdot 21} & L_{1\cdot 22} \end{bmatrix} \underset{\sim}{x},$$

$$\underset{\sim}{y} \to \begin{bmatrix} H_{2\cdot 11}L_{2\cdot 11} & 0 \\ L_{2\cdot 21} & L_{2\cdot 22} \end{bmatrix} \underset{\sim}{y}.$$

Therefore, we may assume

$$\Sigma_{xx} = I_p, \quad \Sigma_{yy} = I_q$$

$$\Sigma_{xy} = P = \begin{bmatrix} P_{11} & P_{12} \\ P_{21} & P_{22} \end{bmatrix} \tag{5.7}$$

with P_{11} in (5.5), $P_{12} = H_{1\cdot 11}L_{1\cdot 11}(\Sigma_{11}L'_{2\cdot 21} + \Sigma_{12}L'_{2\cdot 22})$, $P_{21} = (L_{1\cdot 21}\Sigma_{11} + L_{1\cdot 22}\Sigma_{21})L'_{2\cdot 11}H'_{2\cdot 11}$ and $P_{22} = (L_{1\cdot 21}\Sigma_{11} + L_{1\cdot 22}\Sigma_{21})L'_{2\cdot 21} + (L_{1\cdot 21}\Sigma_{12} + L_{1\cdot 22}\Sigma_{22})L'_{2\cdot 22}$. Let

$$S = \Sigma + \frac{1}{\sqrt{n}} V$$

$$= \begin{bmatrix} I_p & P \\ P' & I_q \end{bmatrix} + \frac{1}{\sqrt{n}} \begin{bmatrix} V_{xx} & V_{xy} \\ V_{yx} & V_{yy} \end{bmatrix}. \quad (5.8)$$

Then r_α is the α-th root of

$$|S_{xy\cdot 11} S^{-1}_{yy\cdot 11} S_{yx\cdot 11} - rS_{xx\cdot 11}| = 0$$

which is equivalent to

$$|P_{11}P'_{11} + \frac{1}{\sqrt{n}} Z_{11} - r^2(I_{p_1} + \frac{1}{\sqrt{n}} V_{xx\cdot 11})| = 0 \quad (5.9)$$

where

$$Z_{11} = \sqrt{n}\{(P_{11} + \frac{1}{\sqrt{n}} V_{xy\cdot 11})(I_{q_1} + \frac{1}{\sqrt{n}} V_{yy\cdot 11})^{-1}$$

$$(P'_{11} + \frac{1}{\sqrt{n}} V_{yx\cdot 11}) - P_{11}P'_{11}\}, \quad (5.10)$$

$S_{xy\cdot 11}$, $V_{xy\cdot 11}$, etc. denote the submatrices of S_{xy} and V_{xy} partitioned as in (5.6). Let Γ_1 and Γ_2 be the orthogonal matrices such that

$$\Gamma_1 P \Gamma'_2 = \tilde{P}: \ p\times q \quad (5.11)$$

where the (α,α) elements of $\tilde{P}$ and $\tilde{\rho}_\alpha$ and other elements are zero. Let

$$\tilde{S} = \begin{bmatrix} \Gamma_1 & 0 \\ 0 & \Gamma_2' \end{bmatrix} S \begin{bmatrix} \Gamma_1' & 0 \\ 0 & \Gamma_2' \end{bmatrix} = \begin{bmatrix} \Gamma_1 S_{xx} \Gamma_1' & \Gamma_1 S_{xy} \Gamma_2' \\ \Gamma_2 S_{yx} \Gamma_2' & \Gamma_2 S_{yy} \Gamma_2' \end{bmatrix}$$

$$= \begin{bmatrix} I_p & P \\ P' & I_q \end{bmatrix} + \frac{1}{\sqrt{n}} \begin{bmatrix} \Gamma_1 V_{xx} \Gamma_1' & \Gamma_1 V_{xy} \Gamma_2' \\ \Gamma_2 V_{yx} \Gamma_1' & \Gamma_2 V_{yy} \Gamma_2' \end{bmatrix}. \tag{5.12}$$

THEOREM 5.1. Let r_α and $\tilde{r}_\alpha$ be the α-th largest canonical correlations between $\underset{\sim}{x}_1$: $p_1 \times 1$ and $\underset{\sim}{y}_1$: $q_1 \times 1$ $(p_1 \le q_1)$ and between $\underset{\sim}{x}$: $p \times 1$ and $\underset{\sim}{y}$: $q \times 1$, based on a sample of size $N - n + 1$ from $N(\underset{\sim}{\mu}, \Sigma)$. Let ρ_α and $\tilde{\rho}_\alpha$ be the corresponding population quantities. Assume that a function $f(d_1, \ldots, d_{p_1})$ satisfies the assumption A1 and the canonical correlations ρ_α and $\tilde{\rho}_\alpha$ $(\alpha = 1, \ldots, p_1)$ are simple. Then

$$\sqrt{n}\{f(d_1, \ldots, d_{p_1}) - f(\delta_1, \ldots, \delta_{p_1})\} \xrightarrow{D} N(0, \sigma^2)$$

as $n \to \infty$, where $d_\alpha = \tilde{r}_\alpha - r_\alpha$, $\delta_\alpha = \tilde{\rho}_\alpha - \rho_\alpha$,

$$\sigma^2 = \frac{1}{2} \operatorname{tr} \left\{ \begin{bmatrix} A_{11} & A_{12} \\ A_{21} & A_{22} \end{bmatrix} \begin{bmatrix} I_p & P \\ P' & I_q \end{bmatrix} \right\}^2$$

$$= \sum_{\alpha,\beta=1}^{p_1} c_\alpha^2 \{(1 - \rho_\alpha^2)^2 + (1 - \tilde{\rho}_\alpha^2)^2\}$$

$$+ \sum_{\alpha,\beta=1}^{p_1} c_\alpha c_\beta (1 - \tilde{\rho}_\beta^2) \{\rho_\alpha \tilde{\rho}_\beta (\gamma_{1\cdot\beta\alpha}^2 + \gamma_{2\cdot\beta\alpha}^2) - 2\gamma_{1\cdot\beta\alpha}\gamma_{2\cdot\beta\alpha}\}$$

and $\Gamma_1 = (\gamma_{1\cdot\alpha\beta})$ and $\Gamma_2 = (\gamma_{2\cdot\alpha\beta})$.

COROLLARY 5.1.1. When ρ_α and $\tilde{\rho}_\alpha$ are simple,

$$\sqrt{n}\{(\tilde{r}_\alpha - r_\alpha) - (\tilde{\rho}_\alpha - \rho_\alpha)\} \xrightarrow{D} N(0, \sigma^2)$$

as $n \to \infty$, where $\sigma^2 = (1-\tilde{\rho}_\alpha^2)^2 + (1-\tilde{\rho}_\alpha^2)$
$[\rho_\alpha\tilde{\rho}_\alpha(\gamma_{1\cdot\alpha\alpha}^2 + \gamma_{2\cdot\alpha\alpha}^2) - 2\gamma_{1\cdot\alpha\alpha}\gamma_{2\cdot\alpha\alpha}]$.

Note: Wijsman (1984) proved the Corollary 5.1.1 in the case of $\alpha = 1$ and $q_2 = 0$ or $p_2 = 0$.

6. FURTHER RESULTS

Let d_α and δ_α be the increases in the α-th largest sample and population eigenvalues in the three cases: (i) principal component analysis, (ii) discriminant analysis, and (iii) canonical correlation analysis. Then in Theorems 3.1, 4.1, and 5.1, we have shown that

$$J = \sqrt{n}\{f(d_1,\ldots,d_{p_1}) - f(\delta_1,\ldots,\delta_{p_1})\} \xrightarrow{D} N(0,\sigma^2). \qquad (6.1)$$

The limiting variance σ^2 depends on unknown Σ for cases (i) and (iii) and unknown Σ and Θ for case (ii). In order to make the formula useful, we need the estimate $\hat{\sigma}^2$ obtained from σ^2 by replacing Σ by S for (i) and (iii), and by replacing Σ and Θ by (1/n)W and (1/n)B, respectively, for (ii). It is easy to see that under the same assumption as in each of the Theorems

$\hat{\sigma} \to \delta$ in probability.

Therefore, it follows from (6.1) that

$$\hat{\sigma}^{-1}J \xrightarrow{D} N(0,1). \qquad (6.2)$$

The formula is useful in constructing an approximate confidence interval for $f(\delta_1,\ldots,\delta_{p_1})$.

It is easy to extend the result for a single function J to the one for several functions. Let

$$J = \sqrt{n}\{f_\alpha(d_1,\ldots,d_{p_1}) - f_\alpha(\delta_1,\ldots,\delta_{p_1})\}$$

for $\alpha = 1,2,\ldots,k$. We assume that f_α's satisfy the assumption A1. Let

$$\underset{\sim}{c}_\alpha = (c_{1,\alpha},\ldots,c_{p_1,\alpha})' = \frac{\partial}{\partial \underset{\sim}{d}} f_\alpha(\underset{\sim}{d})\Big|_{\underset{\sim}{d} = \delta}.$$

Then we can prove that

$$J = (J_1,\ldots,J_k) \overset{D}{\to} N_k(\underset{\sim}{0},Q). \tag{6.3}$$

The limiting covariance matrix $Q = (q_{\alpha\beta})$ is given as follows:

Case (i):

$$q_{\alpha\beta} = 2\ \mathrm{tr}(A_\alpha \Lambda A_\beta \Lambda),$$

where A_α is defined from A in (3.10) by substituting $\underset{\sim}{c}$ into $\underset{\sim}{c}_\alpha$.

Case (ii)

$$q_{\alpha\beta} = 2\ \mathrm{tr}\ A_\alpha^{(1)} A_\beta^{(1)} + 4\ \mathrm{tr}\ A_\alpha^{(2)} A_\beta^{(2)} \Theta,$$

where $A_\alpha^{(1)}$ and $A_\alpha^{(2)}$ are defined from the $A^{(1)}$ and $A^{(2)}$ in (4.9) by substituting $\underset{\sim}{c}$ into $\underset{\sim}{c}_\alpha$.

Case (iii)

$$q_{\alpha\beta} = \frac{1}{4}\ \mathrm{tr}\ A_\alpha \begin{bmatrix} I_p & P \\ P' & I_q \end{bmatrix} A_\beta \begin{bmatrix} I_p & P \\ P' & I_q \end{bmatrix},$$

where A_α is defined from A in (5.14) by substituting $\underset{\sim}{c}$ into $\underset{\sim}{c}_\alpha$. Higher order terms of the joint distribution of $J_1,\ldots,J_k$ can be obtained by using perturbation technique.

Department of Mathematics and Statistics
University of Pittsburgh
Pittsburgh, PA 15260

REFERENCES

[1] ANDERSON, T.W. (1963). Asymptotic theory for principal component analysis. *Ann. Math. Statist.* **34**, 122-148.

[2] CRAMÉR, H. (1946). *Mathematical Methods of Statistics.* Princeton University Press.

[3] FUJIKOSHI, Y. (1982). A test for additional information in canonical correlation analysis. *Ann. Inst. Statist. Math.* **34**, 523-530.

[4] GABRIEL, K.R. (1968). Simultaneous test procedures in multivariate analysis of variance. *Biometrika* **55**, 484-504.

[5] HSU, P.L. (1941a). On the limiting distribution of roots of a determinantal equation. *J. London Math. Soc.* **16**, 183-194.

[6] HSU, P.L. (1941b). On the limiting distribution of the canonical correlations. *Biometrika* **32**, 38-45.

[7] RAO, C.R. (1966). Covariance adjustment and related problems in multivariate analysis. In *Multivariate Analysis* (P.R. Krishnaiah, Editor), 87-103. Academic Press, New York.

[8] RAO, C.R. (1973). *Linear Statistical Inference and Its Applications.* John Wiley, New York.

[9] WIJSMAN, R.A. (1986). Asymptotic distribution of the increase of the largest canonical correlation when one of the vectors is augmented. *J. Multivariate Anal.* **18**, 169-177.

N. C. Giri

ON A LOCALLY BEST INVARIANT AND LOCALLY MINIMAX TEST IN SYMMETRICAL MULTIVARIATE DISTRIBUTIONS

0. INTRODUCTION AND SUMMARY

Let

$$X = (X_{ij}) = \begin{pmatrix} X_1' \\ \vdots \\ X_n' \end{pmatrix}, \quad X_i' = (X_{i1},\ldots,X_{ip}),$$

$$i = 1,\ldots,n,$$

be a $n \times p$ random matrix $(n > p)$ with probability density function

$$f_X(x) = |\Sigma|^{-n/2} q(\mathrm{tr}\Sigma^{-1}(x - e\mu')'(x - e\mu')) \qquad (1)$$

with $x \in \chi = \{x = (x_{ij}) | \text{rank of } x = p\}$,

$\mu = (\mu_1,\ldots,\mu_p)' \in R^p$, $e = (1,\ldots,1)'$, $n \times 1$ and $\Sigma > 0$ ($p \times p$ positive definite matrix). We shall assume throughout that $q \in Q = \{q{:}M(p) \text{ to } [0,\infty)\}$ is convex on $M(p) = \{p \times p \text{ nonnegative definite matrices}\}$ and thrice continuously differentiable. Denote $n\overline{x} = \sum_1^n X_i$,

$S = \sum_1^n (X_i - \overline{X})(X_i - \overline{X})'$. We shall also use the following notations throughout: for any p-vector $b = (b_1,\ldots,b_p)' = (b_{(1)}', b_{(2)}')$, $b_{(1)} = (b_1,\ldots,b_{p_1})'$, $b_{(2)} = (b_{p_1+1},\ldots,b_p)'$ and $b_{[i]} = (b_1,\ldots,b_i)'$ for the

A. K. Gupta (ed.), Advances in Multivariate Statistical Analysis, 63–83.

i-vector containing the first i components of b; for any $p \times p$ matrix

$$c = (c_{ij}) = \begin{pmatrix} c_{(11)} & c_{(12)} \\ c_{(21)} & c_{(22)} \end{pmatrix},$$

$c_{(11)}$ is the $p_1 \times p_1$ upper lefthand corner submatrix of c, $c_{(22)}$ is the $p_2 \times p_2$ lower righthand corner submatrix of c, $p_1 + p_2 = p$ and

$$c_{[ii]} = \begin{pmatrix} c_{11},\ldots,c_{1i} \\ \cdots\cdots\cdots \\ c_{i1},\ldots,c_{ii} \end{pmatrix}.$$

We shall consider here the problem of testing

$$H_0: \ \mu_1 = \ldots = \mu_p = 0$$

against the alternatives H_1: $\mu_1 = \ldots = \mu_{p_1} = 0$, $p_1 < p$ when Σ is unknown. The problem of testing H_0 against H_1 remains invariant under the multiplicative group G of $p \times p$ nonsingular matrices g

$$g = \begin{pmatrix} g_{(11)} & 0 \\ g_{(21)} & g_{(22)} \end{pmatrix} \tag{2}$$

where $g_{(11)}$ is $p_1 \times p_1$. A maximal invariant in the space χ of $(\bar{X},S)$ under G (see Giri (1968)) is

$$\bar{R} = (\bar{R}_1,\bar{R}_2)$$

where

$$\bar{R}_1 = n\bar{X}'_{(1)}(S_{(11)} + n\bar{X}_{(1)}\bar{X}'_{(1)})^{-1}\bar{X}_{(1)}$$

$$= \frac{n\bar{X}'_{(1)}S^{-1}_{(11)}\bar{X}_{(1)}}{1 + n\bar{X}'_{(1)}S^{-1}_{(11)}\bar{X}_{(1)}}, \tag{3}$$

$$\bar{R}_1 + \bar{R}_2 = n\bar{X}'(S + n\bar{X}\bar{X}')^{-1}\bar{X} = \frac{n\bar{X}'S^{-1}\bar{X}}{1 + n\bar{X}'S^{-1}\bar{X}} .$$

A corresponding maximal invariant in the parametric space of (μ,Σ) under the induced group is

$$\Delta = (\bar{\delta}_1, \bar{\delta}_2)$$

where

$$\bar{\delta}_1 = n\mu'_{(1)}\Sigma^{-1}_{(11)}\mu_{(1)},$$

$$\bar{\delta}_1 + \bar{\delta}_2 = n\mu'\Sigma^{-1}\mu. \tag{4}$$

We shall write $\delta = \bar{\delta}_1 + \bar{\delta}_2$. It is obvious that

$$\bar{\delta}_2 = (\mu_{(2)} - \Sigma_{(21)}\Sigma^{-1}_{(11)}\mu_{(1)})'$$

$$\cdot (\Sigma_{(22)} - \Sigma_{(21)}\Sigma^{-1}_{(11)}\Sigma_{(12)})^{-1}$$

$$\cdot (\mu_{(2)} - \Sigma_{(21)}\Sigma^{-1}_{(11)}\mu_{(1)}).$$

For invariant tests under G the problem is reduced to testing H_0: $\bar{\delta}_2 = 0$ against the alternatives H_1: $\bar{\delta}_2 > 0$ when it is given that $\bar{\delta}_1 = 0$. We shall show here that the test which rejects H_0 whenever $\bar{R}_1 + \frac{n - p_1}{p_2}\bar{R}_2 \geq C$, where the constant C depends on the level α of the test, is locally best invariant and locally minimax in the sense of

Giri and Kiefer (1964) as $\overline{\delta}_2 \to 0$ for the family of densities given in (1). In the multivariate normal setup i.e. when

$$q(\mathrm{tr}\Sigma^{-1}(x - e\mu')'(x - e\mu')) = (2\pi)^{-\frac{np}{2}}$$

$$\cdot\ |\Sigma^{-1}|^{\frac{n}{2}}\exp\{-\tfrac{1}{2}\ \mathrm{tr}\Sigma^{-1}(x - e\mu')'(x - e\mu)\},$$

Giri (1968) has shown that this test enjoys the same optimality properties. We shall refer to Kariya and Sinha (1985) and the references contained therein for optimality results in symmetrical distributions. We shall furthermore assume that (i) $\int_{R^{np}} q(\mathrm{tr}z'z)dz = 1$,

(ii) $\int_G (\mathrm{tr}(gg'))^{i/2}|q^{(i)}(\mathrm{tr}gg')|\mu(dg) < \infty$,

(iii) $q^{(3)}(x) \leq 0$ and nondecreasing, where $q^{(3)}(x) = \dfrac{d^3q(x)}{dx^3}$ and $\mu(dg)$ is defined later.

1. LOCALLY BEST INVARIANT TEST

To derive the locally best invariant test (LBI) (Giri (1968)), we need the ratio of the probability density of $\overline{R}$ under H_1 to that of $\overline{R}$ under H_0 when $\overline{\delta}_2 \to 0$. Using Stein's theorem (1956) or Wijsman's representation theorem (1967) the ratio R can be expressed as

$$R = \frac{f_{\overline{R}}(\overline{r}|H_1)}{f_{\overline{R}}(\overline{r}|H_0)}$$

$$= \frac{\int_G p_1(g\overline{x}, gsg')|g_{(11)}g'_{(11)}|^{\frac{n-p_1}{2}} |g_{(22)}g'_{(22)}|^{\frac{n-p}{2}} dg}{\int_G p_0(g\overline{x}, gsg')|g_{(11)}g'_{(11)}|^{\frac{n-p_1}{2}} |g_{(22)}g'_{(22)}|^{\frac{n-p}{2}} dg} \tag{5}$$

where

$$p_1(\overline{x}, s) = |s|^{\frac{n}{2}}|\Sigma^{-1}|^{\frac{n}{2}} q(\mathrm{tr}\Sigma^{-1}(s + n(\overline{x} - \mu)(\overline{x} - \mu)'))$$

$$= |s|^{\frac{n}{2}}|\Sigma^{-1}|^{\frac{n}{2}}$$

$$\cdot\, q(\mathrm{tr}\Sigma^{-1}(s + n\overline{x}\overline{x}') - 2n\overline{x}'\Sigma^{-1}\mu + \delta),$$

$$p_0(\overline{x}, s) = |s|^{\frac{n}{2}}|\Sigma^{-1}|^{\frac{n}{2}} q(\mathrm{tr}\Sigma^{-1}(s + n\overline{x}\overline{x}'))$$

and the range of integration being $-\infty$ to ∞ in each variable. Let A be a matrix in G such that

$$A(s + n\overline{x}\overline{x}')A' = I.$$

Then $A'A = (s + n\overline{x}\overline{x}')^{-1} = s^{-1} - ns^{-1}\overline{x}\overline{x}'s^{-1}(1 + n\overline{x}'s^{-1}\overline{x})^{-1}$ so that $n\overline{x}'A'A\overline{x} = n\overline{x}'s^{-1}\overline{x}(1 + n\overline{x}'s^{-1}\overline{x})^{-1} = \overline{r}_1 + \overline{r}_2$ and $n\overline{x}'_{(1)}A'_{(11)}A_{(11)}\overline{x}_{(1)} = n\overline{x}'_{(1)}s^{-1}_{(11)}\overline{x}_{(11)}(1 + n\overline{x}'_{(1)}s^{-1}_{(11)}\overline{x}_{(1)})^{-1}$ $= \overline{r}_1$. Now writing $y = \sqrt{n}A\overline{x}$ such that $y'_{(1)}y_{(1)} = \overline{r}_1$,

$y'_{(2)}y_{(2)} = \bar{r}_2$, and $gA^{-1} = h \in G$ and noting that the left invariant Haar measure in G is

$$\mu(dg) = |g_{(11)}g'_{(11)}|^{\frac{p_1}{2}}|g_{(22)}g'_{(22)}|^{\frac{p}{2}}dg,$$

we can rewrite (5) as

$$R = \{\int_G q(tr[\sum_{j\le i=1}^{2} h_{(ij)}h'_{(ij)} - 2\sum_{j\le i=1}^{2} \rho_{(i)}y_{(j)}h'_{(ij)}$$

$$+ \bar{\delta}_2])|h_{(11)}h'_{(11)}|^{\frac{n-p_1}{2}}|h_{(22)}h'_{(22)}|^{\frac{n-p}{2}}dh\}/$$

$$\{\int_G q(tr(\sum_{j\le i=1}^{2} h_{(ij)}h'_{(ij)}))|h_{(11)}h'_{(11)}|^{\frac{n-p_1}{2}}$$

$$|h_{(22)}h'_{(22)}|^{\frac{n-p}{2}}dh, \qquad (6)$$

where $\rho'_{(1)}\rho_{(1)} = \bar{\delta}_1$, $\rho'_{(2)}\rho_{(2)} = \bar{\delta}_2$. Obviously $\rho_{(1)} = 0$ under H_0 and H_1. To derive the LBI test we expand the integrand q in the numerator of (6) as

$$q(trhh') + q^{(1)}(trhh')(-2\eta + \bar{\delta}_2) + \frac{q^{(2)}(trhh')}{2}$$

$$(-2\eta + \bar{\delta}_2)^2 + \frac{q^{(3)}(z)}{6}(-2\eta + \bar{\delta}_2)^3 \qquad (7)$$

where

$$\eta = tr\rho'_{(2)}(h_{(21)}Y_{(1)} + h_{(22)}Y_{(2)}),$$

$$z = \mathrm{tr}hh' + (1-\alpha)(-2\eta + \bar{\delta}_2) \quad \text{with} \quad 0 < \alpha < 1,$$

and $q^{(i)}(x) = \dfrac{d^i q(x)}{dx^i}$, and evaluate the integral of each term in (7). To do this we need the following integration results on $O(p)$, the group of $p \times p$ orthogonal matrices with respect to the invariant measure $\tau(dO)$, $O \in O(p)$.

$$\text{(i)} \quad \int_{O(p)} \mathrm{tr}(AOBO')\tau(dO) = \frac{\mathrm{tr}AB}{p}, \tag{8}$$

$$\text{(ii)} \quad \int_{O(p)} (\mathrm{tr}OA)^{\kappa}\tau(dO) = \begin{cases} 0, & \text{if } \kappa \text{ is odd,} \\ \dfrac{\mathrm{tr}A'A}{p}, & \text{if } \kappa = 2. \end{cases} \tag{9}$$

Using (9) the integration of the second term in (7) is

$$\alpha_1 = \bar{\delta}_2 \int q^{(1)}(\mathrm{tr}hh')\nu(dh) \tag{10}$$

where $\nu(dh) = |h_{(11)}h'_{(11)}|^{\frac{n-p_1}{2}} |h_{(22)}h'_{(22)}|^{\frac{n-p}{2}} dh$. To integrate the third term we first observe that given $\rho_{(2)}$ there exist an orthogonal matrix $O \in O(p_2)$ such that

$$O\rho_{(2)} = (\sqrt{\bar{\delta}_2}, 0, \ldots, 0) = \rho^*_{(2)} \quad \text{(say)} \tag{11}$$

or $\rho_{(2)} = O'\rho^*_{(2)}$. Hence

$$\mathrm{tr}\rho'_{(2)}(h_{(21)}Y_{(1)} + h_{(22)}Y_{(2)})$$

$$= \mathrm{tr}(\rho^{*\prime}_2 O(h_{(21)}Y_{(1)} + h_{(22)}Y_{(2)})).$$

Using (9) we get

$$\int_{O(p_2)} [\mathrm{tr}(O(h_{(21)}Y_{(1)} + h_{(22)}Y_{(2)})\rho^{*\prime}_{(2)})]^2 \tau(dO)$$

$$= (\mathrm{tr}(h_{(21)}Y_{(1)} + h_{(22)}Y_{(2)})'$$

$$\cdot (h_{(21)}Y_{(1)} + h_{(22)}Y_{(2)})\rho^{*\prime}_{(2)}\rho^{*}_{(2)})\frac{1}{p_2})$$

$$= \frac{\bar{\delta}_2}{p_2}\, \mathrm{tr}(h_{(21)}Y_{(1)} + h_{(22)}Y_{(2)})'$$

$$(h_{(21)}Y_{(1)} + h_{(22)}Y_{(2)}).$$

Hence the integral of the third term in (7) is

$$\frac{2\bar{\delta}_2}{p_2} \int_G \mathrm{tr}(h_{(21)}Y_{(1)} + h_{(22)}Y_{(2)})'$$

$$(h_{(21)}Y_{(1)} + h_{(22)}Y_{(2)})q^{(2)}(\mathrm{tr}hh')\nu(dh)$$

$$+ \frac{\bar{\delta}_2^2}{2} \int_G q^{(2)}(\mathrm{tr}hh')\nu(dh). \qquad (12)$$

Since the measure $q^{(2)}(\mathrm{tr}hh')\nu(dh)$ is invariant under the sign change $h_{(22)} \to -h_{(22)}$ we conclude that

$$\int_G \mathrm{tr}(Y_{(1)}h_{(21)}h'_{(22)}Y'_{(2)})q^{(2)}(\mathrm{tr}hh')\nu(dh) = 0. \qquad (13)$$

Hence we can rewrite (12) as

$$\frac{2\bar{\delta}_2}{p_2} \int_G \mathrm{tr}(Y'_{(2)}h'_{(22)}h_{(22)}Y_{(2)})q^{(2)}(\mathrm{tr}(hh'))\nu(dh)$$

$$+ \frac{2\bar{\delta}_2}{p_2} \int_G \mathrm{tr}(Y'_{(1)} h'_{(21)} h_{(21)} Y_{(1)}) q^{(2)}(\mathrm{tr}hh')\nu(dh)$$

$$+ \frac{\bar{\delta}_2^2}{2} \int_G q^{(2)}(\mathrm{tr}hh')\nu(dh). \qquad (14)$$

Transforming $Y_{(2)}$ by an orthogonal transformation of the type (11) and integrating with respect to $O(p_2)$, the first term in (14) can be written as

$$\frac{2\bar{\delta}_2\bar{r}_2}{p_2^2} \int_G \mathrm{tr}(h_{(22)} h'_{(22)}) q^{(2)}(\mathrm{tr}hh')\nu(dh). \qquad (15)$$

Let $G_\ell(p)$ be the multiplicative group of $(p \times p)$ nonsingular matrices, $G_T(p)$ be the multiplicative group of $(p \times p)$ nonsingular lower triangular matrices with positive diagonal elements. Obviously $G_\ell(p) = G_T(p) \times O(p)$.

Now writing $h_{(22)} = b_2 O$, with $h_{(22)} \in G_\ell(p_2)$, $b_2 \in G_T(p_2)$, $O \in O(p_2)$, we can get

$$\nu(dh_{(22)}) = \lambda(db_2)\tau(dO) \qquad (16)$$

where τ is the invariant probability measure on $O(p_2)$ and

$$\lambda(db_2) = |b_2 b'_2|^{\frac{n-p_1}{2}} \prod_1^{p_2} (b_{ii}^2)^{-\frac{i}{2}} db_2$$

with $b_2 = (b_{ij})$. Hence we can rewrite (15) as (using (8))

$$\frac{2\bar{\delta}_2\bar{r}_2}{p_2^3} \int_{G_T(p_2)} \mathrm{tr}(b_2 b'_2)\tilde{q}^{(2)}(\mathrm{tr}b_2 b'_2)\lambda(db_2) = \frac{2\bar{\delta}_2\bar{r}_2}{p_2^2}\alpha_2 \qquad (17)$$

where $\tilde{q}^{(2)}(tr(b_2b_2'))$ is the marginal measure of b_2 and is given by

$$\tilde{q}^{(2)}(tr(b_2b_2')) = \int q^{(2)}(tr(h_{(11)}h_{(11)}' + h_{(21)}h_{(21)}' + b_2b_2'))|h_{(11)}h_{(11)}'|^{\frac{n-p_1}{2}} \cdot dh_{(11)}dh_{(12)}, \tag{18}$$

and

$$p_2\alpha_2 = \int tr(b_2b_2')\tilde{q}^{(2)}(tr(b_2b_2'))\lambda(db_2)).$$

We have used the fact that

$$\int(b_2b_2')\tilde{q}^{(2)}(tr(b_2b_2'))\lambda(db_2) = \alpha_2 I.$$

Similarly the second integral in (14) can be reduced to $\frac{2\overline{\delta}_2\overline{r}_1\alpha_3}{p_2}$ where $\alpha_3 p_1 p_3 = \int tr(h_{(21)}h_{(21)}'))q^{*(2)}(tr(h_{(21)}h_{(21)}')) \cdot dh_{(12)}$ and

$$q^{*(2)}(tr(h_{(21)}h_{(21)}'))$$

$$= \int q^{(2)}(tr(h_{(11)}h_{(11)}' + h_{(22)}h_{(22)}' + h_{(21)}h_{(21)}') \cdot |h_{(11)}h_{(11)}'|^{\frac{n-p_1}{2}} |h_{(22)}h_{(22)}'|^{\frac{n-p}{2}} dh_{(11)}dh_{(22)}.$$

To evaluate α_2, α_3 we now consider the marginal density of $h_{(22)}$, $h_{(21)}$ with resepct to the measure

$|h_{(22)}h_{(22)}'|^{\frac{n-p}{2}} dh_{(22)}dh_{(21)}$, as given by,

$$h(\mathrm{tr}(h_{(22)}h'_{(22)} + h_{(21)}h'_{(21)}))$$

$$= \int q^{(2)}(\mathrm{tr}(h_{(22)}h'_{(22)} + h_{(21)}h'_{(21)} + h_{(11)}h'_{(11)})$$

$$\cdot \, |h_{(11)}h'_{(11)}|^{\frac{n-p_1}{2}} dh_{(11)} \tag{19}$$

and write $h_{(22)} = b_2 0$, $p_1 + p_2 = p$. The marginal measure of b_2, $h_{(21)}$ is given by

$$\psi(db_2, dh_{(21)}) = h(\mathrm{tr}(b_2b_2' + h_{(21)}h'_{(21)}))|b_2b_2'|^{\frac{n-p_1}{2}}$$

$$\cdot \prod_1^{p_2} (b_{ii}^2)^{-\frac{i}{2}} db_2 dh_{(21)}.$$

Write

$$h_{(21)} = (h_{ij}),$$

$$L = \mathrm{tr}(b_2b_2' + h_{(21)}h'_{(21)}), \quad e_0 = \mathrm{tr}(h_{(21)}h'_{(21)})/L,$$

$$e_i = b_{ii}^2/L, \quad i = 1,\ldots,p_2$$

$$e_{p_2+i} = b_{i+1,i}^2/L, \quad i = 1,\ldots,p_2 - 1,$$

$$e_{p_2+p_2-1+i} = b_{i+2,i}^2/L, \quad i = 1,\ldots,p_2 - 2,$$

$$\cdots$$

$$e_{\frac{p_2(p_2+1)}{2}} = b_{p_2,1}^2/L. \tag{20}$$

Extending the domain of b_{ii} to $(-\infty,+\infty)$ which simply gives a constant c (say) to the right side of (19), writing

$$K = \int ch(tr(b_2 b_2' + h_{(21)}h_{(21)}'))db_2 dh_{(21)} \tag{21}$$

and observing that $\frac{ch}{K}(tr(b_2 b_2' + h_{(21)}h_{(21)}'))$ is a spherical density of b_{ij}'s and h_{ij}'s we conclude that L and $e = (e_0, e_1, e_2, \ldots, e_{p_2(p_2+1)/2})$ are independent and e obeys (Kariya and Eaton (1977)) Dirichlet distribution $D(\frac{p_1 p_2}{2}, \frac{1}{2}, \frac{1}{2}, \ldots, \frac{1}{2})$. The probability density function of e is given by

$$\frac{\Gamma(\frac{p_2(p_2+1)}{2} + \frac{p_1 p_2}{2})}{\Gamma(\frac{p_1 p_2}{2})(\Gamma(\frac{1}{2}))^{\frac{p_2(p_2+1)}{2}}} \cdot (e_0)^{\frac{p_1 p_2}{2} - 1}$$

$$\cdot \prod_{i=1}^{\frac{p_2(p_2+1)}{2} - 1} (e_i)^{\frac{1}{2}-1} (1 - e_0 - \sum_1^{\frac{p_2(p_2+1)}{2}} e_i)^{\frac{1}{2}-1}.$$

Denote by $N = \int L^{\Sigma_i (n-p_1-i)} ch(L) db_2 dh_{(21)}$. Then

$$\frac{\alpha_3 p_1 p_2}{KN} = E(e_0 \prod_1^{p_2} (e_i)^{\frac{n-p_1-i}{2}})$$

$$\frac{\alpha_2 p_2}{KN} = \sum_{i=1}^{p_2} E(e_i^{\frac{n-p_1+i}{2}+1} \prod_{\substack{j=1 \\ j\neq i}}^{p_2} (e_j^{\frac{n-p_1-j}{}}))$$

$$+ \sum_{j=p_2+1}^{\frac{p_2(p_2+1)}{2}} E(e_j \prod_{i=1}^{p_2} (e_i)^{\frac{n-p_1-i}{2}}) .$$

Hence $\dfrac{\alpha_2 p_2}{\alpha_3 p_1 p_2} = \dfrac{n - p_1}{p_1}$ which implies that $\dfrac{\alpha_2}{\alpha_3} = n - p_1$.

In a straightforward manner we can check that the integration of the last term in (14) and that of the last term in (7) are both $O(\overline{\delta}_2)$ uniformly in $\overline{r}_1$ and $\overline{r}_2$. Thus

$$R = 1 + \frac{\overline{\delta}_2}{D}(\alpha_1 + \frac{2\alpha_3}{p_2}[(\frac{n - p_1}{p_2})\overline{r}_2 + \overline{r}_1]$$

$$+ B(\overline{r}_1, \overline{r}_2, \overline{\delta}_2) \tag{21a}$$

where

$$D = \int_G q(\mathrm{tr}hh')|h_{(11)}h'_{(11)}|^{\frac{n-p_1}{2}} |h_{(22)}h'_{(22)}|^{\frac{n-p}{2}} dh$$

and $B(\overline{r}_1, \overline{r}_2, \overline{\delta}_2) = O(\overline{\delta}_2)$ uniformly in $\overline{r}_1$, $\overline{r}_2$. Hence we get the following theorem.

<u>THEOREM 1</u>. *For testing* H_0: $\overline{\delta}_2 = 0$ *against* H_1: $\overline{\delta}_2 = \lambda$, *given that* $\overline{\delta}_1 = 0$, *the test which rejects* H_0 *whenever* $\overline{r}_1 + \dfrac{n - p_1}{p_2}\overline{r}_2 \geq c$, *is locally best invariant as* $\lambda \to 0$ *for the family of distributions in (1), the constant* C *depends on the level* α *of the test.*

2. LOCALLY MINIMAX TEST OF H_0 AGAINST H_1: $\overline{\delta}_2 = \lambda$

We refer to Giri and Kiefer (1964) for details of locally minimax test. Let $(\overline{\delta}_2, \eta)$, be a typical element in the parametric space Ω where $\overline{\delta}_2 > 0$ and η is a p_2-dimensional vector whose range depends on $\overline{\delta}_2$. Let $p(y, \overline{\delta}_2, \eta)$ denote the probability density function on *(y,b)* with respect to some σ-finite measure. For fixed α, $0 < \alpha < 1$, consider the rejection region of the form

$$R^* = \{y: \ U(y) \geq C_\alpha\} \tag{22}$$

where U is bounded and positive and has a continuous distribution function for each $(\overline{\delta}_2, \eta)$, equicontinuous in $(\overline{\delta}_2, \eta)$ for some $\overline{\delta}_2 < \delta_0$ and that

$$P_{0,\eta}(R^*) = \alpha \tag{23}$$

$$P_{\lambda,\eta}(R^*) = \alpha + h(\lambda) + g(\lambda, \eta) \tag{24}$$

where $g(\lambda, \eta) = o(h(\lambda))$ uniformly in η with $h(\lambda) > 0$ for $\lambda > 0$ and $h(\lambda) = o(1)$. Let also $\xi_{0,\lambda}$, $\xi_{1,\lambda}$ denote apriori probability functions on the set $\{\overline{\delta}_2 = 0\}$ and $\{\overline{\delta}_2 = \lambda\}$ respectively such that

$$\frac{\int p(y,\lambda,\eta)\xi_{1\lambda}(d\eta)}{\int p(y,0,\eta)\xi_{0\eta}(d\eta)} = 1 + h(\lambda)(g(\lambda) + r(\lambda)U(y)) + B(y,\lambda) \tag{25}$$

where $0 < c_1 < r(\lambda) < c_2 < \infty$ for λ sufficiently small and $g(\lambda) = O(1)$, $B(y,\lambda) = o(h(\lambda))$ uniformly in y. If U satisfies (23) and (24) and for sufficiently small λ there exists $\xi_{0\lambda}$, $\xi_{1\lambda}$, satisfying (25) the R^* is locally minimax for testing H_0 against H_1' as $\lambda \to 0$. It is well known that Hunt-Stein theorem can not be applied to G with $p \geq 2$ (Giri and Kiefer (1964a); Giri, Kiefer and Stein (1963)). However this does apply to the subgroup $G_T(p)$. Thus for each λ there is a level α test which is invariant under G_T (Lehmann (1959), p. 225) and which minimizes, among all level α tests, the minimum power under H_1'. In the place of $\overline{R}$, the maximal invariant under G, we obtain here a p-dimensional vector $R = (R_1,\ldots,R_p)$ as the maximal univariant statistic under G_T (Giri (1968)) and is defined by

$$\sum_1^i R_j = n\overline{X}'_{[i]}(S_{[ii]} + n\overline{X}_{[i]}\overline{X}'_{[i]})^{-1}\overline{X}_{[i]},$$
$$i = 1,\ldots,p,$$
$$R_i \geq 0, \quad \sum_1^{p_1} R_j = \overline{R}_1, \quad \sum_1^{p} R_j = \overline{R}_1 + \overline{R}_2.$$

A corresponding maximal invariant in the parametric space of (μ,Σ) under the induced group is $\underline{\sigma} = (\sigma_1,\ldots,\sigma_p)'$, given by

$$\sum_1^i \sigma_j = n\mu'_{[i]}\Sigma^{-1}_{[ii]}\mu_{[i]}, \quad i = 1,\ldots,p,$$

$$\sigma_i \geq 0, \quad \sum_1^{p_1} \sigma_j = \overline{\delta}_1, \quad \sum_1^{p} \delta_j = \overline{\delta}_1 + \overline{\delta}_2.$$

Under both H_0 and H_1' $\sigma_1 = \ldots = \sigma_{p_1} = 0$. The nuisance parameter in this reduced setup is $\eta = (0,\ldots,0,\eta_{p_1+1},\ldots,\eta_p)'$ with $\eta_i = \sigma_i/\overline{\delta}_2$. Let $u' = (\sqrt{r_1},\ldots,\sqrt{r_p})'$, $\underline{\nu} = (0,\ldots,0,\sqrt{\sigma_{p_1+1}},\ldots,\sqrt{\sigma_p})'$. Using Stein's theorem (1956) or Wijsman's representation theorem (1967) the ratio R of the probability density $\underline{R}$ under H_1' to that of $\underline{R}$ under H_0 is given by, with $\overline{g} \in G_T$

$$R = \frac{\int_{G_T} p_1(gx, gsg')\prod_1^p (g_{ii}^2)^{\frac{n-i}{2}} dg}{\int_{G_T} p_0(gx, gsg')\prod_1^p (g_{ii}^2)^{\frac{n-i}{2}} dg} \tag{26}$$

$$= \frac{\int_{G_T} q(tr(\sum_i \sum_{j\le i} g_{ij}^2 - 2\nu' gu + \lambda))\prod_1^p (g_{ii}^2)^{\frac{n-i}{2}} dg}{\int_{G_T} q(\sum_i \sum_{j\le i} q_{ij}^2)\prod_1^p (g_{ii}^2)^{\frac{n-i}{2}} dg}.$$

For simplicity let us write $g = \begin{pmatrix} g_{(11)} & 0 \\ g_{(21)} & g_{(22)} \end{pmatrix}$ with $g_{(11)}, g_{(22)}$ both lower triangular matrices. We can now write

$$R = \int_{G_T} q(tr(gg' - 2\nu'_{(2)}g_{(21)}u_{(1)} - 2\nu'_{(2)}g_{(22)}u_{(2)}))$$

$$\prod_1^p (g_{ii}^2)^{\frac{n-i}{2}} dg / \int_{G_T} q(trgg')\prod_1^p (g_{ii}^2)^{\frac{n-i}{2}} dg \tag{27}$$

As before let us now expand the integrand q in the numerator of (27) as

$$q(trgg') + (-2\eta + \lambda)q^{(1)}(trgg')$$

$$+ (-2\eta + \lambda)^2 q^{(2)}(trgg') + (-2\eta + \lambda)^3 q^{(3)}(z) \tag{28}$$

where $z = trgg' + (1-\alpha)(-2\eta + \lambda)$, $0 < \alpha < 1$, $\eta = tr\nu'_{(2)}(g_{(21)}u_{(1)} + g_{(22)}u_{(2)})$. As in the earlier section the integration of the second term in (28) gives

$$\lambda\alpha_1 = \lambda\int_{G_T} q^{(1)}(\mathrm{tr}gg')\prod_1^p (g_{ii}^2)^{\frac{n-i}{2}} dg. \tag{29}$$

Note we have used the fact that, for $i \neq \ell$, $j \neq k$,

$$\int_{G_T} g_{ij} g_{\ell k} g^{(i)}(\mathrm{tr}gg')\prod_1^p (g_{ii}^2)^{\frac{n-i}{2}} dg = 0,$$

$$i = 1,2,3. \tag{30}$$

To integrate the third term in (28), we first observe that (using (23))

$$\int_{G_T} (\mathrm{tr}\nu'_{(2)} q_{(21)} u_{(1)})^2 q^{(2)}(\mathrm{tr}gg')\prod_1^p (g_{ii}^2)^{\frac{n-i}{2}} dg$$
$$= \int\Big(\sum_{i=p_1+1}^{p} \sigma_i \sum_{j=1}^{p} r_j g_{ij}^2\Big) q^{(2)}(\mathrm{tr}gg')\prod_1^p (g_{ii}^2)^{\frac{n-i}{2}} dg, \tag{31}$$

and

$$\int_{G_T} (\mathrm{tr}\nu'_{(2)} g_{(22)} u_{(2)})^2 q^{(2)}(\mathrm{tr}gg')\prod_1^p (g_{ii}^2)^{\frac{n-i}{2}} dg$$

$$= \int_{G_T} \Big[\sum_{j=p_1+1}^{p} r_j \sum_{j\geq i} (\sigma_i g_{ij}^2 + \ldots + \sigma_j g_{jj}^2)\Big]$$

$$\cdot\ q^{(2)}(\mathrm{tr}(gg'))\prod_1^p (g_{ii}^2)^{\frac{n-i}{2}} dg. \tag{32}$$

Now writing $K = \int q^{(2)}(\mathrm{tr}gg')dg$, $L = \mathrm{tr}gg'$,

$$D = \int q(\operatorname{tr} gg') \sum_1^p (g_{ii}^2)^{\frac{n-i}{2}} dg$$

$$N = \int L^{\sum\frac{(n-i)}{2}} q^{(2)}(L) dg,$$

$$M = E\left(\prod_1^p (e_i)^{\frac{n-i}{2}}\right).$$

We get

$$R = 1 + \frac{\lambda}{D}\Big(\alpha_1 + \frac{2MN}{K}\Big(\overline{r}_1 + \sum_{j=p_1+1}^{p} r_j\Big(\sum_{i>j} \eta_i$$

$$+ (n - j + 1)\eta_j\Big) + B(u,\eta,\lambda)\Big) \qquad (33)$$

where $B(u,\eta,\lambda) = o(\lambda)$ uniformly in u,η.

The set $\{\lambda = 0\}$ is a single point $\eta = 0$. So $\xi_{0\lambda}$ assigns measure 1 to the single point $\eta = 0$. The set $\{\overline{\delta}_2 = \lambda\}$ is a convex $p_2 = (p - p_1)$-dimensional Euclidean set where in each component $\eta_i = O(h(\lambda))$. Any probability measure $\xi_{1\lambda}$ can be replaced by the degenerate measure $\xi_{1\lambda}^*$ which assigns measure 1 to the mean η_i^* of $\xi_{1\lambda}$. Hence

$$\int R\xi_{1\lambda}^*(d\eta) = 1 + \frac{\lambda}{D}\Big(\alpha_1 + \frac{2NN}{K}\Big(\overline{r}_1 + \sum_{j=p_1+1}^{p} r_j$$

$$\cdot\Big(\sum_{i>j}\eta_i^* + (n - j + 1)\eta_j^*\Big) + B(u,\lambda)\Big) \qquad (34)$$

where $B(u,\lambda) = o(h(\lambda))$ uniformly in u. Consider the rejection region

$$c_K = \{x\colon\ U(x) = \overline{r}_1 + K\overline{r}_2 \geq c_\alpha\} \qquad (35)$$

where K is a constant such that (34) is reduced to yield (25) and c_α depends on the level of significance α of the test for the chosen K. Now choose

$$\eta_p^* = \frac{n - p_1}{(n - p + 1)p_2},$$

$$\eta_j^* = \frac{(n - j - 1)\ldots(n - p)}{(n - j + 1)\ldots(n - p + 2)}\left(\frac{n - p_1}{(n - p + 1)p_2}\right),$$

$j = p_1 + 1,\ldots,p$ so that $\sum_{j>i} \eta_i^* + (n - j + 1)\eta_j^* = \frac{n - p_1}{p_2}$, $j = p_1 + 1,\ldots,p$, we can conclude that the test with rejection region

$$c' = \{x: \; U(x) = \overline{r}_1 + \frac{n - p_1}{p_2}\overline{r}_2 \geq c_\alpha\}$$

with $p_{0,\lambda}(c') = \alpha$ satisfies (25) as $\lambda \to 0$. Furthermore any region c_k of the form (35) must have $K = \frac{n - p_1}{p_2}$ to satisfy (25) for some $\xi_{1\lambda}$. From (21a), for any invariant region c', $P_{\lambda,n}(c')$ depends only on λ as $\lambda \to 0$. Hence from (24) $g(\lambda,\eta) = 0$. Since the test with rejection region c' is LBI, Hotelling's test which rejects H_0 whenever $\overline{r}_1 + \overline{r}_2 \geq c$ does not coincide with the LBI test and hence it is locally worse. It can be easily shown that Hotelling's test, whose power depends only on λ, has positive derivative at $\lambda = 0$. Hence the LBI test satisfies the same condition at $\lambda = 0$. Thus $h(\lambda) > 0$. The condition (23) follows from the null robustness of the distribution $\overline{r}_1, \overline{r}_2$. Hence we have

THEOREM 2. *For testing* H_0 *against* H_1', *the test which rejects* H_0 *whenever* $\overline{r}_1 + \frac{n - p_1}{p_2}\overline{r}_2 \geq c$ *is locally minimax as* $\lambda \to 0$ *for the family of distributions in (1).*

Department of Mathematics and Statistics
Université de Montreal
Montreal, Canada

REFERENCES

Giri, N. (1977). *Multivariate Statistical Inference.* Academic Press, New York.

Giri, N. (1968). 'On Tests of the Equality of Two Covariance Matrices.' *Ann. Math. Statist.*, 39, 275-277.

Giri, N. (1969). 'Locally and Asymptotically Minimax Tests of a Multivariate Problem.' *Ann. Math. Statist.* 39, 171-178.

Giri, N. and J. Kiefer (1964). 'Local and Asymptotic Minimax Properties of Multivariate Tests.' *Ann. Math. Statist.*, 35, 21-35.

Giri, N. and J. Kiefer (1964a). 'Minimax character of R^2 test in the simplest case.' *Ann. Math. Statist.*, 35, 1475-1490.

Giri, N., J. Kiefer and C. Stein (1963). 'Minimax Character of T^2 test in the simplest case.' *Ann. Math. Statist.*, 34, 1524-1535.

Kariya, T. and M. Eaton (1977). 'Robust tests for spherical symmetry.' *Ann. Statist.*, 5, 206-215.

Kariya, T. and B. K. Sinha (1985). 'Nonnull and Optimality Robustness of Some Tests.' *Ann. Statist.* 13, 1182-1197.

Lehmann, E. (1959). *Testing Statistical Hypotheses.* Wiley, New York.

Stein, C. (1956). 'Some Problems in Multivariate Analysis.' Technical Report No. 6, Department of Statistics, Stanford University.

Wijsmann, R. A. (1967). 'Cross Section of Orbits and Their Application to Densities of Maximal Invariants.' 5th Berkeley Symposium Vol. 1, University of California Press, 389-400.

Leon Jay Gleser

CONFIDENCE INTERVALS FOR THE SLOPE IN A LINEAR ERRORS-IN-VARIABLES REGRESSION MODEL

1. INTRODUCTION

The linear errors-in-variables model treated in this paper is the following. Independent observations (x_i, y_i), $i = 1, 2, \ldots, N$, are obtained, where

$$\binom{x_i}{y_i} = \binom{u_i}{a+bu_i} + \binom{e_i}{f_i}, \quad i = 1, 2, \ldots, N,$$

$$\binom{e_1}{f_1}, \ldots, \binom{e_N}{f_N}, \text{ conditional on } u_1, \ldots, u_N, \text{ are i.i.d. } N\left(\binom{0}{0}, \Sigma\right) \tag{1.1}$$

The u_i's can be unknown constants (*functional case*) or i.i.d. $N(\mu, \sigma_u^2)$ random variables (*structural case*). The intercept a, slope b, and error covariance matrix

$$\Sigma = \begin{pmatrix} \sigma_{ee} & \sigma_{ef} \\ \sigma_{ef} & \sigma_{ff} \end{pmatrix}$$

are unknown parameters. In the functional case of (1.1) the u_i's are also unknown parameters, while in the structural case μ and σ_u^2 are unknown parameters. Note that in the structural case, the model (1.1) implies that $\{u_i, 1 \leq i \leq N\}$ and $\{(e_i, f_i), 1 \leq i \leq N\}$ are mutually independent collections of random variables.

The model (1.1) can be regarded as a regression model in which the predictor u_i of the dependent variable y_i cannot be observed accurately, but instead is measured with error e_i. Consequently, the model (1.1) is often called an *errors-in-variables regression model*. Alternatively, the model (1.1) can be thought of as modeling bivariate normal observations whose mean vectors fall on a straight line. Under either formulation, the slope parameter b is frequently of primary interest.

Without restrictions on the parameters of (1.1), it is known (Nussbaum, 1976; Gleser, 1983) that the slope b is not identifiable, being confounded with the slope $\beta = \sigma_{ef}\sigma_{ee}^{-1}$ of the linear regression of the errors f_i on the errors e_i, and also with the ratio $\lambda = \sigma_{ff}\sigma_{ee}^{-1}$ of

K. Gupta (ed.), Advances in Multivariate Statistical Analysis, 85–109.

the error variances. Thus, to permit identifiability, it is often assumed that β and λ are known. If this is the case, transforming from y_i to

$$y_i^* = \frac{y_i - \beta x_i}{(\lambda - \beta^2)^{\frac{1}{2}}}$$

yields the model

$$\begin{pmatrix} x_i \\ y_i^* \end{pmatrix} = \begin{pmatrix} u_i \\ a^* + b^* u_i \end{pmatrix} + \begin{pmatrix} e_i \\ f_i^* \end{pmatrix}, \quad i = 1, 2, \ldots, N,$$

$$\begin{pmatrix} e_1 \\ f_1^* \end{pmatrix}, \ldots, \begin{pmatrix} e_N \\ f_N^* \end{pmatrix}, \text{ given } u_1, \ldots u_N, \text{ are i.i.d. } .N(\begin{pmatrix} 0 \\ 0 \end{pmatrix}, \sigma^2 I_2),$$

where $f_i^* = (\lambda - \beta^2)^{-\frac{1}{2}}(f_i - \beta e_i)$ and

$$a^* = (\lambda - \beta^2)^{-\frac{1}{2}} a, \quad b^* = (\lambda - \beta^2)^{-\frac{1}{2}}(b - \beta), \quad \sigma^2 = \sigma_{ee}.$$

Since β and λ are known, the y_i^* are observable and we can use $(x_i, y_i^*), i = 1, 2, \ldots, N$, to estimate a^*, b^* and σ^2. From these estimators, estimators of a, b, σ_{ee} are easily obtained. Consequently, henceforth we act as if the transformation from y_i to y_i^* has been made, and assume that in the model (1.1) we have

$$\Sigma = \sigma^2 I_2, \quad \sigma^2 > 0,$$

Over 100 years ago, Adcock (1878) considered the model (1.1) with $\Sigma = \sigma^2 I_2$, and proposed estimating the parameters a, b by choosing the line L for which the sum of squared distances from the observed points (x_i, y_i) to L along perpendiculars to L is minimized. This approach contrasts with classical least squares methodology where distances from (x_i, y_i) to L are measured along perpendiculars to the x-axis. It is now well known (Kendall and Stuart, 1979; Gleser, 1981; Anderson, 1984) that Adcock's method yields the maximum likelihood estimators $\hat{a}$, $\hat{b}$ of a, b, respectively, in both the functional and the structural cases of the model (1.1). Standard theory shows that $\hat{a}$, $\hat{b}$ are best asymptotic normal $(N \to \infty)$ estimators of a, b, respectively, in the structural case. The fact that $\hat{a}$, $\hat{b}$ have similar large-sample optimality properties in the functional case has recently been shown by Gleser (1983).

Several studies have been made of the finite sample distribution of $\hat{b}$. [See Anderson (1976, 1984) for references.] However, little has been written about the problem of assessing (reporting) the accuracy of $\hat{b}$ as an estimator of b. In practice, perhaps the most common approach to this problem is to report a consistent estimator $\hat{\sigma}_{\hat{b}}$ of the standard deviation $\sigma_{\hat{b}}$ of the large-sample distribution of $\hat{b}$. The usual justification of this practice is that

$$\hat{b} \pm z_{\frac{1}{2}\alpha}\, \hat{\sigma}_{\hat{b}}\, N^{-\frac{1}{2}} \tag{1.2}$$

is a large-sample $100(1-\alpha)\%$ confidence interval for b. (Here, z_ν is the $100(1-\nu)^{th}$ percentile of the $N(0,1)$ distribution.) Unfortunately, the large-sample properties of (1.2) for fixed values of the parameters do not necessarily indicate the properties of this confidence interval when the sample size N is fixed and the parameters vary.

Indeed, the results of Gleser and Huang (1987), when applied to the model (1.1), show that when N is fixed, *every confidence interval for b of finite length* (for example, the interval (1.2)) *has confidence equal to* 0. The analysis in Gleser and Huang isolates a function τ^2 of the parameters which determines the amount of information in the data concerning b. This parameter is defined by

$$\tau^2 = \begin{cases} \frac{1}{N-1} \sum_{i=1}^{N} (u_i - \bar{u})^2/\sigma^2 & \text{in the functional case,} \\ \sigma_u^2/\sigma^2 & \text{in the structural case.} \end{cases} \tag{1.3}$$

When N is fixed, the probability of coverage for any finite confidence interval for b tends to 0 as τ^2 becomes small. (This fact is intuitively obvious since $\tau^2 \to 0$ implies $\sum_{i=1}^{n} (u_i - \bar{u})^2 \to 0$ in the functional case and $\sigma_u^2 \to 0$ in the structural case, and when the u_i's do not vary, it is impossible to fit a unique straight line through the points $(E(y_i), u_i), 1 \leq i \leq N$.)

Consequently, the appropriateness of $\hat{\sigma}_{\hat{b}}$ as a measure of accuracy for $\hat{b}$ and the usefulness of (1.2) as a confidence interval for b depend upon the value of τ^2, as well as the value of N. When τ^2 is sufficiently large, there is some hope that $\hat{\sigma}_{\hat{b}}$ can serve as a measure of accuracy for b. The goal of the present paper is to determine a range of values of N and τ^2 for which (1.2) is approximately a $100(1-\alpha)\%$ confidence interval for b, and $\hat{\sigma}_{\hat{b}}$ can serve as a meaningful index of accuracy for b.

In Section 2, lower bounds for the coverage probability of (1.2) are obtained in both the functional and structural cases of the model (1.1). In Section 3, these lower bounds are tabulated for various choices of N and τ^2. In Section 3 some consideration is also given to modifying the critical constant $z_{\frac{1}{2}\alpha}$ in (1.2) to improve coverage probabilities. Some details of the derivation in Section 2 are given in the Appendix. In addition, it is shown in the Appendix that, contrary to an assertion in Anderson (1976), $\hat{b}$ is not median unbiased for b.

2. BOUNDS FOR COVERAGE PROBABILITIES

Let

$$W = \begin{pmatrix} w_{11} & w_{12} \\ w_{12} & w_{22} \end{pmatrix} = \sum_{i=1}^{n} \begin{pmatrix} x_i - \bar{x} \\ y_i - \bar{y} \end{pmatrix} \begin{pmatrix} x_i - \bar{x} \\ y_i - \bar{y} \end{pmatrix}',$$

where $\bar{x} = N^{-1} \sum_{i=1}^{N} x_i$, $\bar{y} = N^{-1} \sum_{i=1}^{N} y_i$. Let

$$W = G\, D\, G' \tag{2.1}$$

be the spectral decomposition of W, where

$$D = \begin{pmatrix} d_1 & 0 \\ 0 & d_2 \end{pmatrix}, \quad d_1 \geq d_2 \geq 0,$$

is the diagonal matrix of eigenvalues of W and

$$G = \begin{pmatrix} g_{11} & g_{12} \\ g_{21} & g_{22} \end{pmatrix}$$

is the orthogonal matrix whose i^{th} column is the eigenvector of W corresponding to the eigenvalue d_i, $i = 1, 2$.

For the functional case of the model (1.1), Gleser (1981) shows that the maximum likelihood estimator of b is

$$\hat{b} = \frac{g_{21}}{g_{11}} = -\frac{g_{12}}{g_{22}}. \tag{2.2}$$

Hence

$$W = \frac{1}{1+\hat{b}^2}\begin{pmatrix} 1 & -\hat{b} \\ \hat{b} & 1 \end{pmatrix}\begin{pmatrix} d_1 & 0 \\ 0 & d_2 \end{pmatrix}\begin{pmatrix} 1 & \hat{b} \\ -\hat{b} & 1 \end{pmatrix}. \tag{2.3}$$

It also follows from Gleser's (1981) results that if

$$\lim_{N\to\infty} (N-1)^{-1}\sum_{i=1}^{N}(u_i - \bar{u})^2 = \Delta, \qquad \Delta > 0,$$

then

$$\sqrt{N}(\hat{b} - b) \longrightarrow N(0, \sigma_{\hat{b}}^2)$$

in distribution as $N \to \infty$, where

$$\sigma_{\hat{b}}^2 = \frac{1}{(\Delta/\sigma^2)^2} + \frac{1+b^2}{(\Delta/\sigma^2)}.$$

Finally, a consistent estimator of $\sigma_{\hat{b}}^2$ is

$$\hat{\sigma}_{\hat{b}}^2 = \frac{(1+\hat{b}^2)^2 d_1 d_2}{(d_1 - d_2)^2}. \tag{2.4}$$

Hence, a large-sample $100(1-\alpha)\%$ confidence interval for b is

$$\hat{b} \pm (z_{\frac{1}{2}\alpha})\left[\frac{(1+\hat{b}^2)(d_1 d_2)^{\frac{1}{2}}}{(d_1 - d_2)N^{\frac{1}{2}}}\right],$$

or equivalently

$$C = \left\{ b\colon \ \frac{(\hat{b} - b)^2 (d_1 - d_2)^2}{(1+\hat{b}^2)^2 d_1 d_2} \le \frac{1}{N}\chi_{1;\alpha}^2 \right\}, \tag{2.5}$$

where $\chi_{1;\alpha}^2$ is the $100(1-\alpha)^{th}$ percentile of the chi-squared distribution with 1 degree of freedom.

In the structural case of the model (1.1), similar results hold. Thus, $\hat{b}$ defined by (2.2) is the maximum likelihood estimator of b.

The large-sample distribution of $\sqrt{N}(\hat{b}-b)$ is $N(0,\sigma_{\hat{b}}^2)$ with

$$\sigma_{\hat{b}}^2 = \frac{1}{(\sigma_u^2/\sigma^2)^2} + \frac{1+b^2}{(\sigma_u^2/\sigma^2)},$$

and (2.4) is a consistent estimator of $\sigma_{\hat{b}}^2$. Thus, (2.5) also defines a large-sample $100(1-\alpha)\%$ confidence interval for b in the structural case of the model (1.1).

Note that in both the functional and structural cases of the model (1.1),

$$\sigma_{\hat{b}}^2 = \frac{1}{(\tau^2)^2} + \frac{1+b^2}{\tau^2},$$

where in the functional case

$$\tau^2 = \frac{\Delta}{\sigma^2} = \lim_{N\to\infty} \frac{(N-1)^{-1}\sum_{i=1}^{n}(u_i-\bar{u})^2}{\sigma^2}.$$

Instead of finding bounds for the coverage probabilities of C defined by (2.5), we will instead find bounds for coverage probabilities of the more general regions

$$C_k = \left\{ b\colon \frac{(\hat{b}-b)^2(d_1-d_2)^2}{(1+\hat{b}^2)^2 d_1 d_2} \le k \right\}. \tag{2.6}$$

This greater generality allows us in Section 3 to consider choices for k other than $N^{-1}\,\chi^2_{1;\alpha}$ in order to improve coverage probabilities.

Note from (2.2), (2.3), and (2.6) that the confidence region (interval) C_k depends on the data (x_i, y_i), $1 \le i \le N$, only through the matrix W. In the functional case of (1.1), W has a noncentral Wishart distribution with $N-1$ degrees of freedom, covariance matrix parameter $\sigma^2 I_2$, and noncentrality matrix parameter

$$(\sigma^2)^{-1}\sum_{i=1}^{N}(u_i-\bar{u})^2\binom{1}{b}\binom{1}{b}' = (N-1)\tau^2\binom{1}{b}\binom{1}{b}'.$$

In the structural case, W has a central Wishart distribution with $N-1$

degrees of freedom and covariance matrix parameter

$$\sigma^2 I_2 + \sigma_u^2 \begin{pmatrix} 1 \\ b \end{pmatrix} \begin{pmatrix} 1 \\ b \end{pmatrix}' = \sigma^2 \left[I_2 + \tau^2 \begin{pmatrix} 1 \\ b \end{pmatrix} \begin{pmatrix} 1 \\ b \end{pmatrix}' \right].$$

Hence, the coverage probabilities $P\{b \text{ in } C_k\}$ in both cases depend upon the parameters only through b, σ^2 and τ^2. However, if we transform (x_i, y_i) to (cx_i, cy_i), $i = 1, 2, \ldots N$, $c > 0$, then $\hat{b}, b, \tau^2$ and $[(d_1 - d_2)^2/d_1 d_2]$ remain constant under the transformation, while $\sigma^2 \to c^2\sigma^2$. Since C_k depends upon the data only through $\hat{b}$ and $[(d_1 - d_2)^2/d_1 d_2]$, it follows that

$$P_{b,\tau^2,\sigma^2}\{b \text{ in } C_k\} = P_{b,\tau^2,c^2\sigma^2}\{b \text{ in } C_k\}.$$

Letting $c = \sigma^{-1}$ shows that the coverage probabilities for C_k do not depend on the value of σ^2. Hence, we can let

$$H(k, \tau^2, b) = P\{b \text{ in } C_k\}, \tag{2.7}$$

where the dependence of $P\{b \text{ in } C_k\}$ on N is suppressed for notational convenience. (Remember that the sample size N is fixed.) We also can assume that $\sigma^2 = 1$ without loss of generality.

The confidence of the confidence region C_k is the infimum of the coverage probabilities $H(k, \tau^2, b)$ over $-\infty < b < \infty$, $\tau^2 > 0$. Since we are mainly interested in the dependence of the coverage probabilities on τ^2, we will in the following be considering (bounds on) the values of

$$H(k, \tau^2) = \inf_{-\infty<b<\infty} H(k, \tau^2, b) = \inf_{-\infty<b<\infty} P\{b \text{ in } C_k\}. \tag{2.8}$$

We obtain bounds on $H(k, \tau^2)$ for both the functional and structural cases of the model (1.1).

2.1. Distributional Representation

It has been noted already that in the functional case of the model (1.1), W has the noncentral Wishart distribution with $n = N - 1$ degrees of freedom, covariance matrix parameter I_2 (remember we are now assuming $\sigma^2 = 1$), and noncentrality matrix parameter $(N -$

$1)\tau^2(1,b)'(1,b)$. That is,

$$W \sim \mathcal{W}\left(n, I_2, n\tau^2 \begin{pmatrix} 1 \\ b \end{pmatrix} \begin{pmatrix} 1 \\ b \end{pmatrix}'\right).$$

In the structural case of (1.1), W has the central Wishart distribution with n degrees of freedom and covariance matrix parameter $I_2 + \tau^2(1,b)(1,b)'$. Thus,

$$W \sim \mathcal{W}\left(n, I_2 + \tau^2 \begin{pmatrix} 1 \\ b \end{pmatrix} \begin{pmatrix} 1 \\ b \end{pmatrix}'\right).$$

Define the orthogonal matrix

$$\Gamma = (1+b^2)^{-\frac{1}{2}} \begin{pmatrix} 1 & -b \\ b & 1 \end{pmatrix},$$

and let

$$V = \begin{pmatrix} v_{11} & v_{12} \\ v_{12} & v_{22} \end{pmatrix} = \Gamma' W \Gamma. \tag{2.9}$$

Note that W and V have the same eigenvalues d_1, d_2, $|V| = |W| = d_1 d_2$, and

$$V = \frac{1}{1+\hat{b}_0^2} \begin{pmatrix} 1 & -\hat{b}_0 \\ \hat{b}_0 & 1 \end{pmatrix} \begin{pmatrix} d_1 & 0 \\ 0 & d_2 \end{pmatrix} \begin{pmatrix} 1 & \hat{b}_0 \\ -\hat{b}_0 & 1 \end{pmatrix}, \tag{2.10}$$

where

$$\hat{b}_0 = \frac{\hat{b} - b}{1 + \hat{b}b}.$$

Also

$$V \sim \mathcal{W}\left(n, I_2, n\tau^2(1+b^2) \begin{pmatrix} 1 & 0 \\ 0 & 0 \end{pmatrix}\right)$$

in the functional case, and

$$V \sim \mathcal{W}\left(n, I_2 + \tau^2(1+b^2)\begin{pmatrix}1 & 0\\ 0 & 0\end{pmatrix}\right)$$

in the structural case. Finally,

$$\frac{(\hat{b}-b)^2(d_1-d_2)^2}{(1+\hat{b}^2)^2 d_1 d_2} = \frac{(1-b\hat{b}_0)^2\hat{b}_0^2(d_1-d_2)^2}{d_1 d_2(1+\hat{b}_0^2)^2}$$
$$= (1-b\hat{b}_0)^2\frac{v_{12}^2}{|V|},$$

since $v_{12} = (1+\hat{b}_0^2)^{-1}\hat{b}_0(d_1-d_2)$.

Let

$$r = \frac{v_{12}}{(v_{11}v_{22})^{\frac{1}{2}}}, \qquad h = \frac{v_{11}}{v_{22}}.$$

Then

$$\frac{(\hat{b}-b)^2(d_1-d_2)^2}{(1+\hat{b}^2)^2 d_1 d_2} = (1-b\hat{b}_0)^2\frac{v_{12}^2}{|V|} = (1-b\hat{b}_0)^2\frac{r^2}{1-r^2}$$

and it can be shown that

$$\hat{b}_0 = \frac{1-h+[(h-1)^2+4hr^2]^{\frac{1}{2}}}{2rh^{\frac{1}{2}}}. \tag{2.11}$$

Hence,

$$H(k,\tau^2,b) = P\{b \text{ in } C_k\} = P\{\frac{r^2}{1-r^2}(1-b\hat{b}_0)^2 \le k\}, \tag{2.12}$$

where $\hat{b}_0$ is the function of h and r given by (2.11).

Lemma 1. The random variables r and h are statistically independent, with r having the density of the sample correlation coefficient ($\rho = 0$):

$$f_n(r) = \frac{(1-r^2)^{\frac{1}{2}(n-3)}}{B(\frac{1}{2},\frac{1}{2}(n-1))}, \qquad -1 \le r \le 1.$$

In the functional case of (1.1), h has the noncentral F-distribution with numerator and denominator degrees of freedom both equal to $n = N - 1$ and noncentrality parameter

$$\delta^2 = n\tau^2(1 + b^2).$$

Thus,

$$h \sim F_{n,n}(\delta^2).$$

In the structural case of (1.1),

$$h \sim [1 + \tau^2(1 + b^2)]F_{n,n},$$

where F_{ν_1,ν_2} is the central F distribution with ν_1 and ν_2 degrees of freedom.

Proof. Define $z = v_{11}^{-\frac{1}{2}} v_{12}$, $w = v_{22} - z^2$. In the functional case, since $V \sim \mathcal{W}(n, I_2, \delta^2\ (1,0)(1,0)')$, Theorem 2.2 of Gleser (1976) shows that v_{11}, z and w are mutually statistically independent with

$$v_{11} \sim \chi_n^2(\delta^2), \quad z \sim N(0,1), \quad w \sim \chi_{n-1}^2,$$

where $\chi_n^2(\delta^2)$ is the noncentral chi-squared distribution with n degrees of freedom and noncentrality parameter δ^2. In the structural case, since

$$V \sim \mathcal{W}\left(n, \begin{pmatrix} 1 + \tau^2(1 + b^2) & 0 \\ 0 & 1 \end{pmatrix}\right),$$

v_{11}, z, and w are again mutually statistically independent, with

$$v_{11} \sim [1 + \tau^2(1 + b^2)]\chi_n^2, \quad z \sim N(0,1), \quad w \sim \chi_{n-1}^2.$$

However,

$$r = \frac{z}{(w + z^2)^{\frac{1}{2}}}, \quad h = \frac{v_{11}}{w + z^2},$$

and it is well known that r and $w + z^2$ are independent (in either the functional or structural cases). The conclusions of the lemma now follow by standard distributional arguments. □

Observe that the distribution of r does not depend upon the parameters b and τ^2.

2.2. Bounds

For $rb \neq 0$ define

$$q_i(r,b) = \frac{r}{b} + (-1)^i \left|\frac{r}{b}\right| \left(\frac{k(1-r^2)}{r^2}\right)^{\frac{1}{2}}, \quad i = 1,2,$$

and note that when $b \neq 0$,

$$\frac{r^2}{1-r^2}(1-b\hat{b}_0)^2 = \frac{r^2}{1-r^2}(1-(r\hat{b}_0)\left(\frac{b}{r}\right))^2 \leq k$$

if and only if

$$q_1(r,b) \leq r\hat{b}_0 \leq q_2(r,b).$$

Thus, when $b \neq 0$, it follows from (2.12) that

$$H(k,\tau^2,b) = P\{q_1(r,b) \leq r\hat{b}_0 \leq q_2(r,b)\}. \tag{2.13}$$

When $b = 0$, it is easily seen from (2.12) that

$$H(k,\tau^2,0) = P\{\frac{r^2}{1-r^2} \leq k\}.$$

The function $H(k,\tau^2,b)$, regarded as a function of b, is difficult to work with, mainly because of the behavior of $q_1(r,b)$. Consequently, we look for good bounds to $H(k,\tau^2,b)$ that are better behaved as functions of b. To this end, define

$$\psi(k,\tau^2,b) = P\{\frac{r^2}{1-r^2} \leq k, \;\; 0 \leq r\hat{b}_0 \leq q_2(r,b)\}. \tag{2.14}$$

Lemma 2. For all k, τ^2, b,

$$\psi(k,\tau^2,b) \leq H(k,\tau^2,b) \leq \psi(k,\tau^2,b) + \frac{1}{2}P\{\frac{r^2}{1-r^2} > k\}. \tag{2.15}$$

Proof. For fixed $r \neq 0$, it can be seen from (2.11) that

$$r\hat{b}_0 \text{ is strictly decreasing and continuous in } h,$$
$$\lim_{h\to 0} r\hat{b}_0 = \infty, \lim_{h\to\infty} r\hat{b}_0 = 0. \tag{2.16}$$

Consequently, $r\hat{b}_0 \geq 0$.

First consider the case $b = 0$. Note that when $r \neq 0$ is fixed and $(1-r^2)^{-1}r^2 \leq k$, then $\lim_{b\to 0} q_2(r,b) = \infty$. Thus since $r\hat{b}_0 \geq 0$,

$$\psi(k,\tau^2,0) = P\{\frac{r^2}{1-r^2} \leq k, \quad 0 \leq r\hat{b}_0 \leq \infty\}$$
$$= P\{\frac{r^2}{1-r^2} \leq k\} = H(k,\tau^2,0).$$

Thus the left-hand inequality in (2.15) is true as an equality, and the right-hand inequality follows since $P\{(1-r^2)^{-1}r^2 > k\} \geq 0$.

For $b \neq 0$ define the events

$$A = \{q_1(r,b) \leq r\hat{b}_0 \leq q_2(r,b)\}, \quad B = \{\frac{r^2}{1-r^2} \leq k\},$$

and $E = \{rb \geq 0\}$. Note that $H(k,\tau^2,b) = P(A)$. Since $r^2(1-r^2)^{-1} \leq k$ implies that $q_1(r,b) \leq 0 \leq q_2(r,b)$, and since $r\hat{b}_0 \geq 0$,

$$A \cap B = \left\{\frac{r^2}{1-r^2} \leq k, \quad 0 \leq r\hat{b}_0 \leq q_2(r,b)\right\}.$$

Thus,

$$\psi(k,\tau^2,b) = P(A \cap B) \leq P(A) = H(k,\tau^2,b),$$

proving the left-hand inequality in (2.15).

On the other hand,

$$A \cap B^c = (A \cap B^c \cap E) \cup (A \cap B^c \cap E^c).$$

However,

$$A \cap B^c \cap E^c = \{\frac{r^2}{1-r^2} > k, \quad q_1(r,b) \leq r\hat{b}_0 \leq q_2(r,b), \quad rb < 0\}$$

is an impossible event, since $r\hat{b}_0 \geq 0$ while

$$rb < 0, \frac{r^2}{1-r^2} > k \implies q_2(r,b) < 0.$$

Thus

$$A \cap B^c = A \cap B^c \cap E \subset B^c \cap E = \{rb \geq 0, \frac{r^2}{1-r^2} > k\}.$$

Since the density $f_n(r)$ of r is symmetric about 0 (see Lemma 1) and does not depend on b or τ^2,

$$P(A \cap B^c) \leq P\{rb > 0, \frac{r^2}{1-r^2} > k\} = \frac{1}{2}P\{\frac{r^2}{1-r^2} > k\}.$$

Consequently,

$$\begin{aligned} H(k,\tau^2,b) &= P(A) = P(A \cap B) + P(A \cap B^c) \\ &\leq \psi(k,\tau^2,b) + \frac{1}{2}P\{\frac{r^2}{1-r^2} > k\}, \end{aligned}$$

verifying the right-hand inequality in (2.15). □

The desirable properties possessed by $\psi(k,\tau^2,b)$ are the following.

Lemma 3. In both the functional and structural cases of the model (1.1), the function $\psi(k,\tau^2,b)$ is

(i) strictly unimodal in b for fixed k, τ^2, with unique mode at $b=0$,

(ii) strictly decreasing in τ^2 for fixed k, b.

Further, for all fixed k, τ^2,

$$\lim_{b\to\pm\infty} \psi(k,\tau^2,b) = P\{v \leq \frac{\tau^2}{r^4}\left[r + (k(1-r^2))^{\frac{1}{2}}\right]^2\}, \tag{2.17}$$

where

$$\begin{aligned}&v \text{ is statistically independent of } r,\\ &r \text{ has the density } f_n(r) \text{ given in Lemma 1,}\\ &v \sim n^{-1}\chi_n^2, \text{ functional case,}\\ &v \sim F_{n,n}, \text{ structural case.}\end{aligned} \tag{2.18}$$

Proof. See Appendix. □

Let

$$L(k,\tau^2) = \lim_{b\to\pm\infty} \psi(k,\tau^2,b). \tag{2.19}$$

It follows from Lemma 3(i) that

$$L(k,\tau^2) = \inf_{-\infty<b<\infty} \psi(k,\tau^2,b). \tag{2.20}$$

Theorem 1. In both the functional and structural cases of the model (1.1),

$$L(k,\tau^2) \le H(k,\tau^2) \le L(k,\tau^2) + \frac{1}{2}P\{\frac{r^2}{1-r^2} > k\}, \tag{2.21}$$

where

$$H(k,\tau^2) = \inf_{-\infty<b<\infty} P\{b \text{ in } C_k\},$$

and $L(k,\tau^2)$ and r are defined by (2.19), (2.17) and (2.18).

Proof. Since r has distribution independent of b (and τ^2), (2.21) follows from (2.15) and (2.20). □

The quantity $\frac{1}{2}P\{(1-r^2)^{-1}r^2 > k\}$ indicates the size of the error made in approximating $H(k,\tau^2)$ by its lower bound $L(k,\tau^2)$. Since

$$\frac{(n-1)r^2}{1-r^2} \sim F_{1,n-1},$$

this quantity is easily calculated once we are given N and k. For large values of k, $\frac{1}{2}P\{(1-r^2)^{-1}r^2 > k\}$ will be small, and the approximation of $H(k,\tau^2)$ by $L(k,\tau^2)$ will be close. It is worth noting that it follows from (2.14) and (2.20) that

$$L(k,\tau^2) \le P\{\frac{r^2}{1-r^2} \le k\},$$

so that

$$\frac{1}{2}P\{\frac{r^2}{1-r^2} > k\} \le \frac{1}{2}(1 - L(k,\tau^2)). \tag{2.22}$$

This result allows one to bound the size of the error of the approximation $H(k,\tau^2) \approx L(k,\tau^2)$ using only the value of $L(k,\tau^2)$.

3. TABULATION OF THE BOUNDS

Let $G_n(\cdot)$ be the cumulative distribution function of the χ_n^2 distribution and $G_n^*(\cdot)$ be the cumulative distribution function of the $F_{n,n}$ distribution. Then in the functional case of (1.1),

$$L(k,\tau^2) = \int_{-t}^{t} G_n(\frac{n\tau^2}{r^4}[r + (k(1-r^2))^{\frac{1}{2}}]^2) f_n(r)dr$$

and in the structural case

$$L(k,\tau^2) = \int_{-t}^{t} G_n^*(\frac{\tau^2}{r^4}[r + (k(1-r^2))^{\frac{1}{2}}]^2) f_n(r)dr,$$

where $t = [(1+k)^{-1}k]^{\frac{1}{2}}$. These expressions are easily evaluated using Simpson's rule, together with IMSL subroutines for calculating $G_n(\cdot)$ and $G_n^*(\cdot)$. Breaking the interval $[-t,t]$ into $m = 50$ and $m = 100$ intervals yielded the same result (to four decimals) for each choice of N, k and τ^2 considered. Consequently, the results of the calculation of $L(k,\tau^2)$ given below are accurate to at least 3 decimal places.

In Table I appear values of $L(k,\tau^2)$ for $k = n^{-1}\chi^2_{1;\alpha}$, $\alpha =$ .10, .05, .01, $n = N - 1 =$ 10, 12, 15, 25, 30, 50, and $\tau^2 = 0.25$, 0.50, 1.00, 2.00. The values of $L(k,\tau^2)$ provide lower bounds for the minimum coverage probability (over b) of the confidence interval

$$\hat{b} \pm z_{\frac{1}{2}\alpha}\ \hat{\sigma}_{\hat{b}} n^{-\frac{1}{2}}. \tag{3.1}$$

This region is slightly wider than (1.2), but is asymptotically equivalent to (1.2) as $N \to \infty$. Coverage probabilities for the interval (3.1) are thus greater than, but very close to, the coverage probabilities of (1.2).

When reading Table I, it should be kept in mind that the difference between the true minimum coverage probability $H(n^{-1}\chi^2_{1;\alpha}, \tau^2)$ and the lower bound $L(n^{-1}\chi^2_{1;\alpha}, \tau^2)$ can be as large as $\frac{1}{2}P\{n(1-r^2)^{-1}r^2 > \chi^2_{1;\alpha}\}$. A rough idea of this error can be obtained by calculating

$$\frac{1}{2}(1 - L(n^{-1}\chi^2_{1;\alpha}, \tau^2));$$

see Equation (2.22).

One notable feature of Table I is the agreement between corresponding (comparable α, n, τ^2) values of $L(\chi^2_{1;\alpha}, \tau^2)$ in the functional and structural cases. This is not completely unexpected, since coverage probabilities for the structural case can be represented as expected values (over $u_1, \ldots, u_N$) of coverage probabilities in the functional case. The closeness of agreement is, however, surprising.

The decision as to whether the results of Table I support the use of $\hat{\sigma}_{\hat{b}}$ as a measure of accuracy for $\hat{b}$ in practice must be left up to individual judgement. Table I does indicate that the bounds $L(n^{-1}\chi^2_{1;\alpha}, \tau^2)$ are reasonably close to the desired coverage probabilities $1-\alpha$ when $n \geq 25$ and $\tau^2 \geq 1$. Interestingly, the device of using $k = n^{-1}\chi^2_{1;.01}$ to obtain a 95% confidence interval for b works well when $n \geq 25$, $\tau^2 \geq 0.25$.

A possible improvement over the confidence region (1.2) is to use $t_{n-1;\frac{1}{2}\alpha}$ in place of $z_{\frac{1}{2}\alpha}$. Such a substitution is often used in practice to adjust for finite sample size when using a large sample confidence region. Here, $t_{n-1;\frac{1}{2}\alpha}$ is the $100(1-\frac{1}{2}\alpha)^{th}$ percentile of the t distribution with $n-1$ degrees of freedom. Since

$$t^2_{n-1;\frac{1}{2}\alpha} = F_{1,n-1;\alpha},$$

the resulting interval is equivalent to C_k for $k = n^{-1}F_{1,n-1;\alpha}$. It is more convenient instead to use

$$k = (n-1)^{-1} F_{1,n-1;\alpha}$$

since then

$$P\{\frac{r^2}{1-r^2} > k\} = \frac{1}{2}\alpha$$

and the error in approximating $H(k,\tau^2)$ by $L(k,\tau^2)$ is fixed as a function of α. [It has already been noted in Section 2 that $(n-1)r^2(1-r^2)^{-1} \sim F_{1,n-1}$.]

Table II gives values of $L(k,\tau^2)$ for $k = (n-1)^{-1}\ F_{1,n;\alpha}$, $\alpha =$ .10, .05, .01, $n =$ 10, 12, 15, 25, 30, 50 and $\tau^2 =$ 0.25, 0.50, 1.00, 2.00. These values provide lower bounds for the minimum coverage probability (over b) of the interval

$$\hat{b} \pm t_{n-1;\frac{1}{2}\alpha}\ \hat{\sigma}_{\hat{b}}(n-1)^{-\frac{1}{2}}. \tag{3.2}$$

Since it is well-known that

$$\frac{t^2_{n-1;\frac{1}{2}\alpha}}{n-1} = \frac{F_{1,n-1;\alpha}}{n-1} \geq \frac{\chi^2_{1;\alpha}}{n} = \frac{z^2_{\frac{1}{2}\alpha}}{n},$$

the interval (3.2) is wider (has greater coverage probabilities) than the interval (3.1). Again, corresponding values for the functional and structural cases are extremely close. Comparing Tables I and II, we see that use of (3.2) provides substantial improvement in minimal coverage probabilities for small values of n. Consequently, use of the confidence interval (3.2) in place of either the interval (1.2) or (3.1) is recommended, particularly when sample sizes are small. When n is reasonably large, (3.1), (3.2) and (1.2) are nearly identical, since

$$\lim_{n\to\infty}\left(n\left(\frac{F_{1,n-1;\alpha}}{n-1} - \frac{\chi^2_{1;\alpha}}{n}\right)\right) = 0.$$

Use of the confidence interval (3.2) can be recommended in practice, particularly if the investigator has reason to believe that $\tau^2 \geq 1$. (Remember that $\psi(k,\tau^2,b)$, and thus $L(k,\tau^2)$, is strictly increasing in τ^2 so that values of $L(k,\tau^2)$ will be larger than those appearing in Table II when $\tau^2 > 2$.) The assumption that the variability in the u_i's exceeds the error variance σ^2 (that is, $\tau^2 \geq 1$) is quite reasonable.

Indeed, in psychometric practice

$$\rho = \left(\frac{\tau^2}{1+\tau^2} \right)^{\frac{1}{2}}$$

is called the ***reliability*** of the x_i measurements. This index (in the structural case of the model (1.1)) gives the correlation between x_i and an independent replication (with the same value of u_i) of x_i. Measurements (tests, instruments, etc.) with reliabilities of less than .7 are seldom used in psychometric practice. (Reliabilities are usually obtained experimentally when the measurement instrument is constructed.) Note that $\rho = .7$ corresponds approximately to $\tau^2 = 1$.

Although it is subject to the same accuracy problems as the estimator $\hat{b}$ of b, the consistent estimator

$$\hat{\tau}^2 = \frac{d_1 - d_2}{d_2(1 + \hat{b}^2)}$$

can be used to give an internal (from the given data) estimate of τ^2. If $\hat{\tau}^2$ is small (say, $\hat{\tau}^2 < 1$), one should be very careful when using large-sample inference methodology, such as the confidence interval (1.2), particularly when n is also small.

ACKNOWLEDGEMENT

Research on this paper was partially supported by National Science Foundation grant DMS-85-01966. The author wishes to thank Dr. Tai-Fang Lu for her help with the computation of Tables I and II.

APPENDIX

Proof of Lemma 3. To prove part (i) fix k and τ^2. Let

$$M(q; \delta^2, r) = P(\{h\colon\ r\hat{b}_0 \leq q\}),$$

where δ^2 is the parameter of the distribution of h. That is,

$$\delta^2 = \begin{cases} n\tau^2(1+b^2), & \text{in the functional case,} \\ 1 + \tau^2(1+b^2), & \text{in the structural case,} \end{cases}$$

as can be seen from Lemma 1. Using Lemma 1 and (2.14), it can be shown that

$$\psi(k,\tau,b) = \int_{\frac{r^2}{1-r^2}\leq k} M(q_2(r,b);\delta^2,r)f_n(r)dr. \qquad (A.1)$$

Take $(d/db)\psi(k,\tau^2,b)$. Since $M(q;\delta^2,r)$ is bounded between 0 and 1, the derivative can be taken inside the integral sign. However,

$$\begin{aligned}\frac{d}{db}&M(q_2(r,b);\delta^2,r)\\ &= \frac{d}{db}q_2(b,r)\frac{d}{dq}M(q;\delta^2,r)|_{q=q_2(r,b)}\\ &\quad + \left(\frac{d\delta^2}{db}\right)\frac{d}{d\delta^2}M(q_2(r,b);\delta^2,r).\end{aligned}$$

Since $r\hat{b}_0$ is strictly decreasing in h, and since h has strict monotone likelihood ratio in δ^2 (in both the functional and structural cases), it follows that $(d/d\delta^2)M(q;\delta^2,r) < 0$ for all q,δ^2,r. Since $M(q;\delta^2,r)$ is a cumulative distribution function as a function of q, $(d/dq)M(q;\delta^2,r) \geq 0$. Now

$$\frac{d}{db}(\delta^2) = \begin{cases} 2n\tau^2 b, & \text{in the functional case,} \\ 2\tau^2 b, & \text{in the structural case,} \end{cases}$$

while for $rb \neq 0$, $r^2(1-r^2)^{-1} \leq k$,

$$\begin{aligned}\frac{d}{db}q_2(b,r) &= \frac{d}{db}(\frac{r}{b} + |\frac{r}{b}|\left[\frac{r(1-r^2)}{r^2}\right]^{\frac{1}{2}})\\ &= \begin{cases} -(1+[\frac{k(1-r^2)}{r^2}]^{\frac{1}{2}})\frac{r}{b^2}, & \frac{r}{b}>0 \\ ([\frac{k(1-r^2)}{r^2}]^{\frac{1}{2}}-1)\frac{r}{b^2}, & \frac{r}{b}<0 \end{cases}\\ &= \begin{cases} <0, & \text{if } b>0, \\ >0, & \text{if } b<0. \end{cases}\end{aligned}$$

Combining the above results, we see that for all $b \neq 0$, $r \neq 0$,

$$\frac{d}{db} M(q_2(r,b); \delta^2, r) = \begin{cases} < 0, & b > 0, \\ > 0, & b < 0, \end{cases}$$

Thus, for $b \neq 0$, $(d/db)\psi(k, \tau^2, b)$ is < 0 for $b > 0$, and > 0 for $b < 0$. Since it is easy to show that $\psi(k, \tau^2, b) \leq \psi(k, \tau^2, 0)$ for all b, this completes the proof of part (i) of Lemma 3.

To prove part (ii), we can again use (A.1), the fact that h has strict monotone likelihood ratio in δ^2, and the fact that $r\hat{b}_0$ is strictly decreasing in h when $r \neq 0$ is fixed. Consequently, $M(q_2(r,b); \delta^2, r)$ is strictly decreasing in δ^2, and thus in τ^2, when $r \neq 0$ and b are fixed. It then follows immediately from (A.1) that $\psi(k, \tau^2, b)$ is strictly decreasing in τ^2.

Finally, from (A.1) and the Lebesgue dominated convergence theorem

$$\lim_{b \to \pm\infty} \psi(k, \tau^2, b) = \int_{\frac{r^2}{1-r^2} \leq k} \lim_{b \to \pm\infty} M(q_2(r,b); \delta^2, r) f_n(r) dr. \qquad (A.2)$$

Note that for $r \neq 0$, $r^2(1-r^2)^{-1} \leq k$,

$$\begin{aligned} \lim_{|b| \to \infty} b q_2(b,r) &= \lim_{|b| \to \infty} |b| \left(\frac{r}{b} + \frac{r}{b}\left[\frac{k(1-r^2)}{r^2}\right]^{\frac{1}{2}}\right) \\ &= \begin{cases} r + [k(1-r^2)]^{\frac{1}{2}}, & b \to \infty. \\ -r + [k(1-r^2)]^{\frac{1}{2}}, & b \to -\infty. \end{cases} \end{aligned} \qquad (A.3)$$

Also since

$$h \sim F_{n,n}(n\tau^2(1+b^2)), \qquad \text{in the functional case,}$$

$$h \sim [1 + \tau^2(1+b^2)] F_{n,n}, \qquad \text{in the structural case,}$$

in either case it can be shown that

$$\operatorname*{plim}_{|b| \to \infty} h = \infty, \quad \frac{h}{b^2} \to \frac{\tau^2}{v} \text{ in distribution as } b^2 \to \infty, \qquad (A.4)$$

where

$$v \sim n^{-1}\chi^2_n, \text{ in the functional case,}$$

$$v \sim F_{n,n}, \quad \text{in the structural case,}$$

and v is independent of r (since h is). However, for $r \neq 0$,

$$r\hat{b}_0 = \frac{1 - h \pm \sqrt{(h-1)^2 + 4r^2h}}{2h^{\frac{1}{2}}} = \frac{(h-1)}{2h^{\frac{1}{2}}}[-1 + (1 + \frac{4r^2h}{(h-1)^2})^{\frac{1}{2}}]$$
$$= r^2h^{-\frac{1}{2}}(1 + o(1)), h \to \infty.$$

It thus follows from (A.4) that for $r \neq 0$,

$$|b|r\hat{b}_0 \to r^2(\frac{v}{\tau^2})^{\frac{1}{2}} \tag{A.5}$$

in distribution as $b \to \pm\infty$. Since the limiting distribution of v is continuous, it follows from (A.3) and (A.5) that for $r \neq 0$, $r^2(1 - r^2)^{-1} \leq k$,

$$\lim_{b \to \pm\infty} M(q_2(r,b), d^2, r) = P\{v \leq \frac{\tau^2}{r^4}(\pm r + [k(1-r^2)]^{\frac{1}{2}})^2\}. \tag{A.6}$$

However, $f_n(r)$ is symmetric about $r = 0$. Hence, (2.17) directly follows from (A.2) and (A.6). This completes the proof of Lemma 3. □

Proof that b is median biased. Note that

$$P\{\hat{b} > b\} = P\{\frac{\hat{b} - b}{1 + b^2} > 0\} = P\{\frac{\hat{b}_0}{1 - b\hat{b}_0} > 0\}$$
$$= P\{1 - b\hat{b}_0 > 0, \hat{b}_0 > 0\} + P\{1 - b\hat{b}_0 < 0, \hat{b}_0 < 0\} \tag{A.7}$$

Note from (2.11) that $\hat{b}_0$ and r always have the same sign. Further, by the symmetry of $f_n(r)$ about $r = 0$, $P\{r < 0\} = \frac{1}{2}$. Thus, $P\{\hat{b}_0 < 0\} = \frac{1}{2}$. Assume $b > 0$. Then $\hat{b}_0 < 0$ implies $1 - b\hat{b}_0 > 0$. Consequently

(A.7) becomes

$$P\{\hat{b} > b\} = P\{\frac{1}{b} > \hat{b}_0 > 0\} + P\{\hat{b}_0 < 0\} \geq P\{\hat{b}_0 < 0\} = \frac{1}{2}$$

The inequality is strict unless $P\{b^{-1} > \hat{b}_0 > 0\} = 0$, which is easily shown not to be true.

For $b < 0$, similar arguments show

$$P\{\hat{b} > b\} = P\{\hat{b}_0 > 0\} + P\{\frac{1}{b} < 0\} > P\{\hat{b}_0 > 0\} = \frac{1}{2}.$$

Of course, when $b = 0$ $P\{\hat{b} > b\} = P\{\hat{b}_0 > 0\} = \frac{1}{2}$. Thus, $\hat{b}$ is median biased except when $b = 0$.

Department of Statistics
Purdue University
West Lafayette, Indiana 47907
U.S.A.

REFERENCES

ADCOCK, R. J. (1878), 'A problem in least squares', *The Analyst*, **5**, 53–54.

ANDERSON, T. W. (1976), 'Estimation of linear functional relationships: approximate distributions and connection with simultaneous equations in econometrics (with discussion)', *Journal of the Royal Statistical Society, Series B*, **38**, 1–36.

ANDERSON, T. W. (1984), 'Estimating linear statistical relationships'. *Annals of Statistics*, **12**, 1–45.

GLESER, LEON JAY (1976), 'A canonical representation for the Wishart distribution useful for simulation'. *Journal of the American Statistical Association*, **71**, 690–695.

GLESER, LEON JAY (1981), 'Estimation in a multivariate "errors in variables" regression model: Large sample results', *Annals of Statistics*, **9**, 24–44.

GLESER, LEON JAY (1983), 'Functional, structural and ultrastructural errors-in-variables models'. ***1983 Proceedings of the Business and Economic Statistics Section.*** American Statistical Association, Washington, D.C., 57–66.

GLESER, LEON JAY and HWANG, J. T. (1987), 'The nonexistence of $100(1-\alpha)\%$ confidence sets of finite expected diameter in errors-in-variables and related models'. *Annals of Statistics*, to appear.

KENDALL, MAURICE G. and STEWART, A. (1979). *The Advanced*

Theory of Statistics (Vol. II, Fourth Edition), New York: Macmillan.

NUSSBAUM, M. (1976). 'Maximum likelihood and least squares estimation of linear functional relationships'. *Mathematische Operationsforschung und Statistik, Series Statistik*, **7**, 23–49.

Table I

Lower bounds for the minimum (over b) coverage probability for the region C_k when $k = n^{-1}\chi^2_{1;\alpha}$.

		Functional Case			Structural Case		
τ^2	$n = N - 1$	$\alpha = .10$	$\alpha = .05$	$\alpha = .01$	$\alpha = .10$	$\alpha = .05$	$\alpha = .01$
0.25	10	.7471	.8050	.8780	.7430	.8004	.8726
	12	.7610	.8188	.8903	.7578	.8153	.8864
	15	.7759	.8335	.9032	.7734	.8309	.9005
	25	.8033	.8604	.9263	.8019	.8590	.9249
	30	.8114	.8682	.9328	.8103	.8671	.9317
	50	.8303	.8865	.9477	.8297	.8859	.9472
0.50	10	.7677	.8265	.8990	.7642	.8227	.8951
	12	.7812	.8397	.9103	.7783	.8366	.9073
	15	.7954	.8535	.9219	.7931	.8512	.9196
	25	.8209	.8780	.9418	.8197	.8768	.9407
	30	.8282	.8850	.9473	.8272	.8840	.9464
	50	.8449	.9008	.9593	.8444	.9003	.9588
1.00	10	.7855	.8447	.9160	.7823	.8414	.9128
	12	.7983	.8571	.9262	.7958	.8545	.9237
	15	.8118	.8699	.9365	.8098	.8680	.9347
	25	.8353	.8921	.9535	.8343	.8911	.9526
	30	.8418	.8982	.9580	.8410	.8974	.9573
	50	.8565	.9118	.9676	.8561	.9114	.9672
2.00	10	.8002	.8595	.9290	.7975	.8567	.9265
	12	.8125	.8712	.9383	.8103	.8690	.9363
	15	.8251	.8831	.9474	.8235	.8815	.9460
	25	.8467	.9031	.9619	.8459	.9023	.9613
	30	.8526	.9084	.9656	.8520	.9078	.9652
	50	.8655	.9201	.9733	.8652	.9198	.9731

Table II

Lower bounds for the minimum (over b) coverage probability for the region C_k when $k = (n-1)^{-1}F_{1,n;\alpha}$.

τ^2	$n = N - 1$	Functional Case $\alpha = .10$	$\alpha = .05$	$\alpha = .01$	Structural Case $\alpha = .10$	$\alpha = .05$	$\alpha = .01$
0.25	10	.8006	.8598	.9305	.7961	.8546	.9250
	12	.8051	.8636	.9324	.8017	.8598	.9284
	15	.8108	.8687	.9354	.8082	.8659	.9326
	25	.8238	.8807	.9438	.8224	.8793	.9424
	30	.8284	.8850	.9470	.8272	.8838	.9459
	50	.8403	.8963	.9555	.8397	.8957	.9550
0.50	10	.8221	.8812	.9485	.8183	.8773	.9449
	12	.8259	.8843	.9497	.8229	.8812	.9469
	15	.8306	.8883	.9518	.8284	.8860	.9497
	25	.8415	.8980	.9579	.8403	.8968	.9568
	30	.8452	.9014	.9602	.8442	.9005	.9594
	50	.8549	.9103	.9663	.8544	.9099	.9659
1.00	10	.8403	.8988	.9617	.8370	.8955	.9591
	12	.8433	.9011	.9625	.8407	.8985	.9604
	15	.8472	.9043	.9639	.8452	.9023	.9623
	25	.8559	.9118	.9681	.8549	.9108	.9673
	30	.8589	.9144	.9697	.8581	.9136	.9691
	50	.8665	.9211	.9739	.8661	.9207	.9736
2.00	10	.8551	.9125	.9711	.8524	.9099	.9691
	12	.8575	.9143	.9715	.8553	.9122	.9700
	15	.8605	.9166	.9724	.8589	.9151	.9713
	25	.8673	.9222	.9752	.8665	.9215	.9747
	30	.8696	.9242	.9763	.8689	.9236	.9759
	50	.8754	.9292	.9791	.8751	.9289	.9789

A. K. Gupta* and D. K. Nagar

LIKELIHOOD RATIO TEST FOR MULTISAMPLE SPHERICITY

ABSTRACT

This article deals with the null and nonnull distributions of the likelihood ratio criterion for testing multisample sphericity in q multinormal populations. Nonnull moments have been obtained using a simple and shortcut method. The null density has been derived using inverse Mellin transform and the calculus of residues. The nonnull density is given in a series involving zonal polynomials and generalized hypergeometric functions.

1. INTRODUCTION

Let $\underset{\sim}{X}_1,\ldots,\underset{\sim}{X}_q$ be random vectors of order $p \times 1$ which are distributed independently as multivariate normal with mean vectors $\underset{\sim}{\mu}_1,\ldots,\underset{\sim}{\mu}_q$ and covariance matrices $\Sigma_1,\ldots,\Sigma_q$ respectively. Let H denote the hypothesis of multisample sphericity, i.e.

$$H: \ \Sigma_1 = \ldots = \Sigma_q = \sigma^2 I_p \tag{1.1}$$

where $\sigma^2 > 0$ is an unknown constant and I_p is the identity matrix of order p. Such an hypothesis arises in repeated measures designs with two or more repeated factors, where if the assumption of homogeneity of group covariance matrices can not be made a priori, it needs to be tested.

It is easy to see that the modified likelihood ratio criterion for testing H is (see Mendoza (1980))

---

*Research initiated while he was University Grants Commission Visiting Fellow at the University of Rajasthan.

A. K. Gupta (ed.), Advances in Multivariate Statistical Analysis, 111–139.

$$\Lambda^* = \frac{n^{np/2}}{\prod_{i=1}^{q} n_i^{n_i p/2}} \frac{\prod_{i=1}^{q} |A_i|^{n_i/2}}{[\mathrm{tr}\ A/p]^{np/2}} \tag{1.2}$$

where $A = \sum_{i=1}^{q} A_i$, $A_i = \sum_{j=1}^{N_i} (\underset{\sim}{X}_{ij} - \underset{\sim}{X}_{i\cdot})(\underset{\sim}{X}_{ij} - \underset{\sim}{X}_{i\cdot})'$,

$\underset{\sim}{X}_{i\cdot} = \sum_{j=1}^{N_i} \underset{\sim}{X}_{ij}/N_i$, $N_0 = \sum_{i=1}^{q} N_i$, $n_i = N_i - 1$,

$n = \sum_i n_i = N_0 - q$ and $\underset{\sim}{X}_{ij}$ is the j-th $(j = 1,\ldots,N_i)$ independent observation on $\underset{\sim}{X}_i$ $(i = 1,\ldots,q)$.

It may be noted that in the case $q = 1$, (1.1) is the usual Mauchly's sphericity hypothesis. In the case $p = 1$, (1.1) is the Neyman and Pearson hypothesis for testing equality of variances of q univariate normal populations. These two problems have been studied by many authors, e.g. see Pillai and Nagarsenker (1971), Khatri and Srivastava (1971), Gupta (1977), Gupta and Rathie (1982).

The object of this paper is to derive null distribution of a one-to-one function of Λ^* (see Gupta et al. (1975), and Gupta and Tang (1984, 1986)). In Section 2, the nonnull moments of Λ^* are derived in the most general form in terms of Lauricella's hypergeometric functions as well as in multiple series involving zonal polynomials. Various particular cases are also discussed there. In Section 3, the null distribution is derived, using inverse Mellin transform and residue theorem. The nonnull distribution has been studied in Section 4.

2. NONNULL MOMENTS

It is well known that $A_1,\ldots,A_q$ are independent Wishart with parameters $(n_1,\Sigma_1),\ldots,(n_q,\Sigma_q)$ respectively. Since Λ^* is a function of $A_1,\ldots,A_q$, its h-th moment is obtained as

$$E(\Lambda^{*h}) = \left\{\frac{(np)^{np/2}}{\prod_{i=1}^{q} n_i^{n_i p/2}}\right\}^h \prod_{i=1}^{q} \{2^{n_i p/2} |\Sigma_i|^{n_i/2} \Gamma_p(n_i/2)\}^{-1}$$

$$\int_{A_1>0} \cdots \int_{A_q>0} [\mathrm{tr}\, A]^{-nph/2}$$

$$\prod_{i=1}^{q} |A_i|^{hn_i/2+(n_i-p-1)/2}$$

$$\mathrm{etr}\{-\tfrac{1}{2} \sum_{i=1}^{q} A_i \Sigma_i^{-1}\} dA_1 \ldots dA_q \qquad (2.1)$$

where $\mathrm{Re}[hn_i/2 + n_i/2] > (p + 1)/2 - 1$, $i = 1,\ldots,q$, and $\Gamma_p(a) = \{\prod_{j=1}^{p} \Gamma(a - \frac{(j-1)}{2})\}\pi^{p(p-1)/4}$. Replacing $[\mathrm{tr}\, A]^{-nph/2}$ by the equivalent gamma integral, namely

$$\{\Gamma(nph/2)\}^{-1} 2^{-nph/2} \int_0^\infty e^{-x\, \mathrm{tr}A/2} x^{nph/2-1} dx,$$

$$\mathrm{Re}(h) > 0,$$

and changing the order of integration, which is permissible and integrating out $A_1,\ldots,A_q$, we get

$$E(\Lambda^{*h}) = Z(h;p,q;n_i,\Sigma_i^{-1}, i = 1,\ldots,q)$$

$$\int_0^\infty \frac{x^{nph/2-1}}{\Gamma(nph/2)} \prod_{i=1}^{q} |\Sigma_i^{-1}+xI|^{-(\frac{hn_i}{2} + \frac{n_i}{2})} dx \qquad (2.2)$$

where

$$Z(h;p,q;n_i,\Sigma_i^{-1},i = 1,\dots,q) = \left\{\frac{(np)^{np/2}}{\prod_{i=1}^{q} n_i^{n_i p/2}}\right\}^h$$

$$\cdot \prod_{i=1}^{q} \left\{\frac{|\Sigma_i^{-1}|^{n_i/2} \Gamma_p(\frac{n_i h}{2} + \frac{n_i}{2})}{\Gamma_p(n_i/2)}\right\}. \qquad (2.3)$$

2.1 Moments in Terms of Lauricella's Function

Let $\theta_{i1} > \theta_{i2} > \dots > \theta_{ip} > 0$ be the eigenvalues of the p.d. matrix Σ_i, $i = 1,\dots,q$. Then

$$|\Sigma_i^{-1} + xI|^{-(\frac{hn_i}{2} + \frac{n_i}{2})} = (\frac{1 + \eta x}{\eta})^{-p(\frac{hn_i}{2} + \frac{n_i}{2})}$$

$$\cdot \prod_{j=1}^{p} \left[1 - \frac{(1 - \eta\theta_{ij}^{-1})}{(1 + \eta x)}\right]^{-(\frac{hn_i}{2} + \frac{n_i}{2})}. \qquad (2.4)$$

Substituting from (2.4) in the integral in (2.2), and transforming $y = 1/(1 + \eta x)$, we get

$$E(\Lambda*^h) = Z(h;p,q;n_i,\eta\Sigma_i^{-1},i = 1,\dots,q)\{\Gamma(nph/2)\}^{-1}$$

$$\cdot \int_0^1 y^{np/2-1}(1 - y)^{nph/2-1}$$

$$\cdot \prod_{i=1}^{q} \prod_{j=1}^{p} [1 - y(1 - \eta\theta_{ij}^{-1})]^{-(\frac{hn_i}{2} + \frac{n_i}{2})} dy. \qquad (2.5)$$

Now using the definition of Lauricella's function (Mathai

and Saxena (1978, p. 163)), one gets

$$E(\Lambda^{*h}) = Z(h;p,q;n_i,\eta\Sigma_i^{-1},i = 1,\ldots,q)\,\frac{\Gamma(\frac{np}{2})}{\Gamma(\frac{np}{2} + \frac{nph}{2})}$$

$$F_D^{(qp)}\left(\frac{np}{2}; \left\{\frac{hn_i}{2} + \frac{n_i}{2},\ldots,\frac{hn_i}{2} + \frac{n_i}{2}\right\},\right.$$

$$i = 1,\ldots,q;\ \frac{np}{2} + \frac{nph}{2};$$

$$\left.\{1 - \eta\theta_{i1}^{-1},\ldots,1 - \eta\theta_{ip}^{-1}\},\quad i = 1,2,\ldots,q\right)$$

$$\mathrm{Re}(h) > 0,\quad |1 - \eta\theta_{ij}^{-1}| < 1,$$

$$i = 1,2,\ldots,q,\quad j = 1,\ldots,p. \tag{2.6}$$

Various other properties and convergence conditions of $F_D(\ldots)$ are discussed in Exton (1976).

Special Cases. (i) For $p = 1$, $\Sigma_i = \sigma_i^2$, $i = 1,\ldots,q$ and the moment expression given in (2.6) reduces to

$$E(\Lambda^{*h}) = Z(h;1,q;n_i,\eta\sigma_i^{-2},i = 1,\ldots,q)$$

$$\{\Gamma(n/2)/\Gamma[n/2 + nh/2]\}$$

$$F_D^{(q)}\left(\frac{n}{2}; \frac{hn_1}{2} + \frac{n_1}{2},\ldots,\frac{hn_q}{2} + \frac{n_q}{2}; \frac{n}{2} + \frac{nh}{2};\right.$$

$$\left.1 - \eta\sigma_1^{-2},\ldots,1 - \eta\sigma_q^{-2}\right),\quad |1 - \eta\sigma_i^{-2}| < 1,$$

$$i = 1,\ldots,q \tag{2.7}$$

which is the h-th nonnull moment of the modified likelihood ratio statistic for testing equality of variances.

(ii) For $q = 1$, the h-th nonnull moment of the

sphericity criterion is derived as

$$E(\Lambda^{*h}) = Z(h;p,1;n_1,\eta\Sigma_1^{-1}) \frac{\Gamma(n_1p/2)}{\Gamma[n_1p/2+n_1ph/2]}$$

$$F_D^{(p)}\left(\frac{n_1p}{2}; \frac{hn_1}{2} + \frac{n_1}{2},\ldots,\frac{hn_1}{2} + \frac{n_1}{2};\right.$$

$$\left.\frac{n_1p}{2} + \frac{hn_1p}{2}; 1 - \eta\theta_{11}^{-1},\ldots,1 - \eta\theta_{1p}^{-1}\right),$$

$$|1 - \eta\theta_{1j}^{-1}| < 1, \quad j = 1,\ldots,p. \qquad (2.8)$$

Similarly nonnull moments for the cases (iii) $q = 2$, $p = 2$ and $\Sigma_1 = \Sigma_2 = \text{diag}(\theta_1,\theta_2)$ and (iv) $q = 2$, $p = 2$ and $\Sigma_1 = \sigma_1^2 I_2$, $\Sigma_2 = \sigma_2^2 I_2$ can also be derived.

2.2 Moments in Terms of Zonal Polynomials

Alternately expanding $|\Sigma_i^{-1} + xI|^{-\frac{1}{2}n_i(1+h)}$ in terms of zonal polynomials, one gets

$$|\Sigma_i^{-1} + xI|^{-(\frac{hn_i}{2} + \frac{n_i}{2})} = \left(\frac{1 + \eta x}{\eta}\right)^{-p(\frac{hn_i}{2} + \frac{n_i}{2})}$$

$$|I - (I - \eta\Sigma_i^{-1})/(1 + \eta x)|^{-(\frac{n_ih}{2} + \frac{n_i}{2})}$$

$$= \left(\frac{1 + \eta x}{\eta}\right)^{-p(\frac{hn_i}{2} + \frac{n_i}{2})} \sum_{k^{(i)}=0}^{\infty} \sum_{\kappa^{(i)}}$$

$$\frac{\left(\frac{hn_i}{2} + \frac{n_i}{2}\right)_{\kappa^{(i)}}}{k^{(i)}!} \frac{C_{\kappa^{(i)}}(I - \eta\Sigma_i^{-1})}{(1 + \eta x)^{k^{(i)}}} \qquad (2.9)$$

where $\kappa^{(i)} = (k_1^{(i)},\ldots,k_p^{(i)})$, $k_1^{(i)} \geq \ldots \geq k_p^{(i)} \geq 0$, $k_1^{(i)} + \ldots + k_p^{(i)} = k^{(i)}$ and $(a)_m = a(a+1)\ldots(a+m-1)$, $(a)_{\kappa^{(i)}} = \prod_{j=1}^{p} (a-(j-1)/2)_{k_j^{(i)}}$. $C_{\kappa^{(i)}}$ is the zonal polynomial of order $k^{(i)}$. For a discussion of zonal polynomials see James (1964). The series in (2.9) is valid for $\|(I - \eta\Sigma_i^{-1})/(1+\eta x)\| < 1$. Now substituting from (2.9) in (2.2) and integrating out x term by term, one obtains

$$E(\Lambda^{*h}) = Z(h;p,q;n_i,\eta\Sigma_i^{-1},i=1,\ldots,q) \sum_{k^{(1)}=0}^{\infty}\cdots\sum_{k^{(q)}=0}^{\infty}\sum_{\kappa^{(1)}}\cdots\sum_{\kappa^{(q)}} \frac{\left(\frac{hn_1}{2}+\frac{n_1}{2}\right)_{\kappa^{(1)}}}{k^{(1)}!}\cdots\frac{\left(\frac{hn_q}{2}+\frac{n_q}{2}\right)_{\kappa^{(q)}}}{k^{(q)}!}$$

$$C_{\kappa^{(1)}}(I-\eta\Sigma_1^{-1})\ldots C_{\kappa^{(q)}}(I-\eta\Sigma_q^{-1})$$

$$\frac{\Gamma[\frac{np}{2}+k^{(1)}+\ldots+k^{(q)}]}{\Gamma[\frac{np}{2}(1+h)+k^{(1)}+\ldots+k^{(q)}]} \qquad (2.10)$$

Notice that when $p = 1$, the above expression reduces to (2.7) with different notations. When $q = 1$ the above moment sequence reduces to that of sphericity criterion.

3. DENSITY IN SERIES FORM

From the result of the previous section, the h-th null

moment of $V = \Lambda*^{2/n}$, for $n_i = n$, $i = 1,\ldots,q$, is derived as

$$E(V^h) = \frac{(pq)^{pqh}\Gamma(\frac{npq}{2})}{\Gamma[pq(\frac{n}{2} + h)]} \prod_{j=1}^{p} \left\{\frac{\Gamma^q(\frac{n - j + 1}{2} + h)}{\Gamma^q(\frac{n - j + 1}{2})}\right\}. \tag{3.1}$$

Simplifying $\Gamma[pq(\frac{n}{2} + h)]$ by using Gauss-Legendre multiplication formula and writing $\Gamma[\frac{n - 1 + 1}{2} + h]\ldots\Gamma[\frac{n - p + 1}{2} + h]$ in the reverse order as $\Gamma[\frac{n - p + 1}{2} + h]\ldots\Gamma[\frac{n - p + p}{2} + h]$, the above expression is written as

$$E(V^h) = \frac{(2\pi)^{\frac{pq-1}{2}}\Gamma(\frac{npq}{2})}{(pq)^{\frac{npq-1}{2}} \prod_{k=0}^{pq-1} \Gamma(\frac{n}{2} + h + \frac{k}{pq})} \prod_{j=1}^{p} \left\{\frac{\Gamma^q(\frac{n - p + j}{2} + h)}{\Gamma^q(\frac{n - j + 1}{2})}\right\}. \tag{3.2}$$

Now from (3.2) using inverse Mellin transform and substituting $\frac{n - p}{2} + h = t$, we have the density of V as

$$f(v) = K(n,p,q)(2\pi\omega)^{-1} v^{\frac{n-p}{2} - 1} \int_C \Delta(t) v^{-t} dt,$$

$$0 < v < 1 \tag{3.3}$$

where $\omega = (-1)^{\frac{1}{2}}$,

$$K(n,p,q) = (2\pi)^{(pq-1)/2}\Gamma(npq/2)/\left\{(pq)^{\frac{npq-1}{2}} \prod_{j=1}^{p} \Gamma^{q}\left(\frac{n-j+1}{2}\right)\right\} \quad (3.4)$$

$$\Delta(t) = \left\{\prod_{j=1}^{p} \Gamma^{q}(t + j/2)\right\}/\left\{\prod_{k=0}^{pq-1} \Gamma(t + p/2 + k/pq)\right\} \quad (3.5)$$

and C is a suitable contour. A contour C exists for which (3.3) can be represented as a G-function and it can be evaluated as a sum of the residues at the poles of the integrand. Properties of G-function and other details are available in Mathai and Saxena (1973). For the computation of percentage points one needs an explicit representation for $f(v)$. In order to evaluate the density in computable form, that is as a sum of the residues, we identify all the poles of the integrand and their orders. Since the alternate gamma functions in the numerator of (3.4) differ by one and the adjacent ones by half, the poles of the alternate gamma function coincide whereas the poles of adjacent gamma functions do not. So we separate the two types of poles and represent all the poles in two sets. Also some of the gamma functions in the numerator may cancel out with the gamma functions in the denominator, we therefore, for simplicity, consider three cases: p-even; p-odd, q-even and p-odd, q-odd separately.

When p is even the numerator gammas in the integrand are written as

$$\prod_{j=1}^{p} \Gamma^{q}[t + j/2] = \prod_{j=1}^{p/2} \Gamma^{q}(t + j) \prod_{j=1}^{p/2} \Gamma^{q}(t - \tfrac{1}{2} + j)$$

and two gammas in the denominator corresponding to $k = 0$ and $k = pq/2$ cancel out with $\Gamma(t + p/2)$ and $\Gamma(t + p/2 - \tfrac{1}{2})$, and the expression (3.5) is simplified as

$$\Delta_1(t) = [\prod_{j=1}^{p/2-1} \{\Gamma(t+j)\Gamma(t-\tfrac{1}{2}+j)\}^q$$

$$\{\Gamma(t+p/2)\Gamma(t+p/2-\tfrac{1}{2})\}^{q-1}]/$$

$$[\{\prod_{k=1(\neq pq/2)}^{pq-1} \Gamma(t+p/2+k/pq)\}(t+p/2-\tfrac{1}{2})]. \tag{3.6}$$

When p is odd the numerator gamma in the integrand are written as

$$\prod_{j=1}^{p} \Gamma^q(t+j/2) = \prod_{j=1}^{(p-1)/2} \Gamma^q(t+j)$$

$$\cdot \prod_{j=1}^{(p+1)/2} \Gamma^q(t-\tfrac{1}{2}+j)$$

and if q is even two gammas of the denominator corresponding to $k = 0$ and $k = pq/2$ cancel out with $\Gamma(t + p/2)$ and $\Gamma(t + p/2 - \frac{1}{2})$ leaving the factor $(t + p/2 - \frac{1}{2})$ in the denominator. In this case

$$\Delta_2(t) = [\{\prod_{j=1}^{(p-1)/2-1} \Gamma^q(t+j)\}$$

$$\{\prod_{j=1}^{(p-1)/2} \Gamma^q(t-\tfrac{1}{2}+j)\}\{\Gamma(t+\frac{p}{2})$$

$$\Gamma(t+\frac{p-1}{2})\}^{q-1}]/[\{\prod_{k=1(\neq pq/2)}^{pq-1} \Gamma(t+\frac{p}{2}+\frac{k}{pq})\}$$

$$(t+\frac{p-1}{2})]. \tag{3.7}$$

If q is odd only one gamma of the denominator corresponding to $k = 0$ cancels out with $\Gamma(t + p/2)$ of numerator and the expression (3.5) simplifies to

$$\Delta_3(t) = [\prod_{j=1}^{(p-1)/2} \Gamma^q(t - \tfrac{1}{2} + j) \prod_{j=1}^{(p-1)/2} \Gamma^q(t + j)$$

$$\cdot \Gamma^{q-1}(t + p/2)]/[\prod_{k=1}^{pq-1} \Gamma(t + \frac{p}{2} + \frac{k}{pq})]. \qquad (3.8)$$

Now with the help of Δ_1, Δ_2 and Δ_3, one can easily identify poles of the integrand and their order for three different cases. The poles are available by equating to zero each factor of

$$\prod_{j=1}^{\infty} (t + j)^{a_j} \prod_{j=1}^{\infty} (t - \tfrac{1}{2} + j)^{b_j}$$

where a_j and b_j give the orders of the poles at $t = -j$ and $t = -j + \frac{1}{2}$ respectively, which can be easily obtained as follows.

<u>p-even</u>:

$$a_j = \begin{cases} qj & , \quad j = 1,2,\ldots,p/2 - 1, \\ \frac{pq}{2} - 1, & j = p/2, p/2 + 1, \ldots, \end{cases}$$

$$b_j = \begin{cases} qj & , \quad j = 1,2,\ldots,p/2, \\ \frac{pq}{2} - 1, & j = p/2 + 1, \ldots \end{cases} \qquad (3.9)$$

<u>p-odd</u> <u>and</u> <u>q-even</u>:

$$a_j = \begin{cases} qj & , \quad j = 1,2,\ldots,(p - 1)/2, \\ \frac{q(p - 1)}{2} - 1, & j = (p - 1)/2 + 1, \ldots \end{cases} \qquad (3.10)$$

p-odd and q-odd:

$$a_j = \begin{cases} qj & , \quad j = 1,2,\ldots,\frac{p-1}{2}, \\ q(\frac{p-1}{2}), & j = \frac{p-1}{2} + 1,\ldots, \end{cases} \tag{3.11}$$

p-odd:

$$b_j = \begin{cases} qj & , \quad j = 1,2,\ldots,\frac{p-1}{2}, \\ q(\frac{p+1}{2}) - 1, & j = \frac{p+1}{2},\ldots . \end{cases} \tag{3.12}$$

Now, using the residue theorem we get the following result from (3.3).

THEOREM 3.1. *The p.d.f. of* $V = \Lambda^{*2/n}$, *for* $n_i = n$, $i = 1,\ldots,q$, *where* Λ^* *is the modified likelihood ratio criterion for testing* H, *is given by*

$$f(v) = K(n,p,q)v^{\frac{n-p}{2}-1} \sum_{i=1}^{\infty} [R_{1i} + R_{2i}], \quad 0 < v < 1 \tag{3.13}$$

where $K(n,p,q)$ *is defined in (3.4),* R_{1i} *and* R_{2i} *are the residues at the poles* $t = -i$ *and* $t = -i + \frac{1}{2}$, *of orders* a_i *and* b_i *respectively. Also*

$$R_{1i} = \frac{1}{(a_i - 1)!} \lim_{t\to -i} \frac{\partial^{a_i-1}}{\partial t^{a_i-1}} [(t+i)^{a_i}\Delta(t)v^{-t}] \tag{3.14}$$

and

$$R_{2i} = \frac{1}{(b_i - 1)!} \lim_{t\to -i+\frac{1}{2}} \frac{\partial^{b_i-1}}{\partial t^{b_i-1}} [(t-\tfrac{1}{2}+i)^{b_i}\Delta(t)v^{-t}] \tag{3.15}$$

where

$\Delta(t) \equiv \Delta_1(t)$ *and* a_i, b_i *are given by (3.9) if* p *is even,*

$\equiv \Delta_2(t)$ *and* a_i, b_i *are given by (3.10), (3.12) if* p *is odd and* q *is even,*

$\equiv \Delta_3(t)$ *and* a_i, b_i *are given by (3.11), (3.12) if both* p *and* q *are odd.*

Clearly the residues R_{1i} and R_{2i} $(i = 1,2,\ldots)$ will be different for the three cases considered above and hence the density will be different for these cases. Considering the three cases separately one can derive the expressions for residues explicitly. For illustration, we here derive the explicit expressions for the case p-even that is when $\Delta = \Delta_1$ and a_i, b_i are given by (3.9). In this case the expression (3.14) is written as

$$R_{1i} = \frac{1}{(a_i - 1)!} \lim_{t \to -i} \frac{\partial^{a_i - 1}}{\partial t^{a_i - 1}} [A_{1i} v^{-t}] \tag{3.16}$$

where

$$A_{1i} = [\Gamma^{qi}(t + i + 1) \prod_{j=i+1}^{p/2-1} \Gamma^{q}(t + j)$$

$$\prod_{j=1}^{p/2-1} \Gamma^{q}(t - \tfrac{1}{2} + j)\Gamma^{q-1}(t + \tfrac{p}{2})\Gamma^{q-1}(t + \tfrac{p-1}{2})]/$$

$$[\prod_{k=1(\neq pq/2)}^{pq-1} \Gamma(t + p/2 + k/pq)(t + p/2 - \tfrac{1}{2})$$

$$\prod_{j=1}^{i-1} (t + j)^{qj}] \quad \text{for } i = 1,2,\ldots,p/2 - 1,$$

$$= [\Gamma^{\frac{1}{2}pq-1}(t + i + 1) \prod_{j=1}^{p/2-1} \Gamma^{q}(t - \tfrac{1}{2} + j)$$

$$\Gamma^{q-1}(t + \frac{p - 1}{2})]/[\prod_{k=1(\neq pq/2)}^{pq-1} \Gamma(t + p/2 + k/pq)$$

$$(t + \frac{p - 1}{2}) \prod_{j=1}^{p/2-1} (t + j)^{qj}$$

$$\prod_{j=p/2}^{i-1} (t + j)^{pq/2-1}] \quad \text{for} \quad i = p/2, p/2+1, \ldots . \tag{3.17}$$

By using a result from differential calculus it is possible to further develop the expression (3.16), as

$$R_{1i} = \frac{1}{(a_i - 1)!} \lim_{t \to -i} v^{-t} [\frac{\partial}{\partial t} + (-\log v)]^{a_i - 1} A_{1i}$$

$$= \frac{v^i}{(a_i - 1)!} \lim_{t \to -i} \sum_{r=0}^{a_i - 1} \binom{a_i - 1}{r}$$

$$(-\log v)^{a_i - 1 - r} A_{1i}^{(r)}. \tag{3.18}$$

Clearly

$$A_{1i}^{(r)} = \frac{\partial^r}{\partial t^r} A_{1i} = \frac{\partial^{r-1}}{\partial t^{r-1}} \left(\frac{\partial}{\partial t} A_{1i}\right)$$

$$= \frac{\partial^{r-1}}{\partial t^{r-1}} (A_{1i} \frac{\partial}{\partial t} \log A_{1i}) = \frac{\partial^{r-1}}{\partial t^{r-1}} (A_{1i} B_{1i}) \tag{3.19}$$

where

$$B_{1i} = \frac{\partial}{\partial t} \log A_{1i}$$

$$= qi\psi(t + i + 1) + q \sum_{j=i+1}^{p/2-1} \psi(t + j)$$

$$+ q \sum_{j=1}^{p/2-1} \psi(t - \tfrac{1}{2} + j) + (q - 1)\psi(t + \tfrac{p}{2})$$

$$+ (q - 1)\psi(t + \frac{p - 1}{2}) - \sum_{k=1(\neq pq/2)}^{pq-1} \psi(t + \frac{p}{2} + \frac{k}{pq})$$

$$- (t + \frac{p - 1}{2})^{-1} - q \sum_{j=1}^{i-1} j(t + j)^{-1} \quad \text{for}$$

$$i = 1,2,\ldots,p/2 - 1, \tag{3.20}$$

$$= (\frac{pq}{2} - 1)\psi(t + i + 1) + q \sum_{j=1}^{p/2-1} \psi(t - \tfrac{1}{2} + j)$$

$$+ (q - 1)\psi(t + \frac{p - 1}{2}) - \sum_{k=1(\neq pq/2)}^{pq-1} \psi(t + \frac{p}{2} + \frac{k}{pq})$$

$$- (t + \frac{p - 1}{2})^{-1} - q \sum_{j=1}^{p/2-1} j(t + j)^{-1}$$

$$- (\frac{pq}{2} - 1) \sum_{j=p/2}^{i-1} (t + j)^{-1}$$

$$\text{for} \quad i = p/2, p/2 + 1, \ldots \; . \tag{3.21}$$

Consequently all the derivatives of A_{1i} are available from recursive relation

$$A_{1i}^{(r)} = \sum_{m=0}^{r-1} \binom{r - 1}{m} A_{1i}^{(r-1-m)} B_{1i}^{(m)} \tag{3.22}$$

with

$$B_{1i}^{(m)} = \frac{\partial^m}{\partial t^m} B_{1i}$$

$$= (-1)^{m+1} m! [qi\zeta(m + 1, t + i + 1)$$

$$+ q \sum_{j=i+1}^{p/2-1} \zeta(m + 1, t + j)$$

$$+ q \sum_{j=1}^{p/2-1} \zeta(m + 1, t - \tfrac{1}{2} + j)$$

$$+ (q - 1)\zeta(m + 1, t + \tfrac{p}{2})$$

$$+ (q - 1)\zeta(m + 1, t + \frac{p - 1}{2})$$

$$- \sum_{k=1(\neq pq/2)}^{pq-1} \zeta(m + 1, t + \frac{p}{2} + \frac{k}{pq})$$

$$+ (t + \frac{p - 1}{2})^{-1-m} + q \sum_{j=1}^{i-1} j(t + j)^{-1-m}]$$

$$\text{for } i = 1,2,\ldots,p/2 - 1, \qquad (3.23)$$

$$= (-1)^{m+1} m! [(\frac{pq}{2} - 1)\zeta(m + 1, t + i + 1)$$

$$+ q \sum_{j=1}^{p/2-1} \zeta(m + 1, t - \tfrac{1}{2} + j)$$

$$+ (q - 1)\zeta(m + 1, t + \frac{p - 1}{2})$$

$$- \sum_{k=1(\neq pq/2)}^{pq-1} \zeta(m + 1, t + \frac{p}{2} + \frac{k}{pq}) + (t + \frac{p-1}{2})^{-1-m}$$

$$+ q \sum_{j=1}^{p/2-1} j(t + j)^{-1-m}$$

$$+ (\frac{pq}{2} - 1) \sum_{j=p/2}^{i-1} (t + j)^{-1-m}]$$

$$\text{for} \quad i = p/2, p/2 + 1, \ldots . \tag{3.24}$$

From (3.18) it is possible to write

$$R_{1i} = \frac{v^i}{(a_i - 1)!} \sum_{r=0}^{a_i-1} \binom{a_i - 1}{r} (-\log v)^{a_i-1-r} A_{1i0}^{(r)} \tag{3.25}$$

where

$$A_{1i0}^{(r)} = \sum_{m=0}^{r-1} \binom{r - 1}{m} A_{1i0}^{(r-1-m)} B_{1i0}^{(m)} \tag{3.26}$$

with

$$A_{1i0} = [\prod_{j=i+1}^{p/2-1} \Gamma^q(j - i) \prod_{j=1}^{p/2-1} \Gamma^q(j - i - \tfrac{1}{2})$$

$$\cdot \Gamma^{q-1}(\frac{p}{2} - i)\Gamma^{q-1}(\frac{p - 1}{2} - i)]/[\prod_{j=1}^{i-1} (j - i)^{qj}$$

$$\prod_{k=1(\neq pq/2)}^{pq-1} \Gamma(\frac{p}{2} - i + \frac{k}{pq})(\frac{p - 1}{2} - i)]$$

$$\text{for} \quad i = 1,2,\ldots,p/2 - 1, \tag{3.27}$$

$$= [\prod_{j=1}^{p/2-1} \Gamma^q(j - i - \tfrac{1}{2})\Gamma^{q-1}(\frac{p - 1}{2} - i)]/$$

$$[\{\prod_{k=1(\neq pq/2)}^{pq-1} \Gamma(\frac{p}{2} - i + \frac{k}{pq})\}(\frac{p - 1}{2} - i)$$

$$\cdot \prod_{j=1}^{p/2-1} (j - i)^{qj} \prod_{j=p/2}^{i-1} (j - i)^{pq/2-1}]$$

for $i = p/2, p/2 + 1, \ldots,$ (3.28)

$$B_{1i0} = qi\psi(1) + \sum_{j=i+1}^{p/2-1} \psi(j - i)$$

$$+ q \sum_{j=1}^{p/2-1} \psi(j - i - \tfrac{1}{2}) + (q - 1)\psi(\tfrac{p}{2} - i)$$

$$+ (q - 1)\psi(\tfrac{p - 1}{2} - i)$$

$$- \sum_{k=1(\neq pq/2)}^{pq-1} \psi(\tfrac{p}{2} - i + \tfrac{k}{pq}) - (\tfrac{p - 1}{2} - i)^{-1}$$

$$- q \sum_{j=1}^{i-1} j(j - i)^{-1} \quad \text{for} \quad i = 1,2,\ldots,p/2 - 1,$$

(3.29)

$$= (\tfrac{pq}{2} - 1)\psi(1) + q \sum_{j=1}^{p/2-1} \psi(j - i - \tfrac{1}{2})$$

$$+ (q - 1)\psi(\tfrac{p - 1}{2} - i)$$

$$- \sum_{k=1(\neq pq/2)}^{pq-1} \psi(\tfrac{p}{2} - i + \tfrac{k}{pq}) - (\tfrac{p - 1}{2} - i)^{-1}$$

$$- q \sum_{j=1}^{p/2-1} j(j - i)^{-1} - (\tfrac{pq}{2} - 1) \sum_{j=p/2}^{i-1} (j - i)^{-1}$$

for $i = p/2, p/2 + 1, \ldots,$ (3.30)

$$B_{1i0}^{(m)} = (-1)^{m+1} m! [qi\zeta(m + 1,1)$$

$$+ q \sum_{j=i+1}^{p/2-1} \zeta(m + 1, j - i)$$

$$+ q \sum_{j=1}^{p/2-1} \zeta(m + 1, j - i - \tfrac{1}{2})$$

$$+ (q - 1)\zeta(m + 1, \frac{p}{2} - i)$$

$$+ (q - 1)\zeta(m + 1, \frac{p - 1}{2} - i)$$

$$- \sum_{k=1(\neq pq/2)}^{pq-1} \zeta(m + 1, \frac{p}{2} - i + \frac{k}{pq})$$

$$+ (\frac{p - 1}{2} - i)^{-1-m} + q \sum_{j=1}^{i-1} j(j - i)^{-1-m}]$$

$$\text{for} \quad i = 1,2,\ldots,p/2 - 1, \qquad (3.31)$$

$$= (-1)^{m+1} m! [(\frac{pq}{2} - 1)\zeta(m + 1, 1)$$

$$+ q \sum_{j=1}^{p/2-1} \zeta(m + 1, j - i - \tfrac{1}{2})$$

$$+ (q - 1)\zeta(m + 1, \frac{p - 1}{2} - i)$$

$$- \sum_{k=1(\neq pq/2)}^{pq-1} \zeta(m + 1, \frac{p}{2} - i + \frac{k}{pq})$$

$$+ (\frac{p - 1}{2} - i)^{-1-m} + q \sum_{j=1}^{p/2-1} j(j - i)^{-1-m}$$

$$+ (\frac{pq}{2} - 1) \sum_{j=p/2}^{i-1} (j - i)^{-1-m}]$$

$$\text{for} \quad i = p/2, p/2 + 1, \ldots . \tag{3.32}$$

Here the suffix o indicates the value of the function and derivatives at $t = -i$. $\psi(\cdot)$ and $\zeta(\cdot,\cdot)$ are well known psi and zeta functions [see Gradshteyn and Ryzhik (1965)]. Following exactly the same procedure the residue at $t = -i + \frac{1}{2}$ is obtained as

$$R_{2i} = \frac{v^{i-\frac{1}{2}}}{(b_i - 1)!} \sum_{r=0}^{b_i-1} \binom{b_i - 1}{r} (-\log v)^{b_i-1-r} A_{2i0}^{(r)} \tag{3.33}$$

where

$$A_{2i0}^{(r)} = \sum_{m=0}^{r-1} \binom{r-1}{m} A_{2i0}^{(r-1-m)} B_{2i0}^{(m)} \tag{3.34}$$

with

$$A_{2i0} = [\prod_{j=i+1}^{p/2-1} \Gamma^q(j - i) \prod_{j=1}^{p/2-1} \Gamma^q(j + \tfrac{1}{2} - i)$$

$$\cdot\, \Gamma^{q-1}(\tfrac{p+1}{2} - i)\Gamma^{q-1}(\tfrac{p}{2} - i)]/$$

$$[\{\prod_{k=1(\neq pq/2)}^{pq-1} \Gamma(\tfrac{p+1}{2} - i + \tfrac{k}{pq})\}(\tfrac{p}{2} - i)$$

$$\cdot \prod_{j=1}^{i-1} (j - i)^{qj}] \quad \text{for} \quad i = 1,2,\ldots,p/2 - 1, \tag{3.35}$$

$$= [\prod_{j=1}^{p/2-1} \Gamma^q(j + \tfrac{1}{2} - \tfrac{p}{2})\Gamma^{q-1}(\tfrac{1}{2})]/$$

$$[\prod_{k=1(\neq pq/2)}^{pq-1} \Gamma(\tfrac{1}{2} + \tfrac{k}{pq}) \prod_{j=1}^{p/2-1} (j - p/2)^{qj}]$$

$$\text{for} \quad i = p/2, \tag{3.36}$$

$$= \left[\prod_{j=1}^{p/2-1} \Gamma^{q}(j - i + \tfrac{1}{2})\Gamma^{q-1}(\frac{p+1}{2} - i)\right] /$$

$$\left[\prod_{k=1(\neq pq/2)}^{pq-1} \Gamma(-i + \frac{p+1}{2} + \frac{k}{pq}) \prod_{j=1}^{p/2} (j - i)^{qj}\right.$$

$$\left.\prod_{j=p/2+1}^{i-1} (j - i)^{pq/2-1}\right] \quad \text{for} \quad i = p/2 + 1, \ldots, \qquad (3.37)$$

$$B_{2i0} = qi\psi(1) + q\sum_{j=1}^{p/2-1} \psi(j - i + \tfrac{1}{2}) + q\sum_{j=i+1}^{p/2-1} \psi(j-i)$$

$$+ (q - 1)\{\psi(\frac{p+1}{2} - i) + \psi(\frac{p}{2} - i)\}$$

$$- (\frac{p}{2} - i)^{-1} - q\sum_{j=1}^{i-1} j(j - i)^{-1}$$

$$- \sum_{k=1(\neq pq/2)}^{pq-1} \psi(\frac{p+1}{2} - i + \frac{k}{pq})$$

$$\text{for} \quad i = 1,2,\ldots,p/2 - 1, \qquad (3.38)$$

$$= (\frac{pq}{2} - 1)\psi(1) + q\sum_{j=1}^{p/2-1} \psi(j + \tfrac{1}{2} - \frac{p}{2})$$

$$+ (q - 1)\psi(\tfrac{1}{2}) - q\sum_{j=1}^{p/2-1} j(j - \frac{p}{2})^{-1}$$

$$- \sum_{\substack{k=1 \\ \neq \frac{1}{2}pq}}^{pq-1} \psi(\tfrac{1}{2} + \frac{k}{pq}) \quad \text{for} \quad i = p/2, \qquad (3.39)$$

$$= (\frac{pq}{2} - 1)\psi(1) + q\sum_{j=1}^{p/2-1} \psi(j - i + \tfrac{1}{2})$$

$$+ (q - 1)\psi\left(\frac{p + 1}{2} - i\right) - q\sum_{j=1}^{p/2}(j - i)^{-1}j$$

$$- \left(\frac{pq}{2} - 1\right)\sum_{j=p/2+1}^{i-1}(j - i)^{-1}$$

$$- \sum_{k=1(\neq pq/2)}^{pq-1}\psi\left(-i + \frac{p + 1}{2} + \frac{k}{pq}\right)$$

$$\text{for } i = p/2 + 1, \ldots, \tag{3.40}$$

and

$$B_{2i0}^{(m)} = (-1)^{m+1}m![qi\zeta(m + 1,1)$$

$$+ q\sum_{j=1}^{p/2-1}\zeta(m + 1,j - i + \tfrac{1}{2})$$

$$+ q\sum_{j=i+1}^{p/2-1}\zeta(m + 1,j - i)$$

$$+ (q - 1)\zeta\left(m + 1,\frac{p + 1}{2} - i\right)$$

$$+ (q - 1)\zeta\left(m + 1,\frac{p}{2} - i\right) + \left(\frac{p}{2} - i\right)^{-1-m}$$

$$+ q\sum_{j=1}^{i-1}j(j - i)^{-1-m}$$

$$- \sum_{k=1(\neq pq/2)}^{pq-1}\zeta\left(m + 1,-i + \frac{p + 1}{2} + \frac{k}{pq}\right)]$$

$$\text{for } i = 1,2,\ldots,p/2 - 1, \tag{3.41}$$

$$= (-1)^{m+1} m! [(\frac{pq}{2} - 1)\zeta(m + 1,1)$$

$$+ q \sum_{j=1}^{p/2-1} \zeta(m + 1, j - \tfrac{1}{2}p + \tfrac{1}{2})$$

$$+ (q - 1)\zeta(m + 1, \tfrac{1}{2})$$

$$+ q \sum_{j=1}^{p/2-1} j(j - \tfrac{1}{2}p)^{-1-m}$$

$$- \sum_{k=1(\neq pq/2)}^{pq-1} \zeta(m + 1, \tfrac{1}{2} + \frac{k}{pq})]$$

$$\text{for } i = p/2, \tag{3.42}$$

$$= (-1)^{m+1} m! [(\frac{pq}{2} - 1)\zeta(m + 1,1)$$

$$+ q \sum_{j=1}^{p/2-1} \zeta(m + 1, j - i + \tfrac{1}{2})$$

$$+ (q - 1)\zeta(m + 1, \frac{p + 1}{2} - i) + q \sum_{j=1}^{p/2} (j-i)^{-1-m} j$$

$$+ (\frac{pq}{2} - 1) \sum_{j=p/2+1}^{i-1} (j - i)^{-1-m}$$

$$- \sum_{k=1(\neq pq/2)}^{pq-1} \zeta(m + 1, \frac{p + 1}{2} - i + \frac{k}{pq})]$$

$$\text{for } i = p/2 + 1, \ldots . \tag{3.43}$$

Now substituting the expressions for R_{1i} and R_{2i} from (3.25) and (3.33) respectively, in (3.13), we get the following result.

THEOREM 3.2. *When* p *is even, the p.d.f. of* $V = \Lambda^{*2/n}$, *for* $n_i = n$, $i = 1,\ldots,n$, *where* Λ^* *is the modified likelihood ratio criterion for testing* H *is given by*

$$f(v) = K(n,p,q)v^{\frac{n-p}{2}-1} \sum_{i=1}^{\infty} \left[\frac{v^i}{(a_i - 1)!} \sum_{r=0}^{a_i-1} \binom{a_i - 1}{r} (-\log v)^{a_i-1-r} A_{1i0}^{(r)} + \frac{v^{i-\frac{1}{2}}}{(b_i - 1)!} \sum_{r=0}^{b_i-1} \binom{b_i - 1}{r} (-\log v)^{b_i-1-r} A_{2i0}^{(r)} \right],$$

$$0 < v < 1,$$

where a_i, b_i *are given by (3.9), K(n,p,q) is defined in (3.4), and the expressions for* $A_{1i0}^{(r)}$ *and* $A_{2i0}^{(r)}$ *are given by (3.26) and (3.34) respectively.*

The density $f(v)$ for $q = 2$, $p = 2$ is derived as

$$f(v) = K(n,2,2)v^{\frac{n-4}{2}} [\sum_{i=1}^{\infty} v^i A_{1i0} + v^{\frac{1}{2}}(-\log v + B_{210})A_{210} + v^{-\frac{1}{2}} \sum_{i=2}^{\infty} v^i A_{2i0}]$$

where the coefficients A_{1i0}, B_{210}, A_{210} and A_{2i0} are calculated from (3.28), (3.39), (3.36) and (3.37) respectively. Simplifying these coefficients using conversion formula (Mathai (1982, p. 248, Eq. (2.7)) and some properties of psi function the density is finally given by the following corollary.

COROLLARY 3.2. *The density for* $p = 2$, $q = 2$ *is given as*

$$f(v) = K(n,2,2)v^{\frac{n-4}{2}} [4v\Gamma(\tfrac{1}{2})\{\Gamma(\tfrac{1}{4})\Gamma(3/4)\}^{-1}$$

$$\cdot\ {}_3F_2(\tfrac{1}{2},\tfrac{1}{4},3/4;3/2,3/2;v)$$

$$+\ v^{\frac{1}{2}}\{-\log v + 4 \log 2 - 4\}$$

$$\cdot\ 4\Gamma(\tfrac{1}{2})\{\Gamma(3/4)\Gamma(\tfrac{1}{4})\}^{-1}$$

$$+\ v^{3/2}\Gamma(-\tfrac{1}{2})\{\Gamma(-\tfrac{1}{4})\Gamma(\tfrac{1}{4})\}^{-1}$$

$$\cdot\ {}_4F_3\,(\tfrac{5}{4},\tfrac{3}{4},1,1;\tfrac{3}{2},2,2,;v)],\quad 0 < v < 1,$$

where ${}_3F_2$ *and* ${}_4F_3$ *are the hypergeometric functions (Mathai (1982), p. 248).*

4. NONNULL DENSITY

The Lauricella's function F_D used in section two has the following series representation

$$F_D^{(pq)}(a;\{b_{i1},\ldots,b_{ip}\},\ i = 1,\ldots,q;\ c;$$

$$\{v_{i1},\ldots,v_{ip}\},\ i = 1,\ldots,q)$$

$$= \sum_{m_{11}=0}^{\infty} \cdots \sum_{m_{1p}=0}^{\infty} \cdots \sum_{m_{q1}=0}^{\infty} \cdots \sum_{m_{qp}=0}^{\infty}$$

$$\frac{(a)_{m_{11}+\ldots+m_{pq}}}{(c)_{m_{11}+\ldots+m_{pq}}} \prod_{i=1}^{q} \prod_{j=1}^{p} \left\{\frac{(b_{ij})_{m_{ij}} v_{ij}^{m_{ij}}}{m_{ij}!}\right\}$$

for $|v_{ij}| < 1$, $i = 1,\ldots,q$, $j = 1,2,\ldots,p$, where for example $(a)_m = a(a + 1)\ldots(a + m - 1) = \Gamma(a + m)/\Gamma(a)$

whenever $\Gamma(a)$ is defined. Expanding the Lauricella's function given in (2.6) and taking inverse Mellin transform we get the nonnull density of Λ^* as follows:

$$f(\lambda^*) = \lambda^{*-1}\pi^{p(p-1)/4} \prod_{i=1}^{q} \left\{ \frac{|n\Sigma_i^{-1}|^{n_i/2}}{\Gamma_p(n_i/2)} \right\} \sum_{m_{11}=0}^{\infty} \cdots \sum_{m_{pq}=0}^{\infty} \left\{ \prod_{i=1}^{q} \prod_{j=1}^{p} \frac{(1 - n\theta_{ij}^{-1})^{m_{ij}}}{m_{ij}!} \right\}$$

$$\cdot\Gamma\left(\frac{np}{2} + \sum_{i=1}^{q} \sum_{j=1}^{p} m_{ij}\right) H^{2pq,0}_{pq+1,2pq}\left[\lambda^* \prod_{i=1}^{q} \left(\frac{n_i}{np}\right)^{n_i p/2} \middle| \begin{array}{l} \left(\frac{np}{2} + m_{11} + \cdots + m_{pq}, \frac{np}{2}\right), \\ \left\{\left(\frac{n_i}{2} + m_{ij}, \frac{n_i}{2}\right), j = 1,\ldots,p\right\}, \quad i = 1,\ldots,q, \\ \left\{\left(\frac{n_i}{2}, \frac{n_i}{2}\right), \text{repeated } p \text{ times}\right\}, \quad i=1,\ldots,q \\ \left\{\left(\frac{n_i - j + 1}{2}, \frac{n_i}{2}\right), j = 1,\ldots,p\right\}, \ i=1,\ldots,q \end{array}\right]$$

$$0 < \lambda^* < 1,$$

where $H^{2pq,0}_{pq+1,2pq}[\cdot|\cdot]$ is an H-function. For a definition, general expansion and computable representation of H-functions see Mathai and Saxena (1978). One can write this H-function in terms of Meijer's G-function by expanding various gammas by using Gauss-Legendre multiplication formula.

Alternately, using (2.10) and inverse Mellin transform, the nonnull density, in terms of zonal polynomials is derived as

$$f(\lambda^*) = \lambda^{*-1} \prod_{i=1}^{q} \left\{ \frac{|\eta\Sigma_i^{-1}|^{n_i/2}}{\prod_{j=1}^{p} \Gamma(\frac{n_i - j + 1}{2})} \right\}$$

$$\sum_{k^{(1)}=0}^{\infty} \cdots \sum_{k^{(q)}=0}^{\infty} \sum_{\kappa^{(1)}} \cdots \sum_{\kappa^{(q)}}$$

$$\frac{C_{\kappa^{(1)}}(I - \eta\Sigma_1^{-1})}{k^{(1)}!} \cdots \frac{C_{\kappa^{(q)}}(I - \eta\Sigma_q^{-1})}{k^{(q)}!}$$

$$\Gamma(\frac{np}{2} + k^{(1)} + \ldots + k^{(q)})$$

$$H^{pq,0}_{1,pq}\left[\lambda^* \prod_{i=1}^{q} (\frac{n_i}{np})^{n_i p/2} \middle| \begin{array}{l} (\frac{np}{2} + k^{(1)} + \ldots + k^{(q)}, \frac{np}{2}) \\ (\frac{n_i-j+1}{2} + k_j^{(i)}, \frac{n_i}{2}),\ i=1,\ldots,q,\ j=1,\ldots,p \end{array}\right]$$

$$0 < \lambda^* < 1$$

where $H^{pq,0}_{1,pq}[\cdot|\cdot]$ is an H-function.

A. K. Gupta
Department of Mathematics and Statistics
Bowling Green State University
Bowling Green, Ohio 43403
U.S.A.

D. K. Nagar
Department of Statistics
University of Rajasthan
Jaipur - 302014
India

REFERENCES

Exton, H. (1976). *Multiple Hypergeometric Functions.* Wiley, Halsted, New York.

Gradshteyn, I. S. and Ryzhik, I. M. (1965). *Table of Integrals, Series, and Products.* Academic Press, New York.

Gupta, A. K. (1977). 'On the distribution of sphericity test criterion in the multivariate Gaussian distribution'. *Aust. J. Statist.,* 19(3), 202-205.

Gupta, A. K., Chattopadhyay, A. K., and Krishnaiah, P. R. (1975). 'Asymptotic distribution of the determinants of some random matrices'. *Comm. Statist.,* 4, 33-47.

Gupta, A. K. and Rathie, A. K. (1982). 'Distribution of the likelihood-ratio criterion for the problem of k-samples'. *Metron,* 40, 147-156.

Gupta, A. K. and Tang, J. (1984). 'Distribution of likelihood ratio statistic for testing equality of covariance matrices of multivariate Gaussian models'. *Biometrika,* 71, 555-559.

Gupta, A. K. and Tang, J. (1986). 'On testing homogeneity of variances for Gaussian models'. *J. Statist. Comp. Simul.* (to appear).

James. A. T. (1964). 'Distributions of matrix variates and latent roots derived from normal samples'. *Ann. Math. Statist.,* 35, 475-501.

Khatri, C. G. and Srivastava, M. S. (1971). 'On exact nonnull distributions of likelihood ratio criteria for sphericity test and equality of two covariance matrices'. *Sankhyā,* A33, 201-206.

Mendoza, J. L. (1980). 'A significance test for multisample sphericity'. *Psychometrika,* 45(4), 495-498.

Mathai, A. M. (1982). 'On a conjecture in geometric probability regarding asymptotic normality of a random simplex'. *Ann. Prob.,* 10(1), 247-251.

Mathai, A. M. and Saxena, R. K. (1973). *The H-function with Applications in Statistics and Other Disciplines*. Wiley Eastern, New Delhi.

Pillai, K. C. S. and Nagarsenker, B. N. (1971). 'On the distribution of the sphericity test criterion in classical and complex normal populations having unknown covariance matrices'. *Ann. Math. Statist.*, **42**, 764-767.

Shanti S. Gupta and S. Panchapakesan

STATISTICAL SELECTION PROCEDURES IN MULTIVARIATE MODELS

1. INTRODUCTION

Since statistical inference problems were first posed in the now-familiar "selection and ranking" framework over three decades ago, these problems have been studied from several points of view using various goals and formulations. However, selection from multivariate populations is an important topic that has not been adequately studied in the literature. Our interest here is to briefly review developments pertaining to selection from multivariate models. In doing so, we consider: (1) selection from a single multivariate normal population, (2) selection from several multivariate normal populations, (3) selection from a multinomial population, (4) selection from several multinomial populations, and (5) selection from a set of predictor variables in a regression model.

For ranking multivariate populations, usually a scalar function of the unknown parameters has been chosen in all the investigations. This permits a complete order of the populations. The choice of the ranking measure depends, of course, on the specific situations. The selection procedure in these cases depends on a suitably chosen statistic which has a univariate distribution.

Let us consider k independent populations $\pi_1, \ldots, \pi_k$, where π_i has the underlying distribution function F_{θ_i}, $i = 1, \ldots, k$. The θ_i are unknown real-valued parameters; these represent the values of a certain quality characteristic θ for the k populations. The populations are ranked according to their θ-values. To be specific, π_i is defined to be *better than* π_j if $\theta_i \geq \theta_j$. The ordered θ_i are denoted by $\theta_{[1]} \leq \cdots \leq \theta_{[k]}$. It is assumed that there is no prior knowledge regarding the correct pairing of the ordered and the unordered θ_i. Selection problems have been generally studied under one of two formulations, namely, (1) the *indifference-zone* and (2) the *subset selection* formulations.

Considering the basic problem of *selecting the best population* (i.e. the population associated with $\theta_{[k]}$), the indifference-zone formulation of Bechhofer (1954) requires that one of the k populations be

A. K. Gupta (ed.), Advances in Multivariate Statistical Analysis, 141–160.

chosen as the best. A *correct selection* (CS) is said to occur when any population associated with $\theta_{[k]}$ is selected. Any *valid procedure* R must guarantee a specified minimum probability of a correct selection (PCS) whenever the best and the next best populations are sufficiently (to be specified) apart. Let $\delta(\theta_{[k]}, \theta_{[k-1]})$ denote an appropriately chosen measure of the separation between the best and the next best populations, and $P(CS|R)$ denote the PCS using the rule R. Further, let

$$\Omega_{\delta^*} = \{\underset{\sim}{\theta}|\underset{\sim}{\theta} = (\theta_1, \ldots, \theta_k), \delta(\theta_{[k]}, \theta_{[k-1]}) \geq \delta^* > 0\}. \tag{1}$$

Any valid rule R should satisfy

$$P(CS|R) \ \geq \ P^* \text{ whenever } \underset{\sim}{\theta} \in \Omega_{\delta^*}. \tag{2}$$

Both δ^* and $P^* \epsilon (1/k, \ 1)$ are specified by the experimenter in advance. Suppose R is based on samples of size n from each population. Then the problem is to determine the smallest n for which the requirement (2) is satisfied. It should be noted that there is no guarantee to be met when $\underset{\sim}{\theta}$ belongs to $\Omega^c_{\delta^*}$, the complement of Ω_{δ^*}. The region $\Omega^c_{\delta^*}$ is the "indifference-zone" lending its name to the formulation.

In the *subset selection* formulation studied extensively beginning with the pioneering work of Gupta (1956, 1965), the basic problem is to select a nonempty subset of the k populations so that the best population is included in the selected subset with a specified minimum PCS. The size of S, the selected subset, is not determined in advance but by data themselves. Selection of any subset that includes the best population results in a correct selection. Letting Ω denote the entire parameter space, any valid rule R should satisfy

$$P(CS|R) \ \geq P^* \quad \text{for all } \underset{\sim}{\theta} \in \Omega. \tag{3}$$

This requirement (3) is called the ***basic probability requirement***, or the P^*-*condition*. Any configuration $\underset{\sim}{\theta}$ which yields the infimum of PCS over Ω is called a *least favorable configuration* (LFC).

The expected value of $|S|$, the size of S, is a reasonable measure of the performance of a valid rule and has been generally used. Some other possible measures (considered by a few authors) are $E(|S|)/P(CS|R)$ and $E(|S|)-P(CS|R)$, the latter being the expected number of non-best populations included in S.

There are many variations and generalizations of the basic formulation using either of the two approaches described above. There are also related problems such as selecting populations that are better than a standard or a control. A comprehensive survey of the developments encompassing all these aspects with an extensive bibliography is given by Gupta and Panchapakesan (1979). Recently, Gupta and Panchapakesan (1985) have provided a critical review of developments in the subset selection theory with historical perspectives. For a categorized bibliography, see Dudewicz and Koo (1982).

In the present paper, we are concerned with subset selection procedures for multivariate populations. In Section 2, we discuss selection of the best component in a multivariate normal population in terms of the means as well as the variances. Selection from several multivariate normal populations is discussed in Section 3 using different criteria such as the Mahalanobis distance, the generalized variance, and the multiple correlation coefficient. Section 4 deals with selecting the most probable and the least probable cells in a multinomial distribution. Selection from several multinomial populations is discussed in Section 5 using the Shannon entropy function for comparison of the populations. Finally, Section 6 describes subset selection procedures for choosing a best set of predictor variables in a linear regression model.

2. SELECTION FROM A SINGLE MULTIVARIATE NORMAL POPULATION

Consider a p-variate normal population $N_p(\underset{\sim}{\mu}, \Sigma)$ with mean vector $\underset{\sim}{\mu}' = (\mu_1, \ldots, \mu_p)$ and covariance matrix $\Sigma = (\sigma_{ij})$, which is assumed to be positive definite. In this section, we consider ranking the p components according to their means μ_i, and according to their variances σ_{ii}.

2.1. Selection in Terms of the Means

Let $\underset{\sim}{X}' = (X_1, \ldots, X_p)$ be the sample mean based on n independent (vector) observations from the population. We first consider the case of known Σ and assume, without loss of generality, that $\sigma_{ii} = 1$ for $i = 1, \ldots, p$. For selecting the component associated with $\mu_{[p]}$, the largest μ_i, Gnanadesikan (1966) considered the procedure

$$R_1 : \text{ Select the } i\text{th component if and only if } X_i \geq X_{[p]} - \frac{d_1}{\sqrt{n}} \qquad (4)$$

where $X_{[1]} \leq \ldots \leq X_{[p]}$ denote the ordered X_i, and $d_1 = d_1(n, p, \Sigma) > 0$ is the smallest number such that the P^*-condition is satisfied. It is easily shown that

$$\inf_{\Omega} P(CS|R_1) \;=\; \Pr\{Y_p \;\geq\; Y_j - d_1,\;\; j = 1,\ldots,p-1\}, \qquad (5)$$

where $Y_i = \sqrt{n}(X_{(i)} - \mu_{[i]})$, $X_{(i)}$ is the component sample mean associated with $\mu_{[i]}$, and $\Omega = \{\underset{\sim}{\mu} : \;\; -\infty < \mu_i < \infty,\; i = 1,\ldots,p\}$. For evaluating d_1 for which the right-hand side of (5) equals P^*, we need to know $A = (a_{ij})$, the covariance matrix of $\underset{\sim}{Y}' = (Y_1,\ldots,Y_p)$. Even though Σ is known, we do not know the correspondence between the σ_{ij} and the a_{ij} except when $p = 2$. For $p = 2$, the right-hand side of (5) equals $\Phi[d_1/\sqrt{2(1-\sigma_{12})}]$, where $\Phi(\cdot)$ is the cdf of a standard normal random variable; this gives

$$d_1 \;=\; d_1(n, 2, \Sigma) \;=\; \sqrt{2(1-\sigma_{12})}\,\Phi^{-1}(P^*). \qquad (6)$$

For $p > 2$, Gnanadesikan (1966) obtain two different lower bounds for the infimum of PCS. Letting $d_{01} = \min\{d_1/\sqrt{2(1-a_{pj})},\; j = 1,\ldots,p-1\}$, one gets

$$\inf_{\Omega} P(CS|R_1) \;\geq\; \Pr\{Z_j \leq\; d_{01},\; j = 1,\ldots,p-1\} \qquad (7)$$

where $\underset{\sim}{Z}' = (Z_1,\ldots,Z_{p-1})$ has $N_{p-1}(\underset{\sim}{0}, B)$ distribution and B has a known structure with elements being 0, or $[2(1-a_{jp})]^{-\frac{1}{2}}$, or $-[2(1-a_{jp})]^{-\frac{1}{2}}$, $j = 1,\ldots,p-1$. One lower bound for the right-hand side of (7) obtained by Gnanadesikan (1966) is $\Phi^{p-1}(d_{01})$ based on an inequality due to Slepian (1962). The other lower bound is $(2-p) + (p-1)\Phi(d_{01})$ obtained by using a Bonferroni inequality. For $p = 2$, the two bounds coincide. While d_{01}, using either lower bound, is a conservative value for d_1, the computations of Gnanadesikan (1966) show that d_{01} in the former case (Slepian inequality) is closer to the exact value. However, the difference between the two approximate values decreases as P^* increases and is very small for $P^* \;\geq\; .90$.

The determination of the constant d becomes easier when $\sigma_{ij} = \rho > 0$, $i \neq j$. In this case, we get

$$\inf_{\Omega} P(CS|R_1) = \int_{-\infty}^{\infty} \Phi^{p-1}(x + \frac{d}{\sqrt{1-\rho}})d\Phi(x) \qquad (8)$$

and $H = d/\sqrt{2(1-\rho)}$ are tabulated by Gupta (1963a) and by Gupta, Nagel and Panchapakesan (1973) who have also considered the selection problem in this special case.

When the covariance matrix Σ is unknown, let us assume that $\sigma_{ii} = \sigma^2$ for $i = 1, \ldots, p$, and let s_ν^2 denote an estimator of σ^2 on ν degrees of freedom, statistically independent of the X_i. In this case, Gnanadesikan (1966) proposed the procedure

$$R_2: \quad \text{Select the } i\text{th component if and only if } X_i \geq X_{[p]} - \frac{d_2 s_\nu}{\sqrt{n}} \tag{9}$$

where $d_2 = d_2(\nu, p, P^*) > 0$ is the smallest number for which the P^*-condition is satisfied. For this procedure,

$$\begin{aligned} \inf_\Omega P(CS|R_2) &\geq \Pr\{t_i \leq d_{01},\ i = 1, \ldots, p-1\} \\ &\geq 1 - \sum_{i=1}^{p-1} \Pr\{t_i \geq d_{01}\} \end{aligned} \tag{10}$$

where $t_i = Z_i/s_\nu$, $\underset{\sim}{Z}' = (Z_1, \ldots, Z_{p-1})$ has the same distribution as in the known Σ case, $\nu s_\nu^2/\sigma^2$ has a chi-square distribution with ν degrees of freedom, d_{01} is defined as before, and $\Omega = \{(\underset{\sim}{\mu}, \Sigma)\}$. Equating the last member of the inequalities in (10) to P^*, an approximate value of d_{01} is given by

$$(2-p) + (p-1)G_\nu(d_0) = P^* \tag{11}$$

where $G_\nu(\cdot)$ is the cdf of a Student's t variable with ν degrees of freedom. In the special case of $\sigma_{ij} = \rho\sigma^2$, $\rho > 0$, d_{01} can be evaluated as an equicoordinate percentage point of a multivariate t distribution. The d_{01} values are tabulated by Gupta and Sobel (1957), Krishnaiah and Armitage (1966), and Gupta, Panchapakesan and Sohn (1985).

2.2. Selection in Terms of the Variances

We now define the best component as the one associated with the smallest σ_{ii}. A natural procedure is analogous to that of Gupta and Sobel (1962a) in the uncorrelated case. This procedure is

$$R_3: \quad \text{Select the } i\text{th component if } s_{ii} \leq \frac{1}{c} \min_{1 \leq j \leq p} s_{jj} \tag{12}$$

where $c = c(p, n, P^*) \in (0, 1)$ is the largest number for which the P^*-condition is satisfied, and $S = (s_{ij})$ is the sample covariance matrix based on n independent (vector) observations from the population. This procedure has been considered by Frischtak (1973), who has shown that, for $p = 2$, the infimum of PCS is attained when $\sigma_{11} = \sigma_{22}$ and $\sigma_{12} = 0$. Thus c can be obtained from the tables of Gupta and Sobel (1962b).

For $p \geq 3$, Frischtak (1973) obtained only an asymptotic ($n \to \infty$) solution, using the asymptotic normality of $\log(s^2_{(1)}/s^2_{(j)})$, $j = 2, \ldots, p$, after suitable normalization; here $s^2_{(i)}$ is the s_{ii} associated with the ith smallest σ_{ii}. The asymptotic solution c is given by

$$\Pr\{Y_j \leq \sqrt{\frac{n-1}{2}} \log c, \; j = 2, \ldots, p\} = P^* \tag{13}$$

where the Y_j are standard normal random variables with equal correlation 0.5, and can be obtained from the tables of Gupta (1963a) and Gupta, Nagel and Panchapakesan (1973).

3. SELECTION FROM SEVERAL MULTIVARIATE NORMAL POPULATIONS

Let $\pi_1, \ldots, \pi_k$ be k p-variate normal populations, $N_p(\underset{\sim}{\mu}_i, \Sigma_i)$, $i = 1, \ldots, k$, where the $\underset{\sim}{\mu}_i$ are the mean vectors and the Σ_i are positive definite covariance matrices. For defining the best population, several measures have been used such as the generalized variance, Mahalanobis distance, and the multiple correlation coefficient. Also, comparison with a control has been studied using as criteria linear combinations of the elements of the mean vector and those of the covariance matrix. We now discuss these briefly.

3.1. Selection in Terms of Mahalanobis Distance

Let $\lambda_i = \underset{\sim}{\mu}_i' \Sigma_i^{-1} \underset{\sim}{\mu}_i$, the Mahalanobis distance of π_i from the origin. We first assume that the Σ_i are *known*. Let $\underset{\sim}{X}_{ij}$, $j = 1, \ldots, n$, denote n (vector) observations from π_i, $i = 1, \ldots, k$. Define $Y_{ij} = \underset{\sim}{X}_{ij}' \Sigma_i^{-1} \underset{\sim}{X}_{ij}$ and $Y_i = \sum_{j=1}^{n} Y_{ij}$. For selecting a subset containing the population associated with $\lambda_{[k]}$, Gupta (1966) proposed the procedure

$$R_4: \quad \text{Select } \pi_i \text{ if and only if } Y_i \geq c_4 Y_{[k]} \tag{14}$$

where $0 < c_4 = c_4(k, p, n, P^*) < 1$ is to be chosen suitably to meet the P^*-condition. It has been shown [Gupta (1966) and Gupta and Studden (1970)] that the infimum of PCS occurs when $\lambda_1 = \cdots = \lambda_k = 0$. Thus the constant c_4 is given by

$$\int_0^\infty G_\nu^{k-1}(\frac{x}{c_4})dG_\nu(x) = P^* \tag{15}$$

where $G_\nu(x)$ is the cdf of a standardized (i.e unit scale parameter) gamma variable with $\nu = np/2$ degrees of freedom. The values of c are tabulated by Gupta (1963b) and Armitage and Krishnaiah (1964).

An analogous procedure can be defined for selecting the population with the smallest λ_i. In this case, the appropriate constant can be obtained from the tables of Gupta and Sobel (1962b) and Krishnaiah and Armitage (1964).

It should be noted that the procedure R_4 is based on the statistics $Y_i = \sum_{j=1}^n \underset{\sim}{X}'_{ij}\Sigma_i^{-1}\underset{\sim}{X}_{ij}$ rather than $Z_i = \overline{\underset{\sim}{X}}'_i\Sigma_i^{-1}\overline{\underset{\sim}{X}}_i$, where $\overline{\underset{\sim}{X}}_i$ denote the sample mean vector from π_i. If we use Z_i instead of Y_i in R_4, the infimum of PCS and hence the constant c_4 do not depend on n. This makes the procedure unsatisfactory. One can, of course, use a different type of procedure. For example, we can define R' : Select π_i if and only if $Z_i \geq Z_{[k]} - d$, $d > 0$. Such a procedure has not been investigated.

When the Σ_i are *unknown and not necessarily equal*, Gupta and Studden (1970) proposed and studied the rule

$$R_5 : \quad \text{Select } \pi_i \text{ if and only if } T_i \geq c_5 T_{[k]} \tag{16}$$

where $T_i = \overline{\underset{\sim}{X}}'_i S_i^{-1}\overline{\underset{\sim}{X}}_i$, S_i is the usual sample covariance matrix with $(n-1)$ as the divisor, and $0 < c_5 = c_5(k,n,p,P^*) < 1$ is chosen suitably to satisfy the P^*-condition. It has been shown by Gupta and Studden (1970) that

$$\inf_\Omega P(CS|R_5) = \int_0^\infty F_{p,n-p}^{k-1}(\frac{x}{c_5})dF_{p,n-p}(x) \tag{17}$$

where $F_{p,n-p}(x)$ is the cdf of a central F-variable with p and $n-p$ degrees of freedom. The values of c_5 for which the right-hand side of (17) equals P^* have been tabulated by Gupta and Panchapakesan (1969) for various values of k, P^*, p, and n.

Gupta and Studden (1970) also studied the problem of selecting the population associated with the smallest λ_i. Their rule is

$$R'_5: \quad \text{Select } \pi_i \text{ if and only if } T_i \leq \frac{1}{c'_5} T_{[1]} \tag{18}$$

where $0 < c'_5 = c'_5(k, n, p, P^*) < 1$ is to be chosen suitably. In this case,

$$\inf_{\Omega} P(CS|R'_5) = \int_0^{\infty} [1 - F_{p,n-p}(c'_5 x)]^{k-1} dF_{p,n-p}(x). \tag{19}$$

The constant c'_5 for which the right-hand side of (19) equals P^* has been tabulated by Gupta and Panchapakesan (1969) for several combinations of k, P^*, p, and n.

When $\Sigma_1 = \cdots = \Sigma_k = \Sigma$ and Σ is *unknown*, one would define a procedure with $T_i = \overline{\underset{\sim}{X}}'_i S^{-1} \overline{\underset{\sim}{X}}_i$ in R_5, where S is the usual pooled estimator of Σ. This procedure was proposed by Gupta and Studden (1970) and studied later by Chattopadhyay (1981). He has discussed evaluation of the constant in an approximate sense, i.e. the infimum of PCS is approximately P^* but can be on either side of it.

3.2. Selection in Terms of the Generalized Variance

It is meaningful to rank multivariate normal populations according to the amounts of dispersion in them. A frequently used measure of dispersion is the generalized variance which is the determinant of the covariance matrix. Let $\theta_i = |\Sigma_i|$, $i = 1, \ldots, k$. We define the best population as the one associated with the smallest θ_i. Let S_i be the sample covariance matrix based on a sample of size n from π_i, $i = 1, \ldots, k$. Gnanadesikan and Gupta (1970) proposed the rule

$$R_6: \quad \text{Select } \pi_i \text{ if and only if } W_i \leq \frac{1}{c_6} W_{[1]} \tag{20}$$

where $W_i = |S_i|$, and $0 < c_6 = c_6(k, n, p, P^*) < 1$ is to be chosen suitably to satisfy the P^*-condition. It has been shown that

$$\inf_{\Omega} P(CS|R_6) = \Pr\{Y_1 \leq \frac{1}{c_6} Y_j,\ j = 2, \ldots, k\} \tag{21}$$

where $Y_1, \ldots, Y_k$ are independent and identically distributed, each being the product of p independent factors, the rth factor having a chi-square distribution with $(n - r)$ degrees of freedom. An exact solution for c_6 is obtained in the case of $p = 2$, using the fact that $2(n-1)^{p/2}(W_i/\theta_i)^{\frac{1}{2}}$ is then distributed as a chi-square variable with $2(n-2)$ degrees of freedom. The constant c_6 in this case can be obtained from the tables of Gupta and Sobel (1962b) and Krishnaiah and Armitage (1964).

When $p > 2$, one can use Hoel's approximation of the distribution of $Y_i^{1/p}$ by a gamma distribution with scale parameter θ^{-1} and shape parameter m, where $2m = p(n-p)$ and $2\theta = p[1-(2n)^{-1}(p-1)(p-2)]^{1/p}$. Another approximation is that of $p^{-1}\log Y_i$ using the normal approximation of $\log \chi^2$. Gnanadesikan and Gupta (1970) have studied these approximations.

Some alternative procedures have been proposed by Regier (1976). These procedures are R_6' : Select π_i if and only if $W_i \leq a(\prod_{j=1}^{k} W_j)^{1/k}$ and R_6'' : Select π_i if and only if $W_i \leq b\sum_{j=1}^{k} W_j/k$. Again, the evaluation of the constants a and b are based on normal approximation to log χ^2 and the asymptotic distribution of the sample variance, respectively. Regier (1976) has given some numerical comparisons of the three procedures.

3.3. Selection in Terms of Multiple Correlation Coefficient

We now assume that the $\underset{\sim}{\mu}_i$ and Σ_i are unknown. Let ρ_i denote the multiple correlation coefficient between the first variable and the rest in π_i. It is a measure of dependence between the two partitioned sets. Gupta and Panchapakesan (1969) investigated the problem of selecting a subset containing the population associated with $\rho_{[k]}(\rho_{[1]})$. Let R_i denote the multiple correlation coefficient between the first variable and the rest from the sample $\underset{\sim}{X}_{ij}$, $j = 1,\ldots,n$. Two cases arise: (1) the *conditional case* in which the variables 2 to p are fixed, and (2) the *unconditional case* in which all variables are random. Let $R_i^{*^2} = R_i^2/(1-R_i^2), i = 1,\ldots,k$. Gupta and Panchapakesan (1969) proposed the rule

$$R_7: \text{ Select } \pi_i \text{ if and only if } R_i^{*^2} \geq c_7 R_{[k]}^{*^2} \tag{22}$$

for selecting the population associated with $\rho_{[k]}$, and the rule

$$R_7': \text{ Select } \pi_i \text{ if and only if } R_i^{*^2} \leq \frac{1}{c_7'} R_{[1]}^{*^2} \tag{23}$$

for selecting the population associated with $\rho_{[1]}$, where $0 < c_7 = c_7(k,p,n-p,P^*) < 1$ and $0 < c_7' = c_7'(k,p,n-p,P^*) < 1$ are chosen suitably to meet the P^*-condition. The procedures proposed are the same for the conditional as well as the unconditional case. When $\rho_i \neq 0$, the distribution of $R_i^{*^2}$ is different in these two cases. However, the infimum of PCS occurs in either case when $\rho_1 = \cdots = \rho_k = 0$. The distribution of $R_i^{*^2}$ is the same in either case when $\rho_i = 0$. Thus, in either case, the constants c_7 and c_7' are given by

$$\int_0^\infty F_{2q,2m}^{k-1}(\frac{x}{c_7})dF_{2q,2m}(x) = P^* \tag{24}$$

and

$$\int_0^\infty [1 - F_{2q,2m}(c_7'x)]^{k-1} dF_{2q,2m}(x) = P^* \tag{25}$$

where $q = (p-1)/2$, $m = (n-p)/2$, and $F_{2q,2m}(x)$ is the cdf of an F-variable with $2q$ and $2m$ degrees of freedom. The values of c_7 for selected values of k, P^*, m, and q are tabulated by Gupta and Panchapakesan (1969). The values of c_7' can be obtained from the same tables because $c_7'(p,q,m,P^*) = c_7(p,m,q,P^*)$.

3.4. Selection in Terms of Other Measures

Suppose the p variables under consideration are partitioned into two sets consisting of q_1 and $q_2 (q_1 + q_2 = p)$ variables. Let the corresponding partition of Σ_i be denoted by

$$\Sigma_i = \begin{pmatrix} \Sigma_{11}^{(i)} & \Sigma_{12}^{(i)} \\ \Sigma_{21}^{(i)} & \Sigma_{22}^{(i)} \end{pmatrix}, i = 1, \ldots, k.$$

Selection in terms of the conditional generalized variance of the q_2-set given the q_1-set has been considered by Gupta and Panchapakesan (1969). Frischtak (1973) discussed selection in terms $\gamma_i^2 = \frac{|\Sigma_i|}{|\Sigma_{11}^{(i)}||\Sigma_{22}^{(i)}|}$ but has obtained only an asymptotic solution.

For the problem of selecting populations that are better than a control, Krishnaiah (1967) used linear combinations of the elements of the covariance matrices for making comparisons. Krishnaiah and Rizvi (1966) used several linear combinations of the elements of the mean vectors for comparison and studied procedures to select a subset containing good populations (defined through comparison with the control). For more details, reference can also be made to Gupta and Panchapakesan (1979).

4. SELECTION FROM A MULTINOMIAL POPULATION

Let $p_1, \ldots, p_k$ denote the unknown cell probabilities of a k-cell multinomial distribution. The ordered cell probabilities are denoted by $p_{[1]} \leq \cdots \leq p_{[k]}$. Gupta and Nagel (1967) proposed and studied procedures for selecting the most (least) probable cell based on a single sample of size n. Let $X_1, \ldots, X_k$ denote the cell counts. Their procedure for selecting the most probable cell is

$$R_8: \quad \text{Select the } i\text{th cell if and only if } X_i \geq X_{[k]} - D \tag{26}$$

and the procedure for selecting the least probable cell is

$$R_8': \quad \text{Select the } i\text{th cell if and only if } X_i \leq X_{[1]} + C \tag{27}$$

where $D = D(k, n, P^*)$ and $C = C(k, n, P^*)$ are the smallest nonnegative integers for which the P^*-condition is satisfied in each case.

An intereting point about R_8 and R_8' is that, unlike similar analogous rules for normal means, normal variances, etc., the analyses in the maximum and minimum cases do not run parallel. The LFC for either procedure is completely known only when $k = 2$. In this case, it is given by $p_1 = p_2 = \frac{1}{2}$. For $k > 2$, the LFC (in terms of the ordered p_i) is of the type $(0, \ldots, 0, s, p, \ldots, p)$, $s \leq p$, in the case of R_8 and is of the type $(p, \ldots, p, q)$, $p \leq q$, in the case of R_8'. An alternative to R_8 is the inverse sampling selection rule of Panchapakesan (1971, 1973). Observations are made one at a time until the cell count reaches a predetermined integer M in one of the cells. At termination, let $X_1, \ldots, X_k$ be the cell counts (one of them is M). The selection rule is

$$R_9: \quad \text{Select the } i\text{th cell if and only if } X_i \geq M - D \tag{28}$$

where $D (0 \leq D \leq M)$ is the smallest nonnegative integer for which the P^*-condition is satisfied. For R_9, the infimum of PCS occurs when all the cell probabilities are equal.

Again, for selecting the most probable cell, Gupta and Huang (1975) proposed the rule

$$R_{10}: \quad \text{Select the } i\text{th cell if and only if } X_i + 1 \geq cX_{[k]} \tag{29}$$

where $c = c(k, N, P^*)\epsilon(0,1)$ is the largest number for which the P^*-condition is met. The motivation for the rule R_{10} comes from their conditional selection rules for Poisson populations. A conservative value of c can be obtained from their results for Poisson populations.

Recently, Chen (1985) considered an inverse sampling selection rule for selecting a subset containing the least probable cell. For his procedure R_{11} the observations are made one at a time until *either* (1) the count in any cell reaches r, *or* (2) $(k-1)$ cells reach count of at least $r'(1 \leq r' \leq r+1)$. If (1) occurs before (2), the rule R_{11} selects the cells with counts $X_i < r'$. If (2) occurs before (1), then R_{11} selects the cell with count $X_i < r'$. The constants r and r' are to be chosen so as to satisfy the P^*-condition. It has been shown by Chen (1985) that the infimum of $P(CS|R_{11})$ occurs when all the cell probabilities are equal.

Minimax subset selection rules have been investigated by Berger (1979) and Berger and Gupta (1980). For selecting the least probable cell, Berger (1980) investigated a minimax subset selection rule taking as loss the size of the selected subset or the number of non-best cells selected. In another paper, Berger (1982) investigated minimax and admissible subset selection rules for the least probable cell taking as the loss the number of non-best cells selected. His rule, however, satisfies the P^*-condition only if P^* is sufficiently large. For the corresponding procedure for the most probable cell, the P^*-condition has been verified only in certain special cases.

The importance of multinomial selection rules is accented by the fact that they provide distribution-free procedures. Suppose that $\pi_1, \ldots, \pi_k$ have continuous distributions F_{θ_i}, $i = 1, \ldots, k$. We assume that $\{F_\theta\}$ is a stochastically increasing family in θ. Let p_i denote the probability that in a set of k observations, one from each distribution, the observation from π_i is the largest, $i = 1, \ldots, k$. Selecting the stochastically largest (smallest) population is then equivalent to selecting the population associated with the largest (smallest) p_i. If we take observations a vector at a time and note which population yielded the largest observation, the problem can be converted to the multinomial cell problem.

5. SELECTION FROM SEVERAL MULTINOMIAL POPULATIONS

Let $\pi_1, \ldots, \pi_k$ be k multinomial populations each with m cells and let the unknown cell probabilities of π_i be $p_{i1}, \ldots, p_{im}$, $i = 1, \ldots, k$. Let $H_i \equiv H(p_{i1}, \ldots, p_{im}) = -\sum_{j=1}^{m} p_{ij} \log p_{ij}$, the Shannon entropy function associated with π_i. The function is a measure of the uncertainty with regard to the nature of the outcomes from π_i. We want to select the population associated with the largest H_i. For $m = 2$, the problem reduces to that of selecting the binomial population associated with the largest $\psi(\theta_i) = -\theta_i \log \theta_i - (1-\theta_i)\log(1-\theta_i)$, where θ_i is the success probability. In this case, Gupta and Huang (1976) proposed the rule

$$R_{12}: \quad \text{Select } \pi_i \text{ if and only if } \psi(\frac{X_i}{n}) \geq \max_{1\leq j\leq k} \psi(\frac{X_j}{n}) - d_{12} \qquad (30)$$

where X_i is the number of successes in n trials associated with π_i, and $d_{12} = d_{12}(k, n, P^*)$ is the smallest nonnegative constant such that $0 < d \leq \psi([n/2]/n)$ for which the P^*-condition is satisfied. Here $[n/2]$ denotes the largest integer $\leq n/2$. The infimum of $P(CS|R_{12})$ takes place when $\theta_1 = \cdots = \theta_k = \theta$. However, the common value θ for which the infimum takes place is not known. Gupta and Huang (1976) have obtained a conservative value of d using the approach of Gupta, Huang and Huang (1975), who used this approach to obtain a conservative value for the constant defining the procedure of Gupta and Sobel (1960) for selecting the binomial population with the largest success probability. For more details on this, see Gupta and Panchapakesan (1979, 1985).

To discuss the selection procedure of Gupta and Wong (1977) in the case of $m > 2$, let $\underset{\sim}{a} = (a_1, \ldots, a_m)$ and $A_r = \sum_{i=r}^{m} a_{[i]}$, where $a_{[1]} \leq \cdots \leq a_{[m]}$ are the ordered components. Vector $\underset{\sim}{a} = (a_1, \ldots, a_m)$ is said to *majorize* vector $\underset{\sim}{b} = (b_1, \ldots, b_m)$ of the same dimension (written $\underset{\sim}{a} \succ \underset{\sim}{b}$) if $A_r \geq B_r$ for $r = 2, \ldots, m$, and $A_1 = B_1$. Further, a function f is said to be *Schur-concave* if $f(\underset{\sim}{x}) \leq f(\underset{\sim}{x}')$ whenever $\underset{\sim}{x} \succ \underset{\sim}{x}'$.

In our selection problem, we assume that there is a population whose associated vector of cell probabilities is majorized by the associated vector of cell probabilities of any other population. Such a population will have the largest H_i because the entropy function is Schur-concave. Let $\varphi_i = \varphi(\frac{X_{i1}}{n}, \ldots, \frac{X_{im}}{n})$, where φ is a Schur-concave function, and $X_{i1}, \ldots, X_{im}$ are the cell counts based on n independent observations from π_i, $i = 1, \ldots, k$. Gupta and Wong (1977) proposed the rule

$$R_{13}: \quad \text{Select } \pi_i \text{ if and only if } \varphi_i \geq \max_{1\leq j\leq k} \varphi_j - d_{13} \qquad (31)$$

where $d_{13} = d_{13}(k, m, n, P^*)$ is the smallest positive constant for which the P^*-condition is satisfied. Gupta and Wong obtained a conservative value of d using the idea of conditioning as in the paper of Gupta and Huang (1976).

6. SELECTION OF VARIABLES IN LINEAR REGRESSION

In applying regression analysis in practical situations for prediction purposes, we are often faced with a large number of independent variables. In such situations, it may be sufficient to consider a subset of these predictor variables for "adequate" prediction. There arises then a problem of choosing a "good" subset of these variables. Hocking (1976) and Thompson (1978a,b) have reviewed several criteria and techniques that have been used in practice. However, these are ad hoc procedures and are not designed to control the probability of selecting the important variables. McCabe and Arvesen (1974), and Arvesen and McCabe (1975) were the first to formulate this problem in the framework of Gupta-type subset selection.

Consider the standard linear model

$$\underset{\sim}{Y} = X\underset{\sim}{\beta} + \underset{\sim}{\epsilon} \tag{32}$$

where X is an $N \times p$ known matrix of rank $p \leq N$, $\underset{\sim}{\beta}$ is a $p \times 1$ parameter vector, and $\underset{\sim}{\epsilon} \sim N(0, \sigma^2 I_N)$. This model with p independent variables is considered as the "true" model. Now, consider all reduced models that are formed by taking all possible subsets of size $t (< p)$ from the p independent variables. These models are described by

$$\underset{\sim}{Y} = X_i \underset{\sim}{\beta}_i + \underset{\sim}{\epsilon}_i, \quad i = 1, \ldots, k = \binom{p}{t}, \tag{33}$$

where X_i is an $N \times t$ matrix (of rank t), $\underset{\sim}{\beta}_i$ is a $t \times 1$ parameter vector, and $\underset{\sim}{\epsilon}_i \sim N(0, \sigma_i^2 I_N)$. It should be noted that the models in (33) are considered for prediction purposes and must be compared under the true model assumptions. The expectations of residual mean squares in the corresponding ANOVA evaluated under the true model assumption are σ_i^2, $i = 1, \ldots, k$. For the goal of selecting the design X_i (or the corresponding set of independent variables) associated with $\sigma_{[i]}^2$, Arvesen and McCabe (1975) proposed the rule

$$R_{14}: \quad \text{Select the design } X_i \text{ if and only if } SS_i \leq \frac{1}{c_{14}} SS_{[1]} \tag{34}$$

where SS_i is the residual sum of squares in the ANOVA corresponding to the design X_i, and $0 < c_{14} = c_{14}(p, t, N, P^*) < 1$ is to be chosen to satisfy the P^*-condition. An exact evaluation of the constant c_{14} is difficult. Arvesen and McCabe showed that the PCS is asymptotically $(N \to \infty)$ minimized when $\underset{\sim}{\beta} = \underset{\sim}{0}$. The evaluation of c_{14} is not easy even under this asymptotic LFC. An algorithm has been given by McCabe and Arvesen (1974) for determining c_{14} under the asymptotic LFC for given P^* and X, using Monte Carlo methods.

In the above formulation, the size t is arbitrarily fixed. Huang and Panchapakesan (1982) considered a different formulation taking into consideration all possible reduced models. They considered the regression model with $\underset{\sim}{\beta}' = (\beta_0, \ldots, \beta_p)$, and $X = (\underset{\sim}{1}\underset{\sim}{x}_1 \ldots \underset{\sim}{x}_{p-1})$, where $\underset{\sim}{1}' = (1, \ldots, 1)$ and $\underset{\sim}{x}_i' = (x_{i1}, \ldots, x_{iN})$, $i = 1, \ldots, p-1$. For fixed $\alpha \epsilon \{0, 1, \ldots, p-1\}$, consider all the $\binom{p-1}{\alpha}$ subsets of the set of predictor variables $\{x_1, \ldots, x_{p-1}\}$ and the corresponding reduced models obtained from (32). Associated with these reduced models are the multiple correlation coefficients $R_{i\alpha}$, $i = 1, 2, \ldots, \binom{p-1}{\alpha}$. Let $\theta_{i,\alpha} = E(1 - R_{i\alpha}^2)$. Any reduced model with the associated parameter $\theta_{i,\alpha}$ is said to be *inferior* if $\theta_{1,p-1} \le \delta^* \theta_{i,\alpha}$, where $\delta^* \epsilon (0, 1)$ is a specified constant. (The parameter $\theta_{1,p-1}$ is associated with the true model). Huang and Panchapakesan (1982) considered the problem of eliminating all inferior models. A *correct decision* (CD) is selection of any subset of the models such that all inferior models are excluded from the selected subset. They proposed and studied the procedure

$$R_{15}: \quad \text{Exclude a model if and only if } \hat{\theta}_{i,\alpha} \ge \frac{c_{15}}{\delta^*} \hat{\theta}_{1,p-1} \tag{35}$$

where $\hat{\theta}_{i,\alpha} = 1 - R_{i,\alpha}^2$, and the constant $c_{15} = c_{15}(N, p, P^*) > \delta^*$ is determined such that the P^*-condition is satisfied.

The LFC for the rule R_{15} has been established only in the asymptotic $(N \to \infty)$ sense. For evaluating the constant under the asymptotic LFC($\underset{\sim}{\beta} = \underset{\sim}{0}$), Huang and Panchapakesan (1982) used an algorithm similar to that of McCabe and Arvesen (1974).

Hsu and Huang (1982) considered the goal of selecting a subset of the models that contains all the *superior* models, namely, all models for which $\sigma_i^2 \le \Delta \sigma^2$, where $\Delta > 1$ is a specified constant. For this problem, they investigated a sequential procedure.

Gupta, Huang and Chang (1984) studied the problem of eliminating inferior models, using the expected mean squares as the criterion for comparing any model with the true model. Their approach is different from those of the earlier papers in that they use simultaneous tests of a family of hypotheses in constructing their procedure.

Now, for any reduced model, it is known that SS_i/σ_0^2 has (under the full assumption model) a noncentral chi-square distribution with $\nu = N - p + 1$ degrees of freedom and a noncentrality parameter $\lambda_i = (X\beta)'Q_i(X\beta)/2\sigma_0^2$, where $Q_i = I_N - X_i(X_i'X_i)^{-1}X_i'$, and σ_0^2 is the error variance in the full model. Recently, Gupta and Huang (1986) have considered the problem of eliminating *inferior* models, namely, those for which $\lambda_i \geq \Delta > 0$, where Δ is specified in advance. For this problem, they have proposed and investigated a two-stage procedure.

7. CONCLUSION

As we have seen, multivariate selection problems have wider applications. However, in many cases, the existing procedures have not been fully examined in terms of their performances as well as the determination of the LFC. Even the multinomial problems have to be studied more satisfactorily. Also, the criterion employed for ranking multivariate populations usually induce a complete ordering in the space of distributions. However, in many practical problems, there is a need to consider a partial ordering. There has been practically no development in this direction. Also, there has been no work done for distributions other than multivariate normal populations. It will be interesting to consider reliability related models such as increasing failure rate distributions in two or more dimensions.

8. ACKNOWLEDGEMENT

This research was supported by the Office of Naval Research Contract N00014-84-C-0167 at Purdue University. Reproduction in whole or part is permitted for any purpose of the United States Government.

Shanti S. Gupta
Statistics Department
Purdue University
West Lafayette, IN 47907

S. Panchapakesan
Department of Mathematics
Southern Illinois University
Carbondale, IL 62901

REFERENCES

Armitage, J. V. and Krishnaiah, P. R. (1964). 'Tables for the studentized largest chi-square distribution and their applications'. *ARL 64-188*, Aerospace Research Laboratories, Wright-Patterson Air Force Base, Dayton, Ohio.

Arvesen, J. N. and McCabe, G. P., Jr. (1975). 'Subset selection problems of variances with applications to regression analysis'. *Journal of the American Statistical Association*, **70**, 166–170.

Bechhofer, R. E. (1954). 'A single-sample multiple decision procedure for ranking means of normal populations with known variances'. *Annals of Mathematical statistics*, **25**, 16–39.

Berger, R. L. (1979). 'Minimax subset selection for loss measured by subset size'. *Annals of Statistics*, **7**, 1333–1338.

Berger, R. L. (1980). 'Minimax subset selection for multinomial distribution'. *Journal of Statistical Planning and Inference*, **4**, 391–402.

Berger, R. L. (1982). 'A minimax and admissible subset selection rule for the least probable multinomial cell'. *Statistical Decision Theory and Related Topics – III*, Vol. 2 (S. S. Gupta and J. O. Berger, eds.), Academic Press, New York, 143–156.

Berger, R. L. and Gupta, S. S. (1980). 'Minimax subset selection rules with applications to unequal variance (unequal sample size) problems'. *Scandinavian Journal of Statistics*, **7**, 21–26.

Chattopadhyay, A. K. (1981). 'Selecting the normal population with largest (smallest) value of Mahalanobis distance from the origin'. *Communications in Statistics – Theory and Methods*, **A10**, 31–37.

Chen, P. (1985). 'Subset selection for the least probable multinomial cell'. *Annals of the Institute of Statistical Mathematics*, **37**, 303–314.

Dudewicz, E. J. and Koo, J. O. (1982). *The Complete Categorized Guide to Statistical Selection and Ranking Procedures.* Series in Mathematical and Management Sciences, Vol. 6, American Sciences Press, Inc., Columbus, Ohio.

Frischtak, R. M. (1973). *Statistical Multiple Decision Procedures for Some Multivariate Selection Problems.* Ph.D. Thesis (also Technical Report No. 187), Department of Operations Research, Cornell University, Ithaca, New York.

Gnanadesikan, M. (1966). *Some Selection and Ranking Procedures for Multivariate Normal Populations.* Ph.D. Thesis, Department of Statistics, Purdue University, West Lafayette, Indiana.

Gnanadesikan, M. and Gupta, S. S. (1970). 'A selection procedure for multivariate normal distributions in terms of the generalized variances'. *Technometrics*, **12**, 103–117.

Gupta, S. S. (1956). *On A Decision Rule for A Problem in Ranking Means.* Ph.D. Thesis (also Mimeograph Series No. 150), Institute

of Statistics, University of North Carolina, Chapel Hill, North Carolina.
Gupta, S. S. (1963a). 'Probability integrals of the multivariate normal and multivariate t'. *Annals of Mathematical Statistics*, **34**, 792–828.
Gupta, S. S. (1963b). 'On a selection and ranking procedure for gamma populations'. *Annals of the Institute of Statistical Mathematics*, **14**, 199–216.
Gupta, S. S. (1965). 'On some multiple decision (selection and ranking) rules. *Technometrics*, **7**, 225–245.
Gupta, S. S. (1966). 'On some selection and ranking procedures for multivariate normal populations using distance functions'. *Multivariate Analysis* (P. R. Krishnaiah, ed.), Academic Press, New York, 457–475.
Gupta, S. S. and Huang, D.-Y. (1975). 'On subset selection procedures for Poisson populations and some applications to the multinomial selection problems'. *Applied Statistics* (R. P. Gupta, ed.), North-Holland, Amsterdam, 97–109.
Gupta, S. S. and Huang, D.-Y. (1976). 'On subset selection procedures for the entropy function associated with the binomial populations'. *Sankhya*, **38A**, 153–173.
Gupta, S. S. and Huang, D.-Y. (1986). 'Selecting important independent variables in linear regression models'. *Technical Report No. 86-29*, Department of Statistics, Purdue University, West Lafayette, Indiana.
Gupta, S. S., Huang, D.-Y., and Chang, C.-L. (1984). 'Selection procedures for optimal subsets of regression variables'. *Design of Experiments: Ranking and Selection*, (T. J. Santner and A. C. Tamhane, eds.), Marcel Dekker, New York, 67–75.
Gupta, S. S., Huang, D.-Y., and Huang, W.-T. (1975). 'On ranking and selection procedures and tests of homogeneity for binomial populations'. *Essays in Probability and Statistics* (S. Ikeda, T. Hayakawa, H. Hudimoto, M. Okamoto, M. Siatoni and S. Yamamoto, eds.), Shinko Tsusho Co. Ltd., Tokyo, Japan, Chapter 33, 501–533.
Gupta, S. S. and Nagel, K. (1967). 'On selection and ranking procedures and order statistics from the multinomial distribution'. *Sankhya*, **29B**, 1–34.
Gupta, S. S., Nagel, K. and Panchapakesan, S. (1973). 'On the order statistics from equally correlated normal random variables'. *Biometrika*, **60**, 403–413.
Gupta, S. S. and Panchapakesan, S. (1969). 'Some selection and ranking procedures for multivariate normal populations'. *Multivariate Analysis – II* (P. R. Krishnaiah, ed.), Academic Press, New York, 475–505.

Gupta, S. S. and Panchapakesan, S. (1979). *Multiple Decision Procedures: Theory and Methodology of Selecting and Ranking Populations*. John Wiley & Sons, Inc., New York.

Gupta, S. S. and Panchapakesan, S. (1985). 'Subset selection procedures: review and assessment'. *American Journal of Mathematical and Management Sciences*, **5**, 235–311.

Gupta, S. S., Panchapakesan, S. and Sohn, J. K. (1985). 'On the distribution of the studentized maximum of equally correlated normal random variables'. *Communications in Statistics - Simulation and Computation*, **14**, 103–135.

Gupta, S. S. and Sobel, M. (1957). 'On a statistic which arises in selection and ranking problems'. *Annals of Mathematical Statistics*, **28**, 957–967.

Gupta, S. S. and Sobel, M. (1960). 'Selecting a subset containing the best of several binomial populations'. *Contributions to Probability and Statistics* (I. Olkin, S. G. Ghurye, W. Hoeffding, W. G. Madow and H. B. Mann, eds.), Stanford University Press, Stanford, California, Chapter 20, 224–248.

Gupta, S. S. and Sobel, M. (1962a). 'On selecting a subset containing the population with the smallest variance'. *Biometrika*, **49**, 495–507.

Gupta, S. S. and Sobel, M. (1962b). 'On the smallest of several correlated F-statistics'. *Biometrika*, **49**, 509–523.

Gupta, S. S. and Studden, W. J. (1970). 'On a ranking and selection procedure for multivariate populations'. *Essays in Probability and Statistics* (R. C. Bose, I. M. Chakravarti, P. C. Mahalanobis, C. R. Rao and K. J. C. Smith, eds.), University of North Carolina, Chapel Hill, North Carolina, Chapter 16, 327–338.

Gupta, S. S. and Wong, W.-Y. (1977). 'Subset selection procedures for finite schemes in information theory'. *Colloquia Mathematica Societatis János Bolyai, 16: Topics in Information Theory* (I. Csiszár and P. Elias, eds.), 279–291.

Hocking, R. R. (1976). 'The analysis and selection of variables in regression analysis'. *Biometrics*, **32**, 1–49.

Hsu, T.-A. and Huang, D.-Y. (1982). 'Some sequential selection procedures for good regression models'. *Communications in Statistics - Theory and Methods*, **A11**, 411–421.

Huang, D.-Y. and Panchapakesan, S. (1982). 'On eliminating inferior regression models'. *Communications in Statistics - Theory and Methods*, **A11**, 751–759.

Krishnaiah, P. R. (1967). 'Selection procedures based on covariance matrices of multivariate normal populations'. *Blanch Anniversary Volume*, Aerospace Reearch Laboratories, U. S. Air Force Base, Dayton, Ohio, 147–160.

Krishnaiah, P. R. and Armitage, J. V. (1964). 'Distribution of the studentized smallest chi-square, with tables and applications'. *ARL 64-218*, Aerospace Research Laboratories, Wright-Patterson Air Force Base, Dayton, Ohio.

Krishnaiah, P. R. and Armitage, J. V. (1966). 'Tables for multivariate t-distribution'. *Sankhya*, **28B**, 31–56.

Krishnaiah, P. R. and Rizvi, M. H. (1966). 'Some procedures for selection of multivariate normal populations better than a control'. *Multivariate Analysis* (P. R. Krishnaiah, ed.), Academic Press, New York, 477–490.

McCabe, G. P., Jr. and Arvesen, J. N. (1974). 'Subset selection procedures for regression variables'. *Journal of Statistical Computation and Simulation*, **3**, 137–146.

Panchapakesan, S. (1971). 'On a subset selection procedure for the most probable event in a multinomial distribution'. *Statistical Decision Theory and Related Topics* (S. S. Gupta and J. Yackel, eds.), Academic Press, New York, 275–298.

Panchapakesan, S. (1973). 'On a subset selection procedure for the best multinomial cell and related problems'. Abstract. *Bulletin of the Institute of Mathematical Statistics*, **2**, 112–113.

Regier, M. H. (1976). 'Simplified selection procedures for multivariate normal populations'. *Technometrics*, **18**, 483–489.

Slepian, D. (1962). 'On the one-sided barrier problem for Gaussian noise'. *Bell System Technical Journal*, **41**, 463–501.

Thompson, M. L. (1978a). 'Selection of variables in multiple regression: Part I. A review and evaluation'. *International Statistical Review*, **46**, 1–19.

Thompson, M. L. (1978b). 'Selection of variables in multiple regression: Part II. Chosen procedures, computations and examples'. *International Statistical Review*, **46**, 129–146.

C. G. Khatri

QUADRATIC FORMS TO HAVE A SPECIFIED DISTRIBUTION

1. INTRODUCTION

Let $\underset{\sim}{x}' = (x_1, x_2, \ldots, x_n)$ be a random vector and let its joint density be

$$f(\underset{\sim}{x}'\underset{\sim}{x}) = (2\pi)^{-\frac{1}{2}n}\exp(-\underset{\sim}{x}'\underset{\sim}{x}/2) \quad \text{for all} \quad \underset{\sim}{x} \in \mathbb{R}^n, \tag{1.1}$$

where $\mathbb{R}^n$ is a set of n-vectors defined on the real field $\mathbb{R}$. Let us use the transformations: $R = \underset{\sim}{x}'\underset{\sim}{x}$ and $\underset{\sim}{u} = \underset{\sim}{x}/\sqrt{R}$ with $\underset{\sim}{u}'\underset{\sim}{u} = 1$. We shall say that $\underset{\sim}{u} \in O(\tilde{1},\tilde{n})$ {= a set of vectors $\underset{\sim}{u} \in \mathbb{R}^n$ such that $\underset{\sim}{u}'\underset{\sim}{u} = 1$}. On account of the restriction on $\underset{\sim}{u}$, the random variables will be taken as $u_1, u_2, \ldots, u_{n-1}$ while $u_n^2 = (1 - \sum_{i=1}^{n-1} u_i^2)$. Then, it is easy to see that the jacobian of the transformation is

$$J(\underset{\sim}{x} \to R, \underset{\sim}{u}) = R^{\frac{1}{2}n-1}(1 - \sum_{i=1}^{n-1} u_i^2)^{-\frac{1}{2}},$$

and the joint density of R and $\underset{\sim}{u}$ can be obtained from (1.1). This shows that

(i) R and $\underset{\sim}{u}$ are independently distributed,

(ii) R is distributed as Chi-square with n degrees of freedom,

denoted by $R \sim \chi_n^2$ or $R \sim G(\frac{1}{2}n, \frac{1}{2})$ where the density of $G(a,\alpha)$ is

$$(\alpha^a/\Gamma(a))R^{a-1}\exp(-\alpha R) \quad \text{for all} \quad R > 0 \tag{1.2}$$

A. K. Gupta (ed.), Advances in Multivariate Statistical Analysis, 161–190.

with $\alpha > 0$ and $a > 0$, and

(iii) the density of $u_1, u_2, \ldots, u_{n-1}$ is

$$(\Gamma(n/2)/\pi^{\frac{1}{2}n})(1 - \sum_{i=1}^{n-1} u_i^2)^{\frac{1}{2}-1} \quad \text{for all } \underset{\sim}{u} \in O(1,n). \tag{1.3}$$

This distribution of $\underset{\sim}{u}$ will be said to be uniform over the surface $\underset{\sim}{u}'\underset{\sim}{u} = 1$ and will be denoted as $[d\underset{\sim}{u}]$, a unit invariant Haar measure over $O(1,n)$.

Cochran's Theorem (1934) is related to the random normal vector variables $\underset{\sim}{x}$ whose density is given by (1.1). We shall write $\underset{\sim}{x} \sim N(\underset{\sim}{0}, I_n)$. This will be stated as

LEMMA 1. *Let* $\underset{\sim}{x}'A_i\underset{\sim}{x}$ $(i = 1,2,\ldots,k)$ *be* k *quadratic forms and* $\underset{\sim}{x}'A\underset{\sim}{x} = \sum_{i=1}^{k} \underset{\sim}{x}'A_i\underset{\sim}{x}$. *Here,* $A, A_1, \ldots, A_k$ *are symmetric matrices. If* $A = I_n$, *(the identity matrix of order* $n \times n$ *and some times it will be mentioned without the suffix* n *when there is no confusion), then* $\underset{\sim}{x}'A_i\underset{\sim}{x}$ $(i = 1,2,\ldots,k)$ *are independently distributed as Chi-squares with* n_i *degrees of freedom for all* i *if and only if* $\sum_{i=1}^{k} n_i = n$.

Later on the above result was slightly modified by Graybill and Marsaglia (1957) which is given by

LEMMA 2. *Let* $\underset{\sim}{x} \sim N(\underset{\sim}{0}, I_n)$ *and let* $\underset{\sim}{x}'A_i\underset{\sim}{x}$ $(i = 1,2,\ldots,k)$ *be* k *quadratic forms. Let* $\underset{\sim}{x}'A\underset{\sim}{x} = \sum_{i=1}^{k} \underset{\sim}{x}'A_i\underset{\sim}{x}$. *Consider the statements:*

(a) $\underset{\sim}{x}'A_i\underset{\sim}{x} \sim \chi^2_{n_i}$ *for all* i

(b) $\underset{\sim}{x}'A_i\underset{\sim}{x}$ *and* $\underset{\sim}{x}'A_j\underset{\sim}{x}$ *are independently distributed for all* $i \neq j$

(c) $\underset{\sim}{x}'A\underset{\sim}{x} \sim \chi_s^2$

(d) $\text{Rank } A = \sum_{i=1}^{k} \text{Rank } A_i.$

Then (i) any two of (a), (b), (c) imply all the four conditions and (ii) (c) and (d) imply (a) and (b).

Later on, these results were extended to complex normal random vector variables and the singular normal random vector variables. Further, they were extended to matrix variables and the quadratic forms in normal vector variables; (see, for example, Khatri (1959), (1962), (1963), (1977), (1978), (1980a), (1980b), (1982), (1983), (1984), and Shanbhag (1968) and Good (1969)). Recently, Anderson and Fang (1985), Khatri and Mukerjee (1986) and Khatri (1986) extended the results to the elliptical contoured distributions. The random vector $\underset{\sim}{x}$ is said to have an elliptical contoured distribution denoted by $\underset{\sim}{x} \sim E_n(\mu,\Sigma;\phi)$ if its characteristic function is given by

$$\exp(\sqrt{-1}\ \underset{\sim}{t}'\underset{\sim}{\mu})\ \phi(\underset{\sim}{t}'\Sigma\underset{\sim}{t}) \quad \text{for all real} \quad \underset{\sim}{t} \in R^n. \quad (1.4)$$

It has been printed out by Cambanis et al (1981) or Dawid (1977) that if $P((\underset{\sim}{x} - \underset{\sim}{\mu})'\Sigma^-(\underset{\sim}{x} - \underset{\sim}{\mu}) = R > 0) = 1$ where Σ^- is a g-inverse of Σ, say $\Sigma\Sigma^-\Sigma = \Sigma$, then $\underset{\sim}{x} - \underset{\sim}{\mu} \overset{d}{=} \sqrt{R}\ P'\underset{\sim}{u}$ where R and $\underset{\sim}{u}$ are independently distributed, $V = P'P$, $\text{Rank } V = m$ and $\underset{\sim}{u} \sim [d\underset{\sim}{u}]$ for all $\underset{\sim}{u} \in O(1,m)$. Thus, to combine all the above results, we shall develop some results on the quadratic forms $\underset{\sim}{u}'A_i\underset{\sim}{u}$ where $\underset{\sim}{u} \sim [d\underset{\sim}{u}]$. These results are extended to the multivariate situation where $U \sim [dU]$ with $U \in O(p,n)$ {= Stieljes manifold, a set of $U \in R^{p\times n}$ matrices such that $UU' = I_p$} and $[dU]$ denotes the unit Haar measure defined on $O(p,n)$. The explicit expression of $[dU]$ is given by Khatri (1970) for any p. Further, these results are extended to the complex random variables.

2. QUADRATIC FORMS IN $\underset{\sim}{u}$

Let $\underset{\sim}{u}'A\underset{\sim}{u} = z$ be a quadratic form in $\underset{\sim}{u} \sim [d\underset{\sim}{u}]$. Then, since A is an $n \times n$ symmetric matrix of rank r $(< n)$, we can write $A = \Delta D_\lambda \Delta'$ where Δ is an $n \times n$ orthogonal matrix and $D_\lambda = \mathrm{diag}(\lambda_1, \ldots, \lambda_r, 0, \ldots, 0)$ with $\lambda_1, \lambda_2, \ldots, \lambda_r$ being nonzero eigen values of A. Then, taking $\Delta'\underset{\sim}{u} = \underset{\sim}{y}$, we see that $\underset{\sim}{y} \sim [d\underset{\sim}{y}]$ and

$$z = \sum_{i=1}^{r} \lambda_i y_i^2 = \sum_{i=1}^{r} \lambda_i z_i \tag{2.1}$$

where $(z_1, z_2, \ldots, z_r)$ is distributed as $D_r(\frac{1}{2}, \frac{1}{2}, \ldots, \frac{1}{2}; \frac{n-t}{2})$ whose density function is given by

$$\frac{\Gamma(n/2)}{\pi^{r/2}\Gamma(\frac{n-r}{2})} \prod_{i=1}^{r} z_i^{\frac{1}{2}-1}\left(1 - \sum_{i=1}^{r} z_i\right)^{\frac{1}{2}(n-r)-1}$$

$$\text{for all } (z_1, \ldots, z_r) \in \mathcal{D} \tag{2.2}$$

where $\mathcal{D}\{(z_1, \ldots, z_r)$: $0 < z_i < 1$ and $\sum_{i=1}^{r} z_i < 1\}$. Notice that

$$|z| \leq \sum_{i=1}^{r} |\lambda_i| z_i \leq \max_i |\lambda_i| \sum_{i=1}^{r} z_i < \max_i |\lambda_i|.$$

Since the variable z has the finite range, all its moments exist and the distribution is uniquely determined by their moments.

(2A) Moments of $\underset{\sim}{z}$.

Let R_1 and u be independently distributed, $R_1 \sim G(\frac{1}{2}n, 1)$ and $\underset{\sim}{u} \sim [d\underset{\sim}{u}]$. Then, $\underset{\sim}{x} = \sqrt{R_1}\,\underset{\sim}{u} \sim N(0, \frac{1}{2}I_n)$ and

$$E \exp(-tR_1 z) = E \exp(-t\underset{\sim}{x}'A\underset{\sim}{x}) = |I + tA|^{-\frac{1}{2}}$$

for all t such that $(\max_i |\lambda_i|)t < 1$. Hence,

$$\prod_{i=1}^{r} (1 + t\lambda_i)^{-\frac{1}{2}} = E(1 + tz)^{-n/2} = \phi(t), \quad \text{(say)}. \tag{2.3}$$

Let us denote

$$\mu_j = \{\Gamma(\tfrac{n}{2} + j)/\Gamma(\tfrac{n}{2})\}Ex^j \quad \text{for} \quad j = 0,1,2,\ldots .$$

Then

$$\phi(t) = \sum_{i=0}^{\infty} \mu_j(-t)^j/j! = \prod_{i=1}^{r} (1 + t\lambda_i)^{-\frac{1}{2}}$$

for all $|t| < 1/\max_i |\lambda_i|$.

From this, we get

$$\log \phi(t) = \sum_{j=1}^{\infty} K_j(-t)^j/j! = -\tfrac{1}{2} \sum_{i=1}^{r} \log(1 + t\lambda_i)$$

$$= \tfrac{1}{2} \sum_{j=1}^{\infty} \frac{(-t)^j}{j} (\text{tr } A^j).$$

Hence $K_j = (\frac{1}{2}\text{tr}A^j)(j - 1)!$ for $j = 1,2,3,\ldots$. Here, the relations between K_j's and μ_j's are the same as those between cumulants and moments. Thus, we have

$$K_1 = \mu_1, \quad K_2 = \mu_2 - \mu_1^2, \quad K_3 = \mu_3 - 3\mu_2\mu_1 + 2\mu_1^3,$$

$$K_4 + 3K_2^2 = \mu_4 - 4\mu_3\mu_1 + 6\mu_2\mu_1^2 - 3\mu_1^4 \quad \text{etc.}$$

From the above considerations, one can obtain the moments of z.

(2B) Beta distribution of z.

z is said to be distributed as Beta with parameters a and b denoted as $z \sim \beta(a,b)$, when its density function is given by

$$\{B(a,b)\}^{-1} z^{a-1} (1 - z)^{b-1} \quad \text{for} \quad 0 < z < 1 \qquad (2.4)$$

where $a > 0$ and $b > 0$ and $B(a,b) = \Gamma(a)\Gamma(b)/\Gamma(a + b)$. The joint distribution of $z_1, z_2, \ldots, z_r$ is said to be Dirichlet with parameters $a_1, a_2, \ldots, a_r$ and a_{r+1}, denoted as $(z_1, z_2, \ldots, z_r) \; D_r(a_1, a_2, \ldots, a_r;\; a_{r+1})$ if its joint density is given by

$$\{B(a_1, a_2, \ldots, a_r;\; a_{r+1})\}^{-1} \left(\prod_{i=1}^{r} z_i^{a_i - 1} \right)$$

$$\cdot \left(1 - \sum_{i=1}^{r} z_i\right)^{a_{r+1} - 1} \qquad (2.5)$$

for all $(z_1, \ldots, z_r) \in \mathcal{D} = \{(z_1, \ldots, z_r) \mid 0 < z_i < 1$ for all i and $\sum_{i=1}^{r} z_i < 1\}$, where $a_i > 0$ for $i = 1, 2, \ldots, r + 1$, and

$$B(a_1, \ldots, a_r;\; a_{r+1}) = \Gamma\left(\sum_{i=1}^{r+1} a_i \right) / \prod_{i=1}^{r} \Gamma(a_i).$$

When $a_{r+1} = 0$, then $\sum_{i=1}^{r} z_i = 1$ and Dirichlet's distribution is singular and defined on the surface $\{ \sum_{i=1}^{r} z_i = 1$ and $0 < z_i < 1$ for all $i\}$.

Theorem 1. *Let* $\underset{\sim}{u} \sim [d\underset{\sim}{u}]$ *on* $O(1,n)$ *and let* $z = \underset{\sim}{u}' A \underset{\sim}{u}$ *be a quadratic form with* $A = A'$. *Then,* $z \sim \beta(a,b)$ *with* $a > 0$ *and* $b > 0$ *if and only if* $a + b = n/2$, $2a$ *is a positive integer and* A *is idempotent of rank* $(2a)$.

Proof. Since $0 \le z \le 1$, $\underset{\sim}{u}'a\underset{\sim}{u} \ge 0$ and $\underset{\sim}{u}'(I - A)\underset{\sim}{u} \ge 0$ for all $\underset{\sim}{u} \in O(1,\bar{n})$, and hence A and $I - A$ are positive semi-definite. If $\lambda_1 \ge \lambda_2 \ge \cdots \ge \lambda_r > 0$ are the nonzero eigen values of A, then $\lambda_1 \le 1$ and we can write

$$A = \Delta D_\lambda \Delta', \quad \Delta \in O(n),$$

$$D_\lambda = \mathrm{diag}(\lambda_1, \lambda_2, \ldots, \lambda_r, 0, \ldots, 0).$$

Hence, we can write

$$z = \sum_{i=1}^{r} \lambda_i z_i, \quad (z_1, \ldots, z_r) \sim D_r(\tfrac{1}{2}, \ldots, \tfrac{1}{2}; \frac{n - r}{2}).$$

Notice that $z \in [0, \lambda_1]$ while the given range is $[0,1]$. Hence, $\lambda_1 = 1$. We shall obtain the density of z from the joint density of $(z_1, \ldots, z_r)$, given by

$$(\Gamma(n/2)/\pi^{r/2}\Gamma(\frac{n - r}{2})) \prod_{i=1}^{r} z_i^{\frac{1}{2}-1}(1 - \sum_{i=1}^{r} z_i)^{\frac{1}{2}(n-r)-1}$$

for all $(z_1, \ldots, z_r) \in \mathcal{D}$.

Use the domain

$$\mathcal{D}_1 = \{(w_2, w_3, \ldots, w_r) | \sum_{i=\ell}^{r} w_i \le 1, \quad w_i \ge 0 \text{ and}$$

$$z \sum_{i=2}^{r} w_i(1 - \lambda_i)/\lambda_i \le 1 - z\}$$

and note that as $z \to 1$, $\mathcal{D}_1 \to \mathcal{D}_0\{(w_2, \ldots, w_r) | \sum_{i=2}^{r} w_i \le 1$ and $w_i \ge 0\}$ and $\lambda_i \to 1$ for all $i = 2, 3, \ldots, n$. Using the necessary transformations, the density of z is given by

$$z^{\frac{1}{2}r-1}(1-z)^{\frac{n-r}{2}-1} g(z/(1-z)) \quad \text{for} \quad 0 < z < 1$$

where $g(t)$ is given by

$$g(t) = \{\Gamma(n/2)/\Pi^{r/2}\Gamma(\frac{n-r}{2})(\Pi\lambda_i)^{\frac{1}{2}}\}$$

$$\cdot \int_{\mathcal{D}_1} (\prod_{i=\ell}^{r} w_i^{\frac{1}{2}-1})(1 - \sum_{i=2}^{r} w_i)^{\frac{1}{2}-1}$$

$$\cdot (1 - t\sum_{i=\ell}^{r} \frac{1-\lambda_i}{\lambda_i} w_i)^{\frac{1}{2}(n-r)-1} dW_2 \ldots dW_r .$$

Notice that as $t \to \infty$, $\dfrac{t(1-\lambda_i)}{\lambda_i} \to 1$. Hence as $t \to \infty$, $\mathcal{D}_1 \to \mathcal{D}_0$ and $g(t) \to c(\neq 0)$. Hence, we must have

$$z^{\frac{1}{2}r-1}(1-z)^{\frac{n-r}{2}-1} g(z/(1-z)) =$$

$$\{B(a,b)\}^{-1} z^{a-1}(1-z)^{b-1} \quad \text{for all} \quad 0 < z < 1.$$

If $a > \frac{1}{2}r$, then $(1-z)^{\frac{n-r}{2}-1} g(z/(1-z)) = \{B(a,b)\}^{-1} z^{a-r/2}(1-z)^{b-1}$ for $0 \leq z < 1$ and as $z \to 0$, L.H.S. $\to$ constant while R.H.S. $\to$ 0. Hence $a \not> \frac{1}{2}r$. Similarly, we can show that $a \not< \frac{1}{2}r$ and hence, $r = 2a$ must be a positive integer. If $b > \frac{n-r}{2}$, then

$$g(z/(1-z)) = \{B(a,b)\}^{-1}(1-z)^{b-\frac{n-r}{2}} \quad \text{for}$$

$$0 \leq z \leq 1$$

and as noted above, $g(z/(1-z)) \to c\ (\neq 0)$ as $z \to 1$ while R.H.S. $\to 0$, which is impossible. Hence $b \not> \frac{1}{2}(n-r)$. Similarly, we can show that $b \not< \frac{1}{2}(n-r)$. This gives $b = \frac{1}{2}(n-r)$. This shows that $a + b = \frac{1}{2}n$ and $2a = r =$ Rank A. We have already noted that as $z \to 1$, $\mathcal{D}_1 \to \mathcal{D}_0$ and $\lambda_i \to 1$ for all i. Hence, A must be an idempotent matrix of rank r.

This proves the first part of Theorem 1. The converse result is immediate. Thus, Theorem 1 is established.

Note 1. Theorem 1 was established by Khatri and Mukerjee (1986) by a different approach.

Theorem 2. *Let* $\underset{\sim}{u} \backsim [d\underset{\sim}{u}]$ *on* $O(1,n)$ *and let* $z_i = \underset{\sim}{u}'A_i\underset{\sim}{u}$ $(i = 1,2,\ldots,k)$ *be* k *quadratic forms. Then, the joint distribution of* $(z_1,z_2,\ldots,z_k)$ *is* $D_k(a_1,a_2,\ldots,a_k;\ a_{k+1})$ *with* $a_1 > 0,\ldots,a_{k+1} > 0$ *if and only if* $\sum_{i=1}^{k+1} a_i = n/2$, $2a_i$ $(i = 1,2,\ldots,k)$ *are positive integers,* A_i $(i = 1,2,\ldots,k)$ *are idempotent matrices of ranks* $2a_i$ $(i = 1,2,\ldots,k)$ *respectively and* $A_iA_j = 0$ *for all* $i \neq j$, $i,j = 1,2,\ldots,k$.

Note 2. This will be considered as Cochran's Theorem (or Graybill and Marsaglia (1957)) for uniformly distributed random variables defined on $O(1,n)$.

Proof. Theorem 2 due to 'if' part is easy to verify. For the 'only if' part, note that $z_i \backsim \beta(a_i,b_i)$ with $b_i = \sum_{i=1}^{k+1} a_i - a_i$. Then, by Theorem 1, $\sum_{i=1}^{k+1} a_i = n/2$ and A_i is an idempotent matrix of rank $(2a_i)$ and this is true for all i $(i = 1,2,\ldots,k)$. Similarly, $z = \sum_{i=1}^{k} z_i = \underset{\sim}{u}'A\underset{\sim}{u}$ with $A = \sum_{i=1}^{k} A_i \backsim \beta(a,b)$ with $a = \sum_{i=1}^{k} a_i$ and $b = a_{k+1}$, and by Theorem 1, A must be idempotent with rank $(2a)$. Thus, we have proved

$$(a)\ A_i^2 = A_i \text{ for all } i, \quad (c)\ A^2 = A \text{ and}$$

$$(d)\ \text{Rank } A = \sum_{i=1}^{k} \text{Rank } A_i .$$

By Graybill and Marsaglia's result (1957), we get (b) $A_iA_j = 0$ for all $i \neq j$. This proves the required Theorem 2.

<u>Theorem 3</u>. *Let* $\underset{\sim}{u} \sim [d\underset{\sim}{u}]$ *on* $O(1,n)$ *and let* $z_i = \underset{\sim}{u}'A_i\underset{\sim}{u}$ *for* $i = 1,2$ *be two quadratic forms with* $A_i = A_i'$. *If* $m = n/2$ *and*

$$\Gamma(m + i + j)Ez_1^i z_2^j = \Gamma(m + i)\Gamma(m + j)(Ez_1^i)(Ez_2^j)/\Gamma(m) \qquad (2.6)$$

for $i,j = 0,1,2$, *then* $A_1A_2 = A_2A_1 = 0$. *Conversely, if* $A_1A_2 = A_2A_1 = 0$, *then (2.6) holds for all* $i,j = 0,1,2,\ldots$.

<u>*Proof*</u>. Let $w_i = z_i - E(z_i)$ for $i = 1,2$. Then, the given conditions (2.6) are

$$Ew_1^i w_2^j = 0 \text{ for } (i,j) = (1,1),(1,2),(2.1) \qquad (2.7a)$$

and

$$\Gamma(m + 4)E(w_1^2 w_2^2) = (\Gamma(m + 2))^2(Ew_1^2)(Ew_2^2)/\Gamma(m). \qquad (2.7b)$$

Consider $z = t_1w_1 + t_2w_2$ for all t_1 and t_2 and $A = t_1A_1 + t_2A_2$. Then, by the moment relations given in section (2A), we have

$$3\mathrm{tr}(t_1A_1 + t_2A_2)^4 + \tfrac{3}{4}[\mathrm{tr}(t_1A_1 + t_2A_2)^2]^2 =$$

$$\{\Gamma(m + 4)/(\Gamma(m)\}E(t_1w_1 + t_2w_2)^4$$

and

$$\tfrac{1}{2}\mathrm{tr}(t_1A_1 + t_2A_2)^2 = (\Gamma(m + 2)/\Gamma(m))$$

$$E(t_1w_1 + t_2w_2)^2.$$

Now, using (2.7), we get

$$\mathrm{tr}(t_1A_1 + t_2A_2)^4 = t_1^4\mathrm{tr}A_1^4 + t_2^4\mathrm{tr}A_2^4$$

for all t_1 and t_2

and collecting the coefficients of $(t_1t_2)^2$, we get

$$2\mathrm{tr}(A_1^2A_2^2) + \mathrm{tr}(A_1A_2 + A_2A_1)^2 = 0$$

and then it is obvious that $A_1A_2 = A_2A_1 = 0$. The converse result is easy to establish.

(2C) Distribution of simultaneous quadratic forms.

Let $\underset{\sim}{u}$ be distributed uniformly over $O(1,n)$, and $k + p$ be quadratic forms given by $\underset{\sim}{u}'A_i\underset{\sim}{u} = z_i$ $(i = 1,2,\ldots,p+k)$. Suppose that if $\underset{\sim}{x} \overset{d}{=} \underset{\sim}{y}$ denotes that $\underset{\sim}{x}$ and $\underset{\sim}{y}$ have identical distributions, then we shall consider the situation

$$(z_1,z_2,\ldots,z_{p+k}) \overset{d}{=} (w_{11},w_{22},\ldots,w_{pp},w_1,w_2,\ldots,w_k)$$

where $w_{11},w_{22},\ldots,w_{pp}$ are the diagonal elements of a positive definite matrix W and the joint density of $(W,\underset{\sim}{w})$

is given by

$$c|W|^{a_0-\frac{1}{2}(p+1)} \prod_{i=1}^{k} w_i^{a_i-1} \quad \text{for all} \quad (W,\underset{\sim}{w}) \in \mathcal{D}_{p,k} \tag{2.8}$$

where $\mathcal{D}_{p,k} = \{(W,\underset{\sim}{w});\ W > 0,\ \ w_i > 0$ and $\text{tr}\Sigma^{-1}W + \sum_{i=1}^{k} w_i = 1\}$, $\Sigma > 0$ and c is a constant given by

$$c = \Gamma(a_0 p + \sum_{i=1}^{k} a_i)/|\Sigma|^{a_0}\Gamma_p(a_0)\prod_{i=1}^{k}\Gamma(a_i),$$

$$\Gamma_p(a_0) = \pi^{\frac{1}{4}p(p-1)} \prod_{i=1}^{p} \Gamma(a_0 - \frac{i-1}{2}),$$

for $k \geq 1$, while if $k = 0$, the density of W is given by

$$\{\Gamma(a_0 p)/|\Sigma|^{a_0}\Gamma_p(a_0)\}|W|^{a_0-\frac{1}{2}(p+1)} J \quad \text{for all} \quad W \in \mathcal{D} \tag{2.9}$$

where $\mathcal{D} = \{W;\ W > 0$ and $\text{tr}\Sigma^{-1}W = 1\}$ and J denotes the jacobian of the transformation of any positive definite matrix S to W and b by the relations $bW = S$ and $b = \text{tr}\Sigma^{-1}S$ when $b = 1$. The distributions (2.8) and (2.9) are respectively denoted by $D_{p,k}(a_0, \Sigma;\ \underset{\sim}{a})$ and $D_{p,0}(a_0,\Sigma)$ respectively. Theorem 2 gives the results when $(z_1,\ldots,z_k) \sim D_k(\underset{\sim}{a})$ with $\underset{\sim}{a}' = (a_1,\ldots,a_k)$ and $\sum_{i=1}^{k} z_i = 1$. (In the notations of (2.5), $(z_1,\ldots,z_{k-1}) \sim D_{k-1}(a_1,\ldots,a_{k-1};\ a_k)$ which is the same as $(z_1,\ldots,z_k) \sim D_k(\underset{\sim}{a})$ when $\sum_{i=1}^{k} z_i = 1$ and $z_i \geq 0$.)

When $k \geq 1$, $(w_1,w_2,\ldots,w_k) \sim D_k(\underset{\sim}{a};\ pa_0)$, and the density of W is given by

$$\frac{\Gamma(a_0 p + \Sigma a_j)}{\Gamma(\Sigma a_j)\Gamma_p(a_0)} |\Sigma|^{-a_0} |W|^{a_0 - \frac{1}{2}(p+1)} (1 - \mathrm{tr}\Sigma^{-1}W)^{a_i - 1}$$

for all $W > 0$ and $\mathrm{tr}\Sigma^{-1}W < 1$, (2.10)

or $(W, w_1) \sim D_{p,1}(a_0, \Sigma; \sum_{j=1}^{k} a_j)$. If $\Sigma = I_p$, then

$$(w_{11}, w_{22}, \ldots, w_{pp}) \sim D_p(a_0, \ldots, a_0; \Sigma a_i)$$

in the notations of (2.5).

Suppose, Σ is a diagonal matrix with diagonal elements σ_{ii} $(i = 1,2,\ldots,p)$, then

$$(w_{11}/\sigma_{11}, \ldots, w_{pp}/\sigma_{pp}) \sim D_p(a_0, \ldots, a_0; \Sigma a_i). \tag{2.11}$$

Now assume throughout that Σ is a correlation matrix with $\sigma_{ii} = 1$. When $k = 0$ and $p \geq 2$ (or $k \geq 1$ and $p \geq 1$),

$$(w_{ii}/\sigma_{ii}, w_1, \ldots, w_k) \quad D_{k+1}(a_0, a_1, \ldots, a_k; (p-1)a_0)$$

for any $i = 1,2,\ldots,p$. (2.12)

Now, Khatri (1986) has established in a different language the following

Theorem 4. *Let* $\underset{\sim}{u} \sim [d\underset{\sim}{u}]$ *over* $O(1,n)$ *and let* $\underset{\sim}{u}'A_i\underset{\sim}{u} = z_i$ $(i = 1,2,\ldots,p + k)$ *be* $p + k$ *quadratic forms, with* $A_i = A_i'$. *Then,* $(z_1, \ldots, z_{p+k}) \overset{d}{=} (w_{11}, \ldots, w_{pp}, w_1, \ldots, w_k)$ *where* $(W, \underset{\sim}{w}) \sim D_{p,k}(a_0, \Sigma; \underset{\sim}{a})$ *and* $w_{11}, \ldots, w_{pp}$ *are the diagonal elements of* W *with* $a_0 \geq p$ *and* $a_j > 0$ *for* $j = 1,2,\ldots,k$; *if and only if* $2\sum_{j=1}^{k} a_j + 2pa_0 = n$, $2a_j$ $(j = 1,2,\ldots,k)$ *are positive integers* = Rank A_{p+j}, $A_i^2 = A_i$, $A_iA_{p+j} = 0$ *for* $i = 1,2,\ldots,p + k$, $i \neq p + j$, *j* = 1,2,...,k, $A_\alpha A_\beta A_\alpha = \sigma^2_{\alpha\beta} A_\alpha$ *and* $A_\alpha A_\beta A_\gamma A_\alpha = (\sigma_{\alpha\beta}\sigma_{\beta\gamma}\sigma_{\gamma\alpha})A_\alpha$ *for* $\alpha \neq \beta \neq \gamma$, $\alpha,\beta,\gamma = 1,2,\ldots,p$.

<u>*Proof*</u>. Suppose $(z_1,\ldots,z_{p+k}) \stackrel{d}{=} (w_{11},\ldots,w_{pp},w_1,\ldots,w_k)$ and $\sigma_{ii} = 1$ for $i = 1,2,\ldots,p$ are the diagonal elements of Σ. Using (2.12),

$$(z_i,z_{p+1},\ldots,z_{p+k}) \stackrel{d}{=} (w_{ii},w_1,\ldots,w_k)$$

$$D_{k+1}(a_0,a_1,\ldots,a_k;a_0(p-1)).$$

By Theorem 2, we must have

$$n = 2pa_0 + 2\sum_{j=1}^{k} a_j, \quad 2a_j = \text{Rank } A_{p+j} = \text{tr}A_{p+j},$$

$$2a_0 = \text{Rank } A_i, \quad A_i^2 = A_i, \quad A_iA_{p+j} = A_{p+j}A_i = 0,$$

$$A^2_{p+j} = A_{p+j'}A_{p+j}A_{p+j'} = 0$$

for $j \neq j'$, $j,j' = 1,2,\ldots,k$. This is true for any $i = 1,2,\ldots,p$. To establish the other conditions, let R_1 and $\underset{\sim}{u}$ be independently distributed, $R_1 \sim G(pa_0 + \Sigma a_j, 1)$ and $\underset{\sim}{x} = \sqrt{R_1}\,\underset{\sim}{u} \sim N(\underset{\sim}{0},\tfrac{1}{2}I_n)$. Then if $R_1W = S$, then given conditions imply

$$(z_1,\ldots,z_p) \stackrel{d}{=} (w_{11},\ldots,w_{pp}) \text{ and}$$

$$(W,w_1) \sim D_{p,1}(a_0,\Sigma;\ \sum_{j=1}^{k} a_j)$$

and consequently

$$(\underset{\sim}{x}'A_1\underset{\sim}{x},\ldots,\underset{\sim}{x}'A_p\underset{\sim}{x}) \stackrel{d}{=} (s_{11},\ldots,s_{pp}) \text{ and}$$

$$S \sim W_p(a_0, \tfrac{1}{2}\Sigma);$$

that is, the density of S is

$$\{\Gamma_p(a_0)|\Sigma|^{a_0}\}^{-1}|S|^{a_0-\frac{1}{2}(p+1)}\exp(-\mathrm{tr}\Sigma^{-1}S)$$

for $S > 0$.

Then, using the result of Khatri (1980b), we get the required conditions.

Conversely, given the conditions $A_\alpha^2 = A_\alpha$, $A_\alpha A_\beta A_\alpha = \sigma_{\alpha\beta}^2 A_\alpha$, $A_\alpha A_\beta A_\gamma A_\alpha = (\sigma_{\alpha\beta}\sigma_{\beta\gamma}\sigma_{\gamma\alpha})A_\alpha$ and $2a_0 = \text{Rank } A_\alpha$ for $\alpha = 1,2,\ldots,p$, we have

$$(\underset{\sim}{x}'A_1\underset{\sim}{x},\ldots,\underset{\sim}{x}'A_p\underset{\sim}{x}) \overset{d}{=} (s_{11},\ldots,s_{pp})$$

and $S \sim W_p(a_0, \frac{1}{2}\Sigma)$.

Note that R_1 and $\underset{\sim}{u}$ are independently distributed, $R_1 \sim G(pa_0 + \Sigma a_j, 1)$ and (W, w_1) and R_1 are independently with $(W, w_1) \sim D_{p,1}(a_0, \Sigma; \sum_{j=1}^{k} a_j)$, using the condition $2pa_0 + 2\sum_j a_j = n$. Thus, we have

$$R_1(z_1,\ldots,z_p) \overset{d}{=} R_1(w_{11},\ldots,w_{pp})$$

$$\Rightarrow (z_1,\ldots,z_p) \overset{d}{=} (w_{11},\ldots,w_{pp})$$

because $P(z_1 > 0,\ldots,z_p > 0) = 1 = P(w_{11} > 0,\ldots,w_{pp} > 0) = P(R_1 > 0)$. (See Anderson and Fang (1985) or the result A.1 after Lemma 3'). For the other conditions, we use the matrix decomposition Theorem, and get the required result. Thus, Theorem 4 is established.

Lemma 3. *Let* $\underset{\sim}{u} \sim [d\underset{\sim}{u}]$ *on* $O(1,n)$ *and let* $\underset{\sim}{u}'A\underset{\sim}{u} = z$ *be a quadratic form with* $A = A'$. *Then,* $z \overset{d}{=} \sum_{i=1}^{k} \lambda_i z_i$ *where* $(z_1, z_2, \ldots, z_k) \sim D_k(a_1, \ldots, a_k;\ a_{k+1})$, $\lambda_1 \neq \lambda_2 \neq \ldots \neq \lambda_k \neq 0$ *and* $\sum_{i=1}^{k+1} a_i = \frac{1}{2}n$ *if and only if* $\lambda_1, \lambda_2, \ldots, \lambda_k$ *are the distinct eigen values of* A *with respective multiplicities* $2a_j$ $(j = 1,2,\ldots,k)$.

Proof. The 'if' result is obvious. For the 'only if' result, let R_1 and $\underset{\sim}{u}$ be independently distributed and let R_1 and $(z_1, \ldots, z_k)$ are independently distributed with $R_1 \sim G(\frac{1}{2}n, 1)$ and $n = 2\sum_{j=1}^{k} a_j$. Then $\underset{\sim}{x} = \sqrt{R_1}\,\underset{\sim}{u} \sim N(0, \frac{1}{2}I_n)$ and $y_i = R_1 z_i$ $(i = 1,2,\ldots,k)$ are independently distributed as $G(a_i, 1)$ for all i. Further,

$$(R_1 z = \underset{\sim}{x}'A\underset{\sim}{x}) \overset{d}{=} (R_1 \sum_{i=1}^{k} \lambda_i z_i = \sum_{i=1}^{k} \lambda_i y_i).$$

Then, using the c.f. on both sides, we get

$$|I - \sqrt{-1}\, t\, A|^{-\frac{1}{2}} = \prod_{j=1}^{k} (1 - \sqrt{-1}\, t\, \lambda_j)^{-a_j}$$

$$\text{or} \quad |I - t_1 A| = \prod_{j=1}^{k} (1 - t_1 \lambda_j)^{2a_j}$$

for all $t \in \mathbb{R}$ or for all t_1 $(= \sqrt{-1}\, t)$. By analytic continuation this will be true for all real numbers $|t_1| < \rho$, ρ being the radius of convergence. From this it is easy to establish the required result. This proves Lemma 3.

[<u>Conjecture</u>. In the notations of Lemma 3, $z \overset{d}{=} \sum_{i=1}^{k} \lambda_i z_i$

where $(z_1,\ldots,z_k) \sim D_k(a_1,\ldots,a_k;\ a_{k+1})$, and $\lambda_1,\ldots,\lambda_k$ are distinct nonzero real numbers if and only if $2\sum_{j=1}^{k+1} a_j = n$, and $\lambda_1,\ldots,\lambda_k$ are distinct nonzero eigen values of A with respective multiplicities $2a_j$ $(j = 1,2,\ldots,k)$.

This conjecture is established when $k = 1$ through Theorem 1.]

Theorem 5. *Let* $\underset{\sim}{u} \sim [d\underset{\sim}{u}]$ *on* $O(1,n)$ *and* $A_1,A_2,\ldots,A_k$ *and* $A = \sum_{i=1}^{k} A_i$ *be symmetric matrices defined on* $\mathbb{R}$. *Then, consider the following statements:*

(a) $u'A_iu \sim \beta(a_i,b_i)$ *for all* $i = 1,2,\ldots,k$;

(b) $\Gamma(\frac{1}{2}n + \alpha_1 + \alpha_2)E(\underset{\sim}{u}'A_i\underset{\sim}{u})^{\alpha_1}(\underset{\sim}{u}'A_j\underset{\sim}{u})^{\alpha_2} = \Gamma(\frac{1}{2}n + \alpha_1)$

$\cdot\ \Gamma(\frac{1}{2}n + \alpha_2)E(\underset{\sim}{u}'A_i\underset{\sim}{u})^{\alpha_1}E(\underset{\sim}{u}'A_j\underset{\sim}{u})^{\alpha_2}/\Gamma(n/2)$ *for all* $i \neq j$, $i,j = 1,2,\ldots,k$, *and for* $\alpha_1,\alpha_2 = 0,1,2$;

(c) $\underset{\sim}{u}'A\underset{\sim}{u} \sim \beta(a,b)$;

and *(d)* $\text{Rank } A = \sum_{i=1}^{k} \text{Rank } A_i$.

Then (i) any two of (a), (b), (c) imply all the conditions and (ii) (c) and (d) imply (a) and (b) or $(\underset{\sim}{u}'A_1\underset{\sim}{u},\ldots,\underset{\sim}{u}'A_k\underset{\sim}{u}) \sim D_k(a_1,\ldots,a_k;\ a_{k+1})$.

This follows from Theorems 1 to 3. This is a generalization of Lemma 2. A result corresponding to that given by Khatri (1983) is due to conditions: (a), (d) and product of nonzero eigen values of $A = 1 \Rightarrow$ all the conditions of Theorem 5, or $(\underset{\sim}{u}'A_1\underset{\sim}{u},\ldots,\underset{\sim}{u}'A_k\underset{\sim}{u}) \sim D_k(a_1,\ldots,a_k;\ a_{k+1})$.

Theorem 6. *Let* $\underset{\sim}{u} \sim [d\underset{\sim}{u}]$ *on* $O(1,n)$ *and* $A_1,A_2,\ldots,A_k$

and $A = \sum_{i=1}^{k} A_i$ *be symmetric matrices defined on* **R**. *Let* $\underset{\sim}{u}'A\underset{\sim}{u} \backsim \beta(a,b)$, $\underset{\sim}{u}'A_i\underset{\sim}{u} \backsim \beta(a_i,b_i)$ *for* $i = 1,2,\ldots,k-1$ *and* $\underset{\sim}{u}'A_k\underset{\sim}{u} \geq 0$ *for all* $\underset{\sim}{u} \in O(1,n)$. *Then* $(\underset{\sim}{u}'A_1\underset{\sim}{u},\ldots,\underset{\sim}{u}'A_k\underset{\sim}{u}) \backsim D_k(a_1,a_2,\ldots,a_k;\ b)$ *with* $a_k = a - \sum_{i=1}^{k-1} a_i$, *and* $2(b + a) = n$.

This follows by using Theorems 1 and 2 along with the following result on matrix algebra (see, for example, Hogg (1963)):

(a_1) $A_i^2 = A_i$, $i = 1,2,\ldots,k-1$,

(c) $A^2 = A$ with $A = \sum_{i=1}^{k} A_i$ and

(a_2) $A_k \geq 0 \Rightarrow A_i^2 = A_i$ and $A_iA_j = 0$ for all $i \neq j$, $i,j = 1,2,\ldots,k$.

A generalized Theorem 5 can be written as

<u>Theorem 7</u>. *Let* $\underset{\sim}{u} \backsim [d\underset{\sim}{u}]$ *over* $O(1,n)$ *and let* $A_1,A_2,\ldots,A_k$ *and* $\sum_{i=1}^{k} A_i = A$ *be symmetric matrices defined on* **R**. *Then, consider the following conditions:*

(a) $\underset{\sim}{u}'A_i\underset{\sim}{u} \overset{d}{=} \sum_{j=1}^{r} \lambda_j z_{ij}$ *where* $(z_{i1},\ldots,z_{ir}) \backsim D_r(a_{i1},a_{i2},\ldots,a_{ir};\ a_{i,r+1})$ *with* $\sum_{j=1}^{r+1} a_{ij} = n/2$ *for all* $i = 1,2,\ldots,k$ *(when some* $a_{ij} = 0$, *then the corresponding* z_{ij}'s *do not exist) and* $\lambda_1,\ldots,\lambda_r$ *being*

distinct and nonzero.

(b) *The same as mentioned in Theorem 5.*

(c) $\underset{\sim}{u}'A\underset{\sim}{u} \overset{d}{=} \sum_{j=1}^{r} {}_j z_j$ *where*

$(z_1,\ldots,z_r) \sim D_r(a_1,\ldots,a_r;\ a_{r+1})$ *with*

$\sum_{i=1}^{r+1} a_i = n/2$ *and* $\lambda_1,\ldots,\lambda_r$ *being distinct and nonzero.*

(d1) $(r - 1)\ \text{Rank}\ A = \sum_{i=1}^{k} \sum_{\alpha=1}^{r} \text{Rank}(A_i(A + \lambda_\alpha I))$

for $r > 1$.

Then (i) (d1) implies all conditions, (ii) (a1) and (b) imply all conditions, (iii) (b) and (c) imply all conditions and (iv) (a1), (c) and $a_j = \sum_{i=1}^{k} a_{ij}$ *for* $j = 1,2,\ldots,r - 1$ *imply all conditions.*

This follows from the results of Lemma 3, Theorem 3 and Khatri's results on matrices (1982, 1984).

Remark 1. The results of this section 2 are true when $\underset{\sim}{u}$ is a complex random vector such that $\underset{\sim}{u}^*\underset{\sim}{u} = 1$ where $\underset{\sim}{u}^*$ denotes the complex conjugate of $\underset{\sim}{u}$. The space generated by such a vector will be denoted by $\widetilde{CO}(1,n)$ and the uniform distribution over $CO(1,n)$ will be denoted by $[d\underset{\sim}{u}]$, a unit invariant Haar measure. Here, the symmetric matrices will be replaced by Hermitian matrices, and condition $\Sigma a_i = n/2$ will be replaced by $\Sigma a_i = n$; in Theorem 3, $\Gamma(m + i)$ will be changed to $\Gamma(n + i)$; the distribution (2.8) will be changed to

$$C|W|^{a_0-p} \prod_{i=1}^{k} w_i^{a_i-1} \quad \text{for all} \quad (W,\underset{\sim}{w}) \in \tilde{\mathcal{D}}_{p,k}$$

where $\tilde{\mathcal{D}}_{p,k} = \{(W,w);$ W is Hermitian positive definite,

$$\mathrm{tr}\Sigma^{-1}W + \sum_{i=1}^{k} w_i = 1 \quad \text{and} \quad w_i > 0\} \quad \text{and}$$

$$c = \Gamma(a_0 p + \sum_{i=1}^{k} a_i)/|\Sigma|^{a_0} \tilde{\Gamma}_p(a_0) \prod_{i=1}^{k} \Gamma(a_i),$$

$$\tilde{\Gamma}_p(a_0) = \pi^{\frac{1}{2}p(p-1)} \prod_{i=1}^{p} \Gamma(a_0 - i + 1).$$

Similarly, (2.9) can be changed. These distributions will be denoted by $\tilde{D}_{p,k}(a_0,\Sigma;\ a)$ and $\tilde{D}_{p,0}(a_0,\Sigma)$. With these changes, we can modify all the results for complex variables. We only mention Theorems 4 and 3 as

Theorem 4C. *Let* $\underset{\sim}{u} \sim [d\underset{\sim}{u}]$ *over* $CO(1,n)$ *and let* $\underset{\sim}{u}^* A_i \underset{\sim}{u} = z_i$ $(i = 1,2,\ldots,p+k)$ *be* $p + k$ *Hermitian forms with* $A_i = A_i^*$. *Then* $(z_1,\ldots,z_{p+k}) \overset{d}{=} (w_{11},\ldots,w_{pp},w_1,\ldots,w_k)$ *where* $(W,\underset{\sim}{w}) \sim \tilde{D}_{p,k}(a_0,\Sigma;\ \underset{\sim}{a})$ *with* $a_0 \geq p$ *and* $a_j > 0$ *for all* $j = 1,2,\ldots,k$, *if and only if* $pa_0 + \Sigma a_j = n$, a_j $(j = 1,2,\ldots,k)$ *are positive integers* $=$ Rank A_{p+j}, $A_i^2 = A_i$, $A_i A_{p+j} = 0$, $a_0 =$ Rank A, $A_\alpha A_\beta A_\alpha = \sigma_{\alpha\beta}\sigma_{\beta\alpha}A_\alpha$, $A_\alpha A_\beta A_\gamma A_\alpha = (\sigma_{\alpha\beta}\sigma_{\beta\gamma}\sigma_{\gamma\alpha})A_\alpha$ *for all* $i = 1,2,\ldots,p+k$, $j = 1,2,\ldots,k$, $i \neq p + j$, $\alpha \neq \beta \neq \gamma$, $\alpha,\beta,\gamma = 1,2,\ldots,p$.

Theorem 3C. *Let* $\underset{\sim}{u} \sim [d\underset{\sim}{u}]$ *over* $CO(1,n)$ *and let* $z_i = \underset{\sim}{u}^* A_i \underset{\sim}{u}$ *for* $i = 1,2$ *be two Hermitian forms If*

$$\Gamma(n + i + j)Ez_1^i z_2^j = \Gamma(n + i)\Gamma(n + j)Ez_1^i z_2^j/\Gamma(n) \quad \textit{for}$$

$i,j = 0,1,2,$ *then* $A_1A_2 = A_2A_1 = 0$. *Conversely, if* $A_1A_2 = 0 = A_2A_1$, *then the above moment relation is true for all* $i,j = 0,1,2,3,\ldots$.

Remark 2. Let $\underset{\sim}{x} \sim E_{n_0}(\underset{\sim}{\mu},\Sigma;\ \phi)$ and $\underset{\sim}{x} \overset{d}{=} \underset{\sim}{\mu} + \sqrt{R}\, P'\underset{\sim}{u}$ where $P(R > 0) = 1$, R and $\underset{\sim}{u}$ are independently distributed, $\Sigma = P'P$, $\underset{\sim}{u} \sim [d\underset{\sim}{u}]$ over $O(1,n)$, $n =$ Rank Σ and the density of R is

$$\{\Pi^{n/2}/\Gamma(n/2)\}R^{\frac{1}{2}n-1}g(R) \quad \text{for} \quad R > 0. \tag{2.13}$$

The results of section 2 will be true for these variables. They are mentioned below:

Lemma 3'. *Let the joint density of* $y_1, y_2, \ldots, y_k$ *be given through* $y_i = Rz_i$ $(i = 1,2,\ldots,k)$ *where* R *and* $(z_1, z_2, \ldots, z_k)$ *are independently distributed,* $(z_1,\ldots,z_k) \sim D_k(a_1,\ldots,a_k; a_{k+1})$ *and* R *is distributed as (2.13). Let* $y = \underset{\sim}{x}'A\underset{\sim}{x} + 2\underset{\sim}{\ell}'\underset{\sim}{x} + c$ *be a polynomial such that* $y \overset{d}{=} \underset{\sim}{u}'A_1\underset{\sim}{u}R$ *where* $A_1 = PAP'$, $\underset{\sim}{\mu}'A\underset{\sim}{\mu} + 2\underset{\sim}{\ell}'\underset{\sim}{\mu} + c = 0$ *and* $\Sigma(A\underset{\sim}{\mu} + \underset{\sim}{\ell}) = \underset{\sim}{0}$. *Then* $y \overset{d}{=} \sum_{i=1}^{k} \lambda_i y_i$ *where* λ_i*'s are distinct and nonzero real numbers if and only if* $\sum_{i=1}^{k+1} a_i = \frac{1}{2}n$ *and* $\lambda_1, \lambda_2, \ldots, \lambda_k$ *are the distinct nonzero eigen values of* ΣA *with respective multiplicities* $2a_j$ $(j = 1,2,\ldots,k)$.

Proof. This will follow from Lemma 3 if we can show that

$$y = R(\underset{\sim}{u}'A_1\underset{\sim}{u}) \overset{d}{=} R\sum_{j=1}^{k} \lambda_j z_j = \sum_{j=1}^{k} \lambda_j y_j$$

$$\Rightarrow \underset{\sim}{u}'A_1\underset{\sim}{u} \overset{d}{=} \sum_{j=1}^{k} \lambda_j z_j.$$

This will follow from Anderson and Fang's result (1985) given by

A1. Let R and X be independent and R and Y be independent random variables such that $RX \overset{d}{=} RY$. Then, $X \overset{d}{=} Y$ under any one of the following conditions:

(i) $P(R > 0) = P(X > 0) = P(Y > 0) = 1$ and c.f. of $\log R$ is nonzero, or c.f. of $\log X$ (or $\log Y$) equal to some c.f. in an interval implies the validity of the relation throughout the real space. [The later condition on the c.f. will be denoted as the class of c.f.s belonging to $\mathfrak{C}$.]

(ii) $P(R > 0) = 1$ and the c.f. of $(\log X^+ | X > 0)$ belongs to $\mathfrak{C}_+$ and the c.f. of $(\log X^- | X < 0)$ belongs to $\mathfrak{C}$. [Here, $X^+ = X$ if $x > 0$ and $X^+ = 0$ if $x \leq 0$ and $X^- = -X$ if $x < 0$ and $X^- = 0$ if $X \geq 0$.]

(The result A.1(i) is true when X and Y are random vectors.) Notice that if $z = \sum_{j=1}^{k} \lambda_j z_j$, then

$$|z| \leq \sum_{j=1}^{k} |\lambda_j| z_j \leq \max_j |\lambda_j| \sum_{j=1}^{k} z_j < \lambda \quad (= \max_j |\lambda_j|).$$ Thus, $-1 < z/\lambda < 1$, and all the moments of z/λ exist and its distribution is continuous and determined uniquely by the moments. Then, it can be seen that the c.f.s of $(\log(z^+/\lambda) | z > 0)$ and $(\log(z^-/\lambda) | z < 0)$ belongs to $\mathfrak{C}$.

Thus, by A.1(ii), $R(\underset{\sim}{u}'A\underset{\sim}{u}) \overset{d}{=} R \sum_{j=1}^{k} \lambda_j z_j \Rightarrow \underset{\sim}{u}'A\underset{\sim}{u} \overset{d}{=} \sum_{j=1}^{k} \lambda_j z_j$.

In short, for the normal variates, this proves the following important result:

Lemma 4. *Let* $\underset{\sim}{x} \sim N(\underset{\sim}{0}, \Sigma)$ *and let* $\underset{\sim}{x}'A\underset{\sim}{x}$ *be a quadratic form. Then* $\underset{\sim}{x}'A\underset{\sim}{x} \overset{d}{=} Rz$ *where* R *and* z *are independently distributed,* $R \sim G(\frac{1}{2}n, \frac{1}{2})$ *and* $z \sim \beta(a,b)$ *with* $n = \text{Rank}\,\Sigma$, *if and only if* $a + b = \frac{1}{2}n$, $2a = \text{Rank}(\Sigma A \Sigma) = \text{tr}(A\Sigma)$ *and* $\Sigma A \Sigma A \Sigma = \Sigma A \Sigma$.

(Notice the change in the statement of the Lemma 4. If $\frac{1}{2}n = a + b$, then $Rz \sim G(a; \frac{1}{2})$, otherwise Rz may not be distributed as chi-square.)

All the Theorems can be modified except Theorem 3 and consequences of Theorem 3.

Lemma 5. *Under the assumptions mentioned in the beginning, let* $y_i = \underset{\sim}{x}'A_i\underset{\sim}{x} + 2\underset{\sim}{\ell}_i'\underset{\sim}{x} + c_i$ *be second degree polynomials in* $\underset{\sim}{x}$ *such that* $\underset{\sim}{\mu}'A_i\underset{\sim}{\mu} + 2\underset{\sim}{\ell}_i'\underset{\sim}{\mu} + c_i = 0$, $\Sigma(A_i\underset{\sim}{\mu} + \underset{\sim}{\ell}_i) = \underset{\sim}{0}$ *and* $y_i = R\underset{\sim}{u}'A_{(i)}\underset{\sim}{u}$ *for* $i = 1,2,\ldots,k$. *Then (i)*

$(y_1, y_2, \ldots, y_k) \overset{d}{=} R(z_1, \ldots, z_k)$ *where* R *and* $(z_1, \ldots, z_k)$ *are independently distributed,* R *has a density given by (2.13) and* $(z_1, \ldots, z_k) \sim D_k(a_1, \ldots, a_k;\ a_{k+1})$ *if and only if*

$n = 2\sum_{j=1}^{k+1} a_j, \quad \Sigma A_i \Sigma A_i \Sigma = \Sigma A_i \Sigma, \quad \Sigma A_i \Sigma A_j \Sigma = 0$ *and*

$\text{Rank}(\Sigma A_i \Sigma) = \text{tr}A_i \Sigma = 2a_i, \quad i \neq j, \quad i,j = 1,2,\ldots,k.$

Further (ii) if $y_i \overset{d}{=} Rz_i \ (i = 1,2,\ldots,k-1)$, $y_k \geq 0$ *for all* $\underset{\sim}{x}$ *and* $\sum_{i=1}^{k} y_i \overset{d}{=} Rz$ *where* R *and* z_i, *and* R *and* z *are independently distributed, the density of* R *is given by (2.13),* $z_i \sim \beta(a_i, b_i)$, $i = 1,2,\ldots,k-1$ *and* $z \sim \beta(a,b)$, *then* $a + b = a_i + b_i = n/2$ *for all* $i = 1,\ldots,k-1$, *and* $\Sigma A_i \Sigma A_i \Sigma = \Sigma A_i \Sigma$, $\Sigma A_i \Sigma A_j \Sigma = 0$ *and* $\text{Rank}(\Sigma A_i \Sigma) = 2a_i$ *for all* $i \neq j$, $i,j = 1,2,\ldots,k-1$, *with* $a = \sum_{i=1}^{k} a_i$.

<u>Lemma 6</u>. *As before, let* $y_i = \underset{\sim}{x}'A_i\underset{\sim}{x} + 2\underset{\sim}{\ell}_i'\underset{\sim}{x} + c_i = R\underset{\sim}{u}'A_{(i)}\underset{\sim}{u}$ *for* $i = 1,2,\ldots,k$, *where* R *is distributed as (2.13),* $\underset{\sim}{u} \sim [d\underset{\sim}{u}]$ *on* $O(1,n)$ *and* R *and* $\underset{\sim}{u}$ *are independent. Then,* $y_i \overset{d}{=} R\sum_{j=1}^{r} \lambda_j z_{ij}$, $(z_{i1}, \ldots, z_{ir}) \sim D_r(a_{i1}, \ldots, a_{ir};\ a_{i,r+1})$ *and* $y = \sum_{i=1}^{k} y_i \overset{d}{=} R\sum_{j=1}^{k} \lambda_j z_j$, $(z_1, \ldots, z_r) \sim D_r(a_1, \ldots, a_r;\ a_{r+1})$ *where* R *and* $(z_1, \ldots, z_r)$, *and* R *and* $(z_{i1}, \ldots, z_{ir})$ *are independently distributed for all* i, *and* $a_j = \sum_{i=1}^{k} a_{ij}$ *for* $j = 1,2,\ldots,r-1$ $(r > 1)$ *if and only if*

$$(r-1)\ \text{Rank}(\Sigma A \Sigma) = \sum_{i=1}^{k} \sum_{\alpha=1}^{r} \text{Rank}(\Sigma A_i(\Sigma A \Sigma + \lambda_\alpha \Sigma)).$$

(This will follow from Theorem 7 after using A.1(ii).)

The result corresponding to Theorem 4 can be written down in the same way, and it is left to the readers to write down.

Remark 3. Let $\underset{\sim}{x}$ be a complex random vector and let $\underset{\sim}{x} \overset{d}{=} \underset{\sim}{\mu} + \sqrt{R}\, P^*\underset{\sim}{u}$ where $P^*P = \Sigma$ is a given matrix, R and $\underset{\sim}{u}$ are independently distributed, $\underset{\sim}{u} \sim [d\underset{\sim}{u}]$ over $CO(1,n)$ and the density of R is $\{\pi^n/\Gamma(n)\}R^{n-1}g(R)$ for $R > 0$. Such a distribution of $\underset{\sim}{x}$ is to be complex elliptical and denoted as $\underset{\sim}{x} \sim CE_n(\underset{\sim}{\mu}, \Sigma; \tilde{\phi})$. Then, with the help of the modifications mentioned in Remark 1, we can obtain the corresponding results mentioned in Remark 2 for such variates. Explicit statements are left to the readers.

3. MULTIVARIATE EXTENSION

In section 2, we have considered $\underset{\sim}{u} \sim [d\underset{\sim}{u}]$ on $O(1,n)$. Let $O(p,n)$ be the Steiljes manifold, (that is, a set of $p \times n$ matrices $U \in \mathbb{R}^{p\times n}$ such that $UU' = I_p$ $(n \geq p)$) and let $[dU]$ be the unit invariant Haar measure defined on $O(p,n)$. For the study of this distribution, one can refer to Khatri (1970). If $U = (U_1, U_2)$ where U_i is a $p \times n_i$ $(i = 1,2)$ matrix, then the distribution of $U_1U_1' = B$ is known as the matrix Beta distribution and its density function can be given provided $n_1 \geq p$ and $n_2 \geq p$. Under these conditions, the density of B is given by

$$\{B_p(n_i/2, n_2/2)\}^{-1}|B|^{-(n_1-p-1)}|I - B|^{\frac{1}{2}(n_2-p-1)}$$

$$\text{for } 0 < B < I. \tag{3.1}$$

We shall say $B \sim \beta_p(a,b)$ with $a = n_1/2$ and $b = n_2/2$, and write $0 < B < I$ where $B > 0$ and $I - B > 0$ (positive definite). Here $B_p(a,b) = \Gamma_p(a)\Gamma_p(b)/\Gamma_p(a + b)$.

Notice that if $\underset{\sim}{a} \in \mathbb{R}^p$ is any nonzero given p-vector, then $\underset{\sim}{a}'U/\sqrt{\underset{\sim}{a}'\underset{\sim}{a}} = \underset{\sim}{u} \sim [d\underset{\sim}{u}]$ over $O(1,n)$. Notice the above matrix distribution is connected with positive integers n_1 and n_2. We shall say that $B \sim \beta_p(a,b)$ if $2a \geq p$ and

$2b \geq p$, and its density function is given by (3.1) after replacing $n_1/2$ by a, and $n_2/2$ by b. In this situation, it is easy to see that $\underset{\sim}{a}'B\underset{\sim}{a}/\underset{\sim}{a}'\underset{\sim}{a} \sim \beta(a,b)$ for for any nonzero vector $\underset{\sim}{a} \in \mathbb{R}^p$.

Let the density function of X be

$$f(XX') \quad \text{for} \quad X \in \mathbb{R}^{p\times n} \tag{3.2}$$

and if $n \geq p$, then the density of S $(= XX')$ is given by

$$\{\Pi^{\frac{1}{2}pn}/\Gamma_p(\tfrac{1}{2}n)\}|S|^{\frac{1}{2}(n-p-1)}f(S) \quad \text{for} \quad S > 0. \tag{3.3}$$

Let $S = TT'$ where T is a lower triangular matrix (or an upper triangular matrix) with positive diagonal elements, or $T > 0$. Then, it is easy to see that $X = TU$ (see for example, Perlman (1969) or Khatri (1970)) where T and U are independently distributed, the density of $S = TT'$ is given by (3.2) and $U \sim [dU]$ over $O(p,n)$.

Let $\underset{\sim}{a} \in \mathbb{R}^p$ be any non-null vector and let

$$\underset{\sim}{x} = X'\underset{\sim}{a} \overset{d}{=} U'T'\underset{\sim}{a} \overset{d}{=} \underset{\sim}{u}\sqrt{R} \tag{3.4}$$

where $R = \underset{\sim}{a}'S\underset{\sim}{a}$ and $\underset{\sim}{u}$ are independently distributed, the density of R can be obtained from (3.3) and $\underset{\sim}{u} \sim [d\underset{\sim}{u}]$ over $O(1,n)$. For the proof of the above result, let $A = (T'\underset{\sim}{a}/\sqrt{R}, A_1)$ be a $p \times p$ orthogonal matrix. Then, given T, U and $A'U$ are identically distributed as $[dU]$ over $O(p,n)$, and hence $U'T'\underset{\sim}{a}/\sqrt{R} = \underset{\sim}{u} \sim [d\underset{\sim}{u}]$ over $O(1,n)$. This proves the required result (3.4). Now, the extension of Theorem 1 to the matrix variates is given by

<u>Theorem</u> <u>1E</u>. *Let* X *have a density function given by (3.2). Then* $XAX' \overset{d}{=} TWT'$ *where* $A = A'$, T *and* W *are independently distributed,* $S = TT'$ *is distributed as (3.3) and* $W \sim \beta_p(a,b)$, *if and only if* $2a + 2b = n$, *and* A *is idempotent of* $\text{rank}(2a)$.

We shall define the matrix Dirichlet distribution $M_pD_k(\underset{\sim}{a};\, a_{k+1})$ through the density function

$$\{\Gamma_p(\sum_{i=1}^{k+1} a_i)/\prod_{i=1}^{k+1}\Gamma_p(a_i)\}(\prod_{i=1}^{k}|W_i|^{a_i-\frac{1}{2}(p+1)})$$

$$\cdot\ |I - \sum_{i=1}^{k} W_i|^{a_{k+1}-\frac{1}{2}(p+1)}$$

for all $0 < W_i < I$ and $\sum_{i=1}^{k} W_i < I$. Here, for the existence of the density, it is assumed at $a_i \geq p$ for all $i = 1,2,\ldots,k+1$. Even though these conditions are not satisfied, observe that for any $\underset{\sim}{\mathfrak{c}} \in \mathbb{R}^p$ with $\underset{\sim}{\mathfrak{c}}'\underset{\sim}{\mathfrak{c}} = 1$

$$(\underset{\sim}{\mathfrak{c}}'W_1\underset{\sim}{\mathfrak{c}},\ldots,\underset{\sim}{\mathfrak{c}}'W_k\underset{\sim}{\mathfrak{c}}) \quad D_k(\underset{\sim}{a};\ a_{k+1}),$$

or $M_1D_k(\underset{\sim}{a};\ a_{k+1}) = D_k(\underset{\sim}{a};\ a_{k+1})$ with $a_i > 0$ for all i. If a_i's are half integers, this can be proved through $U \sim O(p,n)$ and taking $U = (U_1,\ldots,U_k,U_{k+1})$ and $W_i = U_iU_i'$ for $i = 1,2,\ldots,k$. Now, Theorem 2 can be mentioned as

Theorem 2E. *Let* X *have a density function given by (3.2), and let* $A_1,\ldots,A_k,A = \sum_{i=1}^{k} A_i$ *be symmetric matrices. If*

$$(XA_iX',\ i = 1,2,\ldots,k) \overset{d}{=} (TW_1T',TW_2T',\ldots,TW_iT'),$$

where T *and* $(W_1,W_2,\ldots,W_k)$ *are independently distributed,* $S = TT'$ *is distributed as (3.3) and* $(W_1,W_2,\ldots,W_k) \sim M_pD_k(\underset{\sim}{a};\ a_{k+1})$, *if and only if* $A_i^2 = A_i$, $A_iA_j = 0$ *for all* $i \neq j$, $i,j = 1,2,\ldots,k$,

$\mathrm{tr}A_i = \mathrm{Rank}\ A_i = 2a_i$ *and* $2\sum_{i=1}^{k+1} a_i = n$.

Proof. This will follow Theorem 1E. Note that $XA_iX' \overset{d}{=} TW_iT'$ and $W_i \sim \beta_p(a_i,\ \mathfrak{c} - a_i)$ with $\mathfrak{c} = \sum_{i=1}^{k+1} a_i$

and $X(A_i + A_j)X' \overset{d}{=} T(W_i + W_j)T'$ with $W_i + W_j \sim \beta_p(a_i + a_j, b - a_i - a_j)$ for $i \neq j$, $i,j = 1,2,\ldots,p$. From these we get the required result.

[The Theorem 2E can be reworded as: *Consider the following conditions:*

(a) $XA_iX' \overset{d}{=} TW_iT'$, $W_i \sim \beta_p(a_i,\mathfrak{C}_i)$ *for all* $i = 1,2,\ldots,k$

(c) $XAX' \overset{d}{=} TWT'$, $W \sim \beta_p(a,\mathfrak{C})$

(d) $\text{Rank } A = \sum_{i=1}^{k} \text{Rank } A_i$

(b) $A_iA_j = 0$ *for all* $i \neq j$, $i,j = 1,2,\ldots,k$.

Then, (i) any two of (a), (b), (c) imply all conditions and (ii) (c) and (d) imply all conditions.]

The extension of Lemma 3 is given by

Theorem 3E. *Let* X *have a density given by (3.2), and let* A *be a symmetric matrix. Then* $XAX' \overset{d}{=} \sum_{i=1}^{k} \lambda_i TW_iT'$ *where* $\lambda_1,\ldots,\lambda_k$ *are distinct real numbers,* T *and* $(W_1,W_2,\ldots,W_k)$ *are independently distributed,* $S = TT'$ *is distributed as (3.3) and* $(W_1,W_2,\ldots,W_k) \sim M_pD_k(a_1,\ldots,a_k;\ a_{k+1})$ *with* $2\sum_{i=1}^{k+1} a_i = n$ *if and only if* $\lambda_1,\lambda_2,\ldots,\lambda_k$ *are the distinct nonzero eigen values of* A *with respective multiplicities* $2a_j$ $(j = 1,2,\ldots,k)$.

This follows from Lemma 3 after using (3.4).

For the density of X given by (3.2), we mention below the extensions of Lemma 6 as

Lemma 6E. *Let* $A_1,A_2,\ldots,A_k$ *and* $A = \sum_{i=1}^{k} A_i$ *be symmetric matrices. Then,* $XA_iX' \overset{d}{=} \sum_{j=1}^{k} \lambda_j T W_{ij} T'$, T *and* $(W_{i1},\ldots,W_{ir})$ *are independent,* $S = TT'$ *is distributed as (3.3),* $(W_{i1}, W_{i2},\ldots,W_{ir}) \sim M_pD_r(a_{i1},\ldots,a_{ir};\ a_{ir+1})$ *with* $\sum_{j=1}^{r+1} a_{ij} = n/2$ *for* $i = 1,2,\ldots,k$, *and* $XAX' \overset{d}{=} \sum_{j=1}^{r} \lambda_j T W_j T'$, T *and* $(W_1,\ldots,W_r)$ *are independent, and* $(W_1,\ldots,W_r) \sim M_pD_k(a_1,\ldots,a_r;\ a_{r+1})$ *with* $\sum_{i=1}^{k+1} a_i = n/2$, *and*

$$a_j = \sum_{i=1}^{k} a_{ij} \quad \textit{for} \quad j = 1,2,\ldots,r-1, \quad \textit{if and only if}$$

$$(r-1)\text{Rank } A = \sum_{i=1}^{k} \sum_{\alpha=1}^{r} \text{Rank}(A_i(A + \lambda_\alpha I)) \quad \textit{with} \quad r > 1.$$

The above results can be extended for the complex random matrix variate $X \in \mathbb{C}^{p\times n}$, but we shall not mention them explicitly. The changes similar to Remark 1 are necessary for this situation. This is left to the readers.

C. G. Khatri
Department of Statistics
Gujarat University
Ahmedabad, Gujarat, India

REFERENCES

Anderson, T. W. and Fang, Kai-Tai (1985). 'Cochran's Theorem for elliptically contoured distributions.' Department of Statistics, Stanford University (unpublished).

Cambanis, S., Huang, S., and Simons, G. (1981). 'On the theory of elliptically contoured distributions.' *J. Multivariate Analysis*, **11**, 368-385.

Cochran, W. G. (1934). 'The distribution of quadratic forms in a normal system, with applications to the analysis of covariance.' *Proc. Cambridge Philosoph. Soc.*, 30, 178-191.

Dawid, A. P. (1977). 'Spherical matrix distributions and a multivariate model.' *J. Royal Statist. Soc., Series B*, 39, 254-261.

Good, I. J. (1969). 'Conditions for a quadratic form to have a Chi-square distribution.' *Biometrika*, 56, 215-216; Correction (1970), *Biometrika*, 57, 225.

Graybill, F. A. and Marsaglia, G. (1957). 'Idempotent matrices and quadratic forms in general linear hypothesis.' *Ann. Math. Statist.*, 28- 678-686.

Hogg, R. V. (1963). 'On the independence of certain Wishart variables.' *Ann. Math. Statist.*, 34, 935-939.

Khatri, C. G. (1959). 'On conditions for the forms of the type XAX' to be distributed independently or to obey Wishart distribution.' *Bull. Calcutta Statist. Assoc.*, 8, 162-168.

Khatri, C. G. (1962). 'Conditions for Wishartness and independence of second degree polynomials in normal vectors.' *Ann. Math. Statist.*, 33, 1002-1007.

Khatri, C. G. (1963). 'Further contributions to Wishartness and independence of second degree polynomials in normal vectors.' *J. Indian Statist. Assoc.*, 1, 61-70.

Khatri, C. G. (1970). 'A note on Mitra's paper 'A density free approach to the matrix variate Beta distribution'.' *Sankhyā Series A*, 32, 311-318.

Khatri, C. G. (1977). 'Quadratic forms and extension of Cochran's theorem to normal vector variables.' *Multivariate Analysis - IV* (ed. P. R. Krishnaiah), 79-94. North Holland, New York.

Khatri, C. G. (1978). 'A remark on the necessary and sufficient conditions for a quadratic form to be distributed as Chi-squares.' *Biometrika*, 65, 239-240.

Khatri, C. G. (1980a). 'The necessary and sufficient conditions for dependent quadratic forms to be distributed as Multivariate Gamma.' *J. Multivariate Analysis*, **10**, 233-242.

Khatri, C. G. (1980b). 'Quadratic forms.' *Handbook of Statistics* (ed. P. R. Krishnaiah), Chapter 8, 443-468. North Holland, New York.

Khatri, C. G. (1982). 'A theorem on quadratic forms for normal variables.' *Essays in Statistics and Probability* (ed. P. R. Krishnaiah, et al.). North Holland, New York.

Khatri, C. G. (1983a). 'A generalization of Lavoie's inequality concerning the sum of idempotent matrices.' *Linear Algebra and its Applications*, **54**, 97-108.

Khatri, C. G. (1983a). 'A theorem on the decomposition of matrices.' *Matrices Today*, **1**, 55-61.

Khatri, C. G. (1984). 'Some results on decomposition of matrices.' *Proceedings of Indian Statistical Institute Golden Jubilee International Conference on "Statistics Applications and New Directions"* (eds. J. K. Ghosh and J. Roy). Publishing Society, Calcutta, 313-325.

Khatri, C. G. (1986). 'Quadratic forms and null robustness for elliptical distributions' (to be presented in the Second International Tampere Conference in Statistics, June 1-4, 1987).

Khatri, C. G. and Mukerjee, Rahul (1986). 'Characterization of normality within the class of elliptical distributions' (unpublished).

Perlman, M. D. (1969). 'One sided problems in multivariate analysis.' *Ann. Math. Statist.*, **40**, 549-567.

Shanbhag, D. N. (1968). 'Some remarks concerning Khatri's result on quadratic forms.' *Biometrika*, **55**, 593-596.

S. Kocherlakota, K. Kocherlakota and N. Balakrishnan

ASYMPTOTIC EXPANSIONS FOR ERRORS OF MISCLASSIFICATION: NONNORMAL SITUATIONS

SUMMARY

A unified development of the expansions available for the asymptotic distribution of the errors of misclassification is given in the nonnormal situations. Several examples are presented.

A detailed study is undertaken in the case of outliers in sampling from the normal distribution. It is shown that outliers with large variances lead to inflation in the errors of misclassification.

1. INTRODUCTION

The classification rule for classifying an observation X into one of two normal populations Π_1, Π_2, when the variances are assumed equal, is to classify X as

(i) Belonging to Π_1, if

$$X \le \frac{1}{2}(\bar{X}_1+\bar{X}_2) \quad \text{for} \quad \bar{X}_1 < \bar{X}_2$$

or

$$X > \frac{1}{2}(\bar{X}_1+\bar{X}_2) \quad \text{for} \quad \bar{X}_1 \ge \bar{X}_2$$

(ii) Belonging to Π_2 if

$$X > \frac{1}{2}(\bar{X}_1+\bar{X}_2) \quad \text{for} \quad \bar{X}_1 < \bar{X}_2$$

or

$$X \le \frac{1}{2}(\bar{X}_1+\bar{X}_2) \quad \text{for} \quad \bar{X}_1 \ge \bar{X}_2.$$

For this rule, the errors of misclassification are

$$\begin{aligned} e_{12}(\bar{X}_1,\bar{X}_2) &= P\{X \ge \tfrac{1}{2}(\bar{X}_1+\bar{X}_2) \mid X \in \Pi_1\} \text{ for } \bar{X}_1 < \bar{X}_2 \\ &= P\{X < \tfrac{1}{2}(\bar{X}_1+\bar{X}_2) \mid X \in \Pi_1\} \text{ for } \bar{X}_1 \ge \bar{X}_2 \end{aligned}$$

A. K. Gupta (ed.), Advances in Multivariate Statistical Analysis, 191–211.

and a similar expression for $e_{21}(\bar{X}_1,\bar{X}_2)$. In the above, we have assumed that the means are unknown and estimated by $\bar{X}_1,\bar{X}_2$.

Considerable work has been done by the present authors in connection with the distribution of these errors and the expected values of the errors, when the underlying distribution is nonnormal. Here a unified development is given of the distribution of e_{12} (and of e_{21}) when the distribution sampled from has the pdf $f_1(x)$ under Π_1, $f_2(x)$ under Π_2. In particular, several situations are identified for the expansion of the error rates. The basic results have been developed in Chinganda and Subrahmaniam (1979) and Subrahmaniam and Chinganda (1978). Much detail is also given in Chinganda (1976).

2. DISTRIBUTION FUNCTIONS OF e_{12} AND e_{21}

Let the pdf and the cdf of X under Π_i be f_i and F_i, respectively. Then, writing $\delta = (\bar{X}_1 + \bar{X}_2)$,

$$e_{12}(\bar{X}_1,\bar{X}_2) = \int_{\delta/2}^{\infty} f_1(x)dx \quad \text{for } \bar{X}_1 < \bar{X}_2$$

$$= \int_{-\infty}^{\delta/2} f_1(x)dx \quad \text{for } \bar{X}_1 \geq \bar{X}_2$$

or

$$e_{12}(\bar{X}_1,\bar{X}_2) = 1 - F_1(\tfrac{\delta}{2}) \quad \text{for } \bar{X}_1 < \bar{X}_2$$

$$= F_1(\tfrac{\delta}{2}) \quad \text{for } \bar{X}_1 \geq \bar{X}_2.$$

Also

$$e_{21}(\bar{X}_1,\bar{X}_2) = F_2(\tfrac{\delta}{2}) \quad \text{for } \bar{X}_1 < \bar{X}_2$$

$$= 1 - F_2(\tfrac{\delta}{2}) \quad \text{for } \bar{X}_1 \geq \bar{X}_2.$$

This yields for the distribution function of e_{12}

$$G_1(z) = P\{e_{12}(\bar{X}_1,\bar{X}_2) \leq z\}$$

$$= P\{F_1(\tfrac{\delta}{2}) \leq z,\ \bar{X}_1 \geq \bar{X}_2\}$$

$$+ P\{1 - F_1(\tfrac{\delta}{2}) \leq z,\ \bar{X}_1 < \bar{X}_2\}$$

or

$$G_1(z) = P\{F_1(\tfrac{\delta}{2}) \le z,\ \bar{X}_1 \ge \bar{X}_2\} + P\{F_1(\tfrac{\delta}{2}) \ge 1-z,\ \bar{X}_1 < \bar{X}_2\}, \qquad (2.1)$$

with a similar representation for $G_2(z)$, the distribution function of $e_{21}(\bar{X}_1,\bar{X}_2)$.

In what follows, the joint distribution of $U = (\bar{X}_1+\bar{X}_2)$, $V = (\bar{X}_1-\bar{X}_2)$ is considered for the notation

$$E(X_i) = \mu_i,\quad V(X_i) = \sigma_i^2 \quad \text{for } i = 1,2,$$

$$\mu_u = \mu_1 + \mu_2,\quad \mu_v = \mu_1 - \mu_2$$

$$\sigma_u^2 = \sigma_v^2 = (\sigma_1^2/n_1) + (\sigma_2^2/n_2)$$

and

$$\sigma_u\sigma_v\rho_{uv} = (\sigma_1^2/n_1) - (\sigma_2^2/n_2)\ .$$

2.1 $F_1^{-1}(p)$ Known Explicitly

If $F_1^{-1}(p)$, $0 < p < 1$, is known and can be found explicitly, then we can write

$$\begin{aligned} G_1(z) &= P\{F_1(\tfrac{\delta}{2}) \le z,\ \bar{X}_1-\bar{X}_2 \ge 0\} \\ &\quad + P\{F_1(\tfrac{\delta}{2}) \ge 1-z,\ \bar{X}_1-\bar{X}_2 < 0\} \\ &= P\{U \le 2F_1^{-1}(z),\ V \ge 0\} \\ &\quad + P\{U \ge 2F_1^{-1}(1-z),\ V < 0\}. \end{aligned} \qquad (2.2)$$

If $h(u,v)$ is the joint pdf of U and V, then

$$G_1(z) = \int_0^{\infty}\int_{-\infty}^{2F_1^{-1}(z)} h(u,v)\,du\,dv + \int_{-\infty}^{0}\int_{2F_1^{-1}(1-z)}^{\infty} h(u,v)\,du\,dv. \qquad (2.3)$$

If, in particular, (U,V) are known to have the bivariate normal distribution, then

$$G_1(z) = H[\frac{\mu_u - t_1}{\sigma_u}, -\frac{\mu_v}{\sigma_v}; -\rho_{uv}] + H[\frac{t_2 - \mu_u}{\sigma_u}, \frac{\mu_v}{\sigma_v}; -\rho_{uv}], \quad (2.4)$$

with

$$t_1 = 2F_1^{-1}(1-z), \quad t_2 = 2F_1^{-1}(z).$$

In (2.4), $H(\cdot,\cdot;\rho)$ is the distribution function of the SBVN(ρ) distribution.

Examples. (i) Let X be the truncated normal so that under Π_i, $Z_i = \frac{X_i - \mu_i}{\sigma_i} \sim$ standard TN(a,b), i = 1,2. If $\Phi(\cdot)$ is the distribution function of the standard normal variate, then it can be seen that

$$t_1 = 2[\mu_1 + \sigma_1\Phi^{-1}\{\Phi(a) + z[\Phi(b) - \Phi(a)]\}]$$

and

$$t_2 = 2[\mu_1 + \sigma_1\Phi^{-1}\{\Phi(b) - z[\Phi(b) - \Phi(a)]\}].$$

(ii) Johnson's system. Let $Y = \gamma(X) \sim N(\mu_i, \sigma^2)$ under Π_i, i = 1,2. Then, if $u = F_1^{-1}(z)$,

$$z = F_1(u) = P\{\gamma(X) \le \gamma(u)\} = \Phi[\frac{\gamma(u) - \mu_1}{\sigma}]$$

or

$$u = \gamma^{-1}[\mu_1 + \sigma\Phi^{-1}(z)].$$

This yields

$$t_1 = 2\gamma^{-1}[\mu_1 - \sigma\Phi^{-1}(z)]$$

$$t_2 = 2\gamma^{-1}[\mu_1 + \sigma\Phi^{-1}(z)],$$

as found in Chinganda and Subrahmaniam (1979).

(iii) Dichotomous and Normal Variables. Let X conditional on Y = y be $N(\mu_i^{(y)}, \sigma^{(y)^2})$, i = 1,2. Here Y is taken to be Bernoulli with probability θ. Then it can be shown that

$$e_{12}^{(y)} = \Phi\left\{\left[\frac{\bar{X}_1^{(y)}+\bar{X}_2^{(y)}}{2} - \mu_1^{(y)}\right]/\sigma^{(y)}\right\} \text{ if } \bar{X}_1^{(y)} > \bar{X}_2^{(y)}$$

$$= \Phi\left\{-\left[\frac{\bar{X}_1^{(y)}+\bar{X}_2^{(y)}}{2} - \mu_1^{(y)}\right]/\sigma^{(y)}\right\} \text{ if } \bar{X}_1^{(y)} \le \bar{X}_2^{(y)}.$$

Considering the conditional distribution of (U,V) for $Y = y$, we can write the distribution function of e_{12} as

$$G_1(z) = \sum_{y=0}^{1} \theta^y (1-\theta)^{1-y} G_1^{(y)}(z)$$

where

$$G_1^{(y)}(z) = H\left[\frac{\mu_u^{(y)}-t_1^{(y)}}{\sigma_u^{(y)}}, -\frac{\mu_v^{(y)}}{\sigma_v^{(y)}}; -\rho_{uv}^{(y)}\right] + H\left[\frac{t_2^{(y)}-\mu_u^{(y)}}{\sigma_u^{(y)}}, \frac{\mu_v^{(y)}}{\sigma_v^{(y)}}; -\rho_{uv}^{(y)}\right].$$

From the preceding development

$$t_1^{(y)} = 2F_1^{(y)^{-1}}(1-z)$$

and

$$t_2^{(y)} = 2F_1^{(y)^{-1}}(z)$$

or

$$t_1^{(y)} = 2[\mu_1^{(y)} - \sigma^{(y)}\Phi^{-1}(z)],$$

$$t_2^{(y)} = 2[\mu_1^{(y)} + \sigma^{(y)}\Phi^{-1}(z)].$$

These results are given in Balakrishnan et al (1986a).

2.2 $F^{-1}(p)$ Cannot Be Determined Explicitly

If the quantity $F^{-1}(p)$, $0 < p < 1$, cannot be found explicitly, then we can consider the expansion of

$$e_{12}(\bar{X}_1,\bar{X}_2) = 1 - F_1(\tfrac{\delta}{2}), \quad \bar{X}_1 < \bar{X}_2$$

$$= F_1(\tfrac{\delta}{2}) \quad , \quad \bar{X}_1 \ge \bar{X}_2$$

in terms of $\delta = (\bar{X}_1+\bar{X}_2)$. Thus

$$F_1(\tfrac{\delta}{2}) \cong F_1(0) + \delta\{\tfrac{1}{2} f_1(0)\}$$

and

$$\begin{aligned} e_{12}(\bar{X}_1,\bar{X}_2) &\cong \kappa_1 - \kappa_2\delta \qquad & \bar{X}_1 < \bar{X}_2 \\ &\cong \kappa_1' + \kappa_2\delta \qquad & \bar{X}_1 \geq \bar{X}_2 \end{aligned}$$

where

$$\kappa_1 = 1 - F_1(0),\ \kappa_1' = F_1(0),\ \kappa_2 = f_1(0)/2.$$

From this we find the asymptotic distribution function of e_{12} as

$$G_1(z) = P\left\{U \geq \frac{\kappa_1 - z}{\kappa_2},\ V < 0\right\} + P\left\{U < \frac{z-(1-\kappa_1)}{\kappa_2},\ V \geq 0\right\}.$$

Two cases arise here: (i) The exact distribution of (U,V) is known to be h(u,v); (ii) The asymptotic distribution of (U,V) is BVN distribution.

In case (i) we can write

$$G_1(z) = \int_{-\infty}^{0} \int_{(\kappa_1-z)/\kappa_2}^{\infty} h(u,v)\,du\,dv + \int_{0}^{\infty} \int_{-\infty}^{[z-(1-\kappa_1)]/\kappa_2} h(u,v)\,du\,dv \qquad (2.5)$$

while in case (ii)

$$G_1(z) = H[-z_{11}, z_{12}; -\rho_{uv}] + H[z_{21}, -z_{22}; -\rho_{uv}] \qquad (2.6)$$

where

$$z_{11} = \left\{\frac{\kappa_1 - z}{\kappa_2} - \mu_u\right\}/\sigma_u,\quad z_{12} = -\mu_v/\sigma_v,$$

$$z_{21} = \left\{\frac{z-(1-\kappa_1)}{\kappa_2} - \mu_u\right\}/\sigma_u,\quad z_{22} = -\mu_v/\sigma_v.$$

Examples. (i) Mixtures of Normals. Let Π_i be defined by the density function

$$f_i(x) = p\phi(x;\mu_i,\sigma^2) + (1-p)\phi(x;\mu_i+a\sigma,\sigma^2)$$

for i = 1,2. In this case it is known that the exact distribution of (U,V) is

$$g^*(u,v) = \sum_{rs}\sum \binom{n_1}{r}\binom{n_2}{s} p^{r+s}(1-p)^{n-(r+s)}$$

$$h^*[u,v;\lambda_1+\lambda_2,\lambda_1-\lambda_2,\tau^2,\tau^2,\rho]$$

where h^* is the bivariate normal density function with the parameters as indicated and

$$\lambda_1 = \{n_1\mu_1 + (n_1-r)a\sigma\}/n_1,$$

$$\lambda_2 = \{n_2\mu_2 + (n_2-s)a\sigma\}/n_2,$$

$$\tau^2 = \sigma^2(\frac{1}{n_1} + \frac{1}{n_2}),\ n = n_1 + n_2,\ \rho = (n_2-n_1)/n.$$

Since $\kappa_1 = 1 - F_1(0)$ we get

$$\kappa_1 = p\Phi(\frac{\mu_1}{\sigma}) + (1-p)\Phi(a + \frac{\mu_1}{\sigma}),$$

$$1-\kappa_1 = p\Phi(-\frac{\mu_1}{\sigma}) + (1-p)\Phi[-(a + \frac{\mu_1}{\sigma})].$$

From this we can see that

$$G_1(z) = \sum_{rs}\sum c(r,s)\{H(a_1,b_1;-\rho) + H(a_2,b_2;-\rho)\} \tag{2.7}$$

where

$$a_1 = [(\lambda_1+\lambda_2) + (z-\kappa_1)/\kappa_2]/\sigma_u,\ b_1 = -b_2 = (\lambda_2-\lambda_1)/\sigma_v,$$

$$a_2 = [\{z - (1-\kappa_1)\}/\kappa_2 - (\lambda_1+\lambda_2)]/\sigma_u.$$

Equation (2.7) agrees with the result of Balakrishnan and Kocherlakota (1985).

(ii) Edgeworth Series Distribution. It has been shown by Subrahmaniam and Chinganda (1978) that in this case

$$G_1(z) = H\left[\frac{z-\kappa_1}{\sigma_u\kappa_2}, \frac{\tau}{\sigma_u}; \rho\right] + H\left[\frac{z-(1-\kappa_1)}{\sigma_u\kappa_2}, -\frac{\tau}{\sigma_u}; \rho\right]$$

where

$$\kappa_1 = F_1(\tau),\ \kappa_2 = f(\tau),\ \sigma_u^2 = \sigma^2\left(\frac{1}{n_1} + \frac{1}{n_2}\right),$$

$$\rho = (n_2-n_1)/(n_1+n_2),\ \tau = (\mu_2-\mu_1)/2.$$

This result is obtained from (2.6), after a slight modification of the arguments.

3. OUTLIERS

Let us consider sampling from normal populations under Π_i, $i = 1,2$, with s_i outliers in each case. That is, for $i = 1,2$,

$$X_{i,j} \sim N(\mu_i,\sigma^2),\quad j = 1,2,\ldots,n_i-s_i$$

and

$$X_{i,j} \sim N(\mu_i+a_i\sigma, b_i^2\sigma^2),\ j = n_i-s_i+1,\ldots,n_i.$$

Then it can be seen that the sample mean

$$\bar{X}_i \sim N(\mu_i+p_ia_i\sigma, \frac{\sigma^2}{n_i}[1 + p_i(b_i^2-1)])$$

independently, for $i = 1,2$. Here $p_i = s_i/n_i$, the proportion of outliers in the i-th sample. Using the usual classification rule for the observation X, the errors of misclassification are found to be:

$$e_{12}(\bar{X}_1,\bar{X}_2) = 1 - \Phi\left[\frac{\bar{X}_1+\bar{X}_2-2\mu_1}{2\sigma}\right] \quad \text{if } \bar{X}_1 < \bar{X}_2$$

$$= \Phi\left[\frac{\bar{X}_1+\bar{X}_2-2\mu_1}{2\sigma}\right] \quad \text{if } \bar{X}_1 \geq \bar{X}_2$$

and

$$e_{21}(\bar{X}_1,\bar{X}_2) = \Phi\left[\frac{\bar{X}_1+\bar{X}_2-2\mu_2}{2\sigma}\right] \quad \text{if } \bar{X}_1 < \bar{X}_2$$

$$= 1 - \Phi\left[\frac{\bar{X}_1-\bar{X}_2-2\mu_2}{2\sigma}\right] \quad \text{if } \bar{X}_1 \geq \bar{X}_2.$$

Considering the outliers as described above,

$$\begin{pmatrix} \bar{X}_1+\bar{X}_2 \\ \bar{X}_1-\bar{X}_2 \end{pmatrix} \sim \mathrm{BVN}\left[\begin{pmatrix} \nu_1 \\ \nu_2 \end{pmatrix}, \begin{pmatrix} \tau^2 & \rho\tau^2 \\ & \tau^2 \end{pmatrix}\right]$$

where

$$\nu_1 = \mu_1 + \mu_2 + \sigma(p_1a_1 + p_2a_2),$$

$$\nu_2 = \mu_1 - \mu_2 + \sigma(p_1a_1 - p_2a_2),$$

$$\tau^2 = \sigma^2\left[\frac{1+p_1(b_1^2-1)}{n_1} + \frac{1+p_2(b_2^2-1)}{n_2}\right]$$

and

$$\rho\tau^2 = \sigma^2\left[\frac{1+p_1(b_1^2-1)}{n_1} - \frac{1+p_2(b_2^2-1)}{n_2}\right].$$

3.1 Distribution Function of e_{12}

Using the results derived earlier for the general situation, we can write the asymptotic distribution function of e_{12} as

$$G_1(z) \cong H\left[\frac{\nu_1-t_1}{\tau}, -\frac{\nu_2}{\tau}; -\rho\right] + H\left[\frac{t_2-\nu_1}{\tau}, \frac{\nu_2}{\tau}; -\rho\right] \quad (3.1)$$

where

$$t_1 = 2F_1^{-1}(1-z), \quad t_2 = 2F_1^{-1}(z).$$

Since under Π_1, $X \sim N(\mu_1,\sigma^2)$,

$$F_1^{-1}(1-z) = \mu_1 - \sigma\Phi^{-1}(z)$$

and

$$F_1^{-1}(z) = \mu_1 + \sigma\Phi^{-1}(z).$$

Hence

$$G_1(z) = H\left[\frac{\nu_1 - 2\{\mu_1 - \sigma\Phi^{-1}(z)\}}{\tau}, -\frac{\nu_2}{\tau}; -\rho\right] + H\left[\frac{2\{\mu_1 + \sigma\Phi^{-1}(z)\} - \nu_1}{\tau}, \frac{\nu_2}{\tau}; -\rho\right].$$

Substituting for ν_1, ν_2 and writing $\delta = (\mu_2 - \mu_1)/\sigma$, the standardized distance between the two populations, we have

$$G_1(z) = H\left[\frac{\delta + (a_1p_1 + a_2p_2) + 2\Phi^{-1}(z)}{\xi}, \frac{\delta - (a_1p_1 - a_2p_2)}{\xi}; -\rho\right] + H\left[\frac{2\Phi^{-1}(z) - \delta - (a_1p_1 + a_2p_2)}{\xi}, \frac{-\delta + (a_1p_1 - a_2p_2)}{\xi}; -\rho\right] \tag{3.2}$$

where

$$\xi^2 = \frac{1 + p_1(b_1^2 - 1)}{n_1} + \frac{1 + p_2(b_2^2 - 1)}{n_2}$$

and

$$\rho = \left\{\frac{1 + p_1(b_1^2 - 1)}{n_1} - \frac{1 + p_2(b_2^2 - 1)}{n_2}\right\}/\xi^2.$$

Similarly, for e_{21},

$$G_2(z) = H\left[\frac{2\Phi^{-1}(z) + \delta - (a_1p_1 + a_2p_2)}{\xi}, \frac{\delta - (a_1p_1 - a_2p_2)}{\xi}; \rho\right] + H\left[\frac{2\Phi^{-1}(z) - \delta + (a_1p_1 + a_2p_2)}{\xi}, \frac{-\delta + (a_1p_1 - a_2p_2)}{\xi}; \rho\right]. \tag{3.3}$$

It should be noted that if $p_1 = p_2 = 0$, that is when there are no outliers, the distribution function $G_1(z)$ in (3.2) reduces to the form given in equation (13) of John (1961).

3.2 Density Function

It is readily seen that

$$\frac{d}{dz} H[u,k(z);\rho] = k'(z)\phi\{k(z)\}\Phi\left[\frac{u-\rho k(z)}{(1-\rho^2)^{\frac{1}{2}}}\right]$$

from which we have

$$g_1(z) = \frac{2}{\xi\phi[\Phi^{-1}(z)]}\left[\phi\left(\frac{\delta_1^*}{\xi}\right)\Phi\left(\frac{\rho\delta_1^*+\delta'}{\xi(1-\rho^2)^{\frac{1}{2}}}\right) + \phi\left(\frac{\delta_2^*}{\xi}\right)\Phi\left(\frac{-\rho\delta_2^*-\delta'}{\xi(1-\rho^2)^{\frac{1}{2}}}\right)\right] \quad (3.4)$$

where

$$\delta_1^* = \delta + (a_1p_1 + a_2p_2) + 2\Phi^{-1}(z),$$
$$\delta_2^* = \delta + (a_1p_1 + a_2p_2) - 2\Phi^{-1}(z)$$

and

$$\delta' = \delta - (a_1p_1 - a_2p_2).$$

Similarly

$$g_2(z) = \frac{2}{\xi\phi[\Phi^{-1}(z)]}\left[\phi\left(\frac{\delta_3^*}{\xi}\right)\Phi\left(\frac{\delta'-\rho\delta_3^*}{\xi(1-\rho^2)^{\frac{1}{2}}}\right) + \phi\left(\frac{\delta_4^*}{\xi}\right)\Phi\left(\frac{-\delta'+\rho\delta_4^*}{\xi(1-\rho^2)^{\frac{1}{2}}}\right)\right] \quad (3.5)$$

where

$$\delta_3^* = \delta - (a_1p_1 + a_2p_2) + 2\Phi^{-1}(z)$$
$$\delta_4^* = \delta - (a_1p_1 + a_2p_2) - 2\Phi^{-1}(z).$$

3.3 Expected Values

The expected value of e_{12} is of interest as it is the unconditional probability of misclassification. From the definition of e_{12} it can be readily seen that

$$E[e_{12}] = P\{\bar{X}_1+\bar{X}_2 < 2X,\ \bar{X}_1-\bar{X}_2 < 0 | X \in II_1\}$$
$$+ P\{\bar{X}_1+\bar{X}_2 \geq 2X,\ \bar{X}_1-\bar{X}_2 \geq 0 | X \in II_1\}. \quad (3.6)$$

Fixing X, standardizing the variables $(\bar{X}_1+\bar{X}_2, \bar{X}_1-\bar{X}_2)$ we can write the first term as

$$P\{U < \frac{2x-\mu_u}{\sigma_u},\ V < -\frac{\mu_v}{\sigma_v} \Big| X = x\}$$
$$= H\left[\frac{2x-\mu_u}{\sigma_u},\ -\frac{\mu_v}{\sigma_v};\ \rho\right]$$

and hence

$$P\{\bar{X}_1+\bar{X}_2 < 2X,\ \bar{X}_1-\bar{X}_2 < 0 | X \in II_1\}$$
$$= \int_{-\infty}^{\infty} H\left[\frac{2x-\mu_u}{\sigma_u},\ -\frac{\mu_v}{\sigma_v};\rho\right] \frac{1}{\sqrt{2\pi}\sigma} \exp - \frac{(x-\mu_1)^2}{2\sigma^2}\, dx. \quad (3.7)$$

Upon rearranging the terms and simplifying, the right hand side reduces to

$$\int_{-\infty}^{\infty} \left\{ \int_{-\infty}^{-\nu_2/\tau} \Phi\left[\frac{2x-\nu_1-u\rho\tau}{\tau(1-\rho^2)^{\frac{1}{2}}}\right] \phi(u)du \right\} \frac{1}{\sqrt{2\pi}\sigma} \exp - \frac{(x-\mu_1)^2}{2\sigma^2}\, d$$
$$= \int_{-\infty}^{-\nu_2/\tau} \Phi\left[\frac{2\mu_1-\nu_1-u\rho\tau}{\{4\sigma^2+\tau^2(1-\rho^2)\}^{\frac{1}{2}}}\right] \phi(u)du$$

which further reduces to

$$H\left[\frac{2\mu_1-\nu_1}{(\tau^2+4\sigma^2)^{\frac{1}{2}}},\ -\frac{\nu_2}{\tau};\ \frac{\rho\tau}{(\tau^2+4\sigma^2)^{\frac{1}{2}}}\right].$$

Similarly, the second term in (3.6) is evaluated stepwise to be

$$\int_{-\infty}^{\nu_2/\tau} \Phi\left[\frac{\nu_1-2\mu_1-\rho\tau u}{\{\tau^2(1-\rho^2)+4\sigma^2\}^{\frac{1}{2}}}\right]\phi(u)du$$

$$= H\left[\frac{\nu_1-2\mu_1}{(\tau^2+4\sigma^2)^{\frac{1}{2}}}, \frac{\nu_2}{\tau}; \frac{\rho\tau}{(\tau^2+4\sigma^2)^{\frac{1}{2}}}\right].$$

Combining these results, with $\tau^{*2} = \tau^2+4\sigma^2$,

$$E[e_{12}] = H\left[\frac{2\mu_1-\nu_1}{\tau^*}, -\frac{\nu_2}{\tau}; \frac{\rho\tau}{\tau^*}\right]$$

$$+ H\left[-\frac{\nu_1-2\mu_1}{\tau^*}, \frac{\nu_2}{\tau}; \frac{\rho\tau}{\tau^*}\right].$$

Substituting for ν_1, ν_2, τ^2 we have, writing $\delta^* = \delta + (a_1p_1+a_2p_2)$ and $\xi^{*2} = \xi^2+4$,

$$E[e_{12}] = H\left[-\frac{\delta^*}{\xi^*}, \frac{\delta'}{\xi}; \frac{\rho\xi}{\xi^*}\right]$$

$$+ H\left[\frac{\delta^*}{\xi^*}, -\frac{\delta'}{\xi}; \frac{\rho\xi}{\xi^*}\right]. \qquad (3.7)$$

As before, if $p_1 = p_2 = 0$ (i.e., no outliers in either sample the quantity on the right-hand side reduces to that of (19) in John (1961).

Similarly, with $\delta^{**} = \delta - (a_1p_1+a_2p_2)$,

$$E[e_{21}] = H\left[-\frac{\delta^{**}}{\xi^*}, \frac{\delta'}{\xi}; \frac{-\rho\xi}{\xi^*}\right]$$

$$+ H\left[\frac{\delta^{**}}{\xi^*}, -\frac{\delta'}{\xi}; \frac{-\rho\xi}{\xi^*}\right]. \qquad (3.8)$$

3.4 Numerical Results

The distribution function $G_1(z)$ and the expected value $E[e_{12}]$ have been evaluated for a variety of values of the parameters. A small selection of the numerical results obtained are reproduced here. While $G_1(z)$ is depicted graphically, the expected values are tabled. We discuss these individually.

<u>$G_1(z)$:</u> The graphs provided here are for the parameter combinations $n_1 = n_2 = 20$, $a_1 = a_2 = 1$,

$b_1 = b_2 = 5$, 10 and $\delta = 0.5, 1.0$. The outlier numbers (s_1, s_2) are (0,0), (0,1), (1,1), (0,2), (2,2). It should be noted that the first one represents the normal situation. The presence of the outliers affects the outcome through its effect on the distribution of $\bar{X}$: For $X \sim N(\mu, \sigma^2)$, with s outliers out of n observations

$$\bar{X} \sim N[\mu + ap\sigma, \frac{\sigma^2}{n} \{1 + p(b^2-1)\}]$$

where $p = s/n$, the outliers having $N(\mu+a\sigma, b^2\sigma^2)$ distribution.

The distribution function $G_1(z|N)$, in the normal case, remains unaltered for all the rest of the parameters considered except δ. It is seen that $G_1(z)$ for the outliers is less than $G_1(z|N)$ when $z \le E(e_{12})$ but tends to be flatter for $z > E(e_{12})$. This makes the probability of the error rates being large much higher in the presence of outliers than in the normal case. The effect of the outliers in <u>both</u> samples is much more pronounced than in the case of <u>only one</u> having outliers. This is illustrated by comparing the graphs where $s_1 = 0$ with those in which $s_1 \neq 0$. Thus when $s_1 = 0$, the error distribution is close to that of the normal case while when $s_1 \neq 0$ this distribution function is extremely flat. The latter case leads to very high probability of the error being large. These results are illustrative of a general behaviour pattern.

<u>$E(e_{12})$</u>: This table re-enforces, in a summary fashion, the behaviour discussed above. In this respect they are not as informative as the distribution function. If $b_1 = b_2 = 1$ we find that the error rates with outliers are on the average smaller than those without outliers. This seems to be the result of the variance being independent of the outliers. The distributions are far apart. If, however, $b = 5$ or 10 the distributions of $\bar{X}$ under Π_1 and Π_2 have larger variances. This leads to a larger overlap in the distributions of $\bar{X}$ under the two populations. Hence we see a larger $E(e_{12})$ when both s_1 and s_2 are not equal to zero. On the other hand, if $s_1 = 0$ while $s_2 \neq 0$, the populations are once again moved apart (farther than in the case of $s_1 = s_2 = 0$. This accounts for the smaller expectation of e_{12} (than for the normal situation).

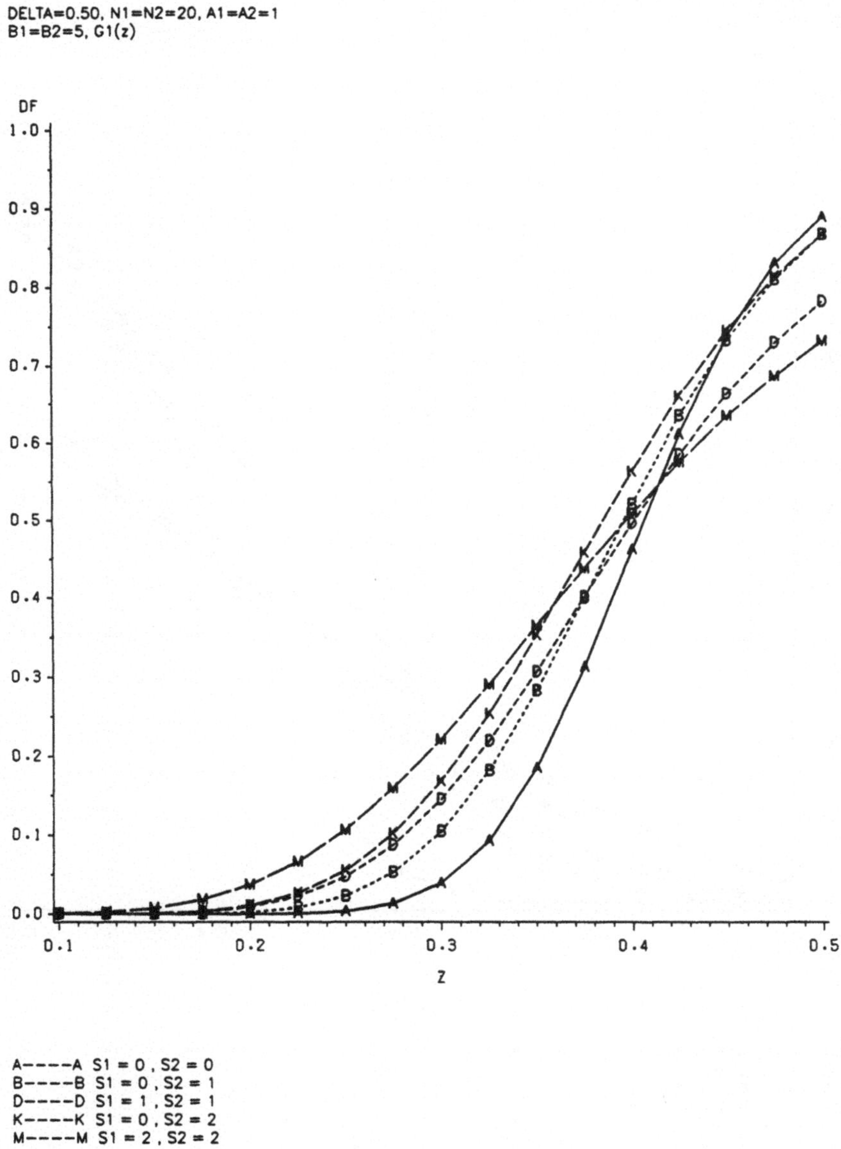
DELTA=0.50, N1=N2=20, A1=A2=1
B1=B2=5, G1(z)
DF
1.0
0.9
0.8
0.7
0.6
0.5
0.4
0.3
0.2
0.1
0.0
0.1
0.2
0.3
0.4
0.5
Z
A----A S1 = 0 , S2 = 0
B----B S1 = 0 , S2 = 1
D----D S1 = 1 , S2 = 1
K----K S1 = 0 , S2 = 2
M----M S1 = 2 , S2 = 2

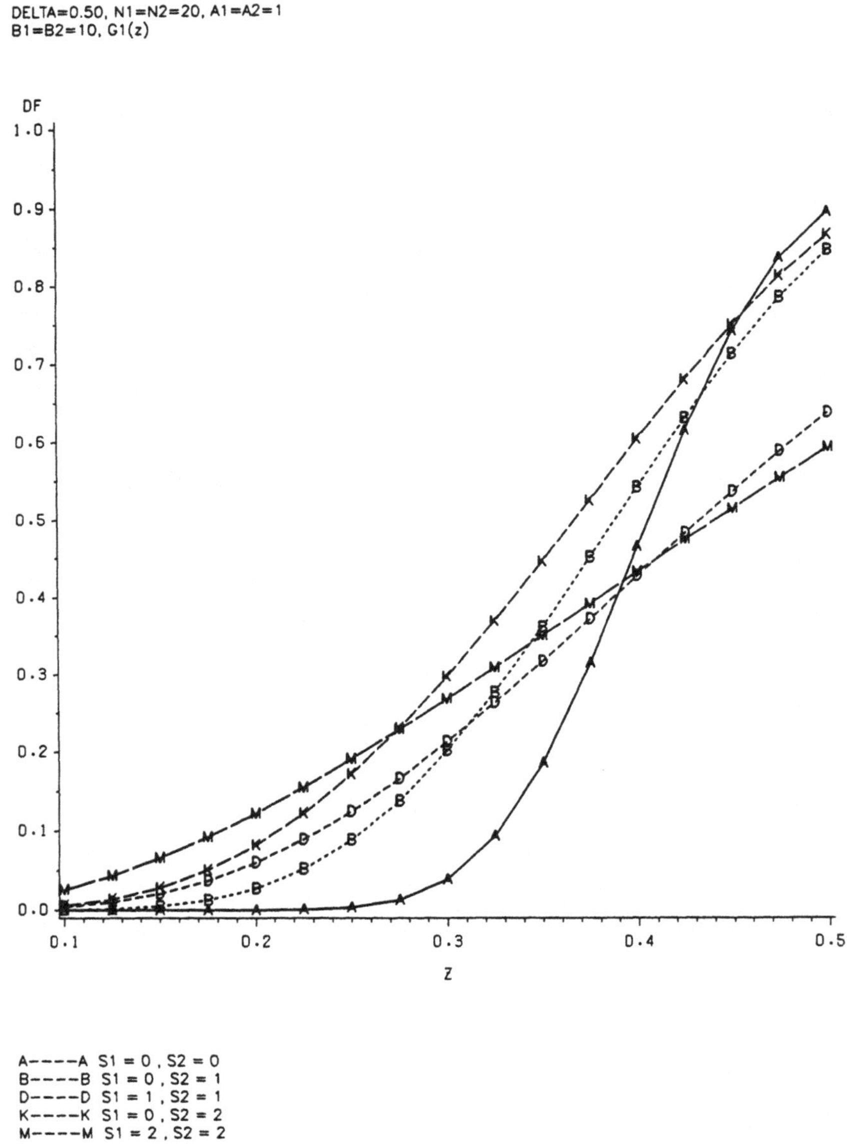
DELTA=0.50, N1=N2=20, A1=A2=1
B1=B2=10, G1(z)
DF
1.0
0.9
0.8
0.7
0.6
0.5
0.4
0.3
0.2
0.1
0.0
0.1
0.2
0.3
0.4
0.5
Z
A----A S1 = 0 , S2 = 0
B----B S1 = 0 , S2 = 1
D----D S1 = 1 , S2 = 1
K----K S1 = 0 , S2 = 2
M----M S1 = 2 , S2 = 2

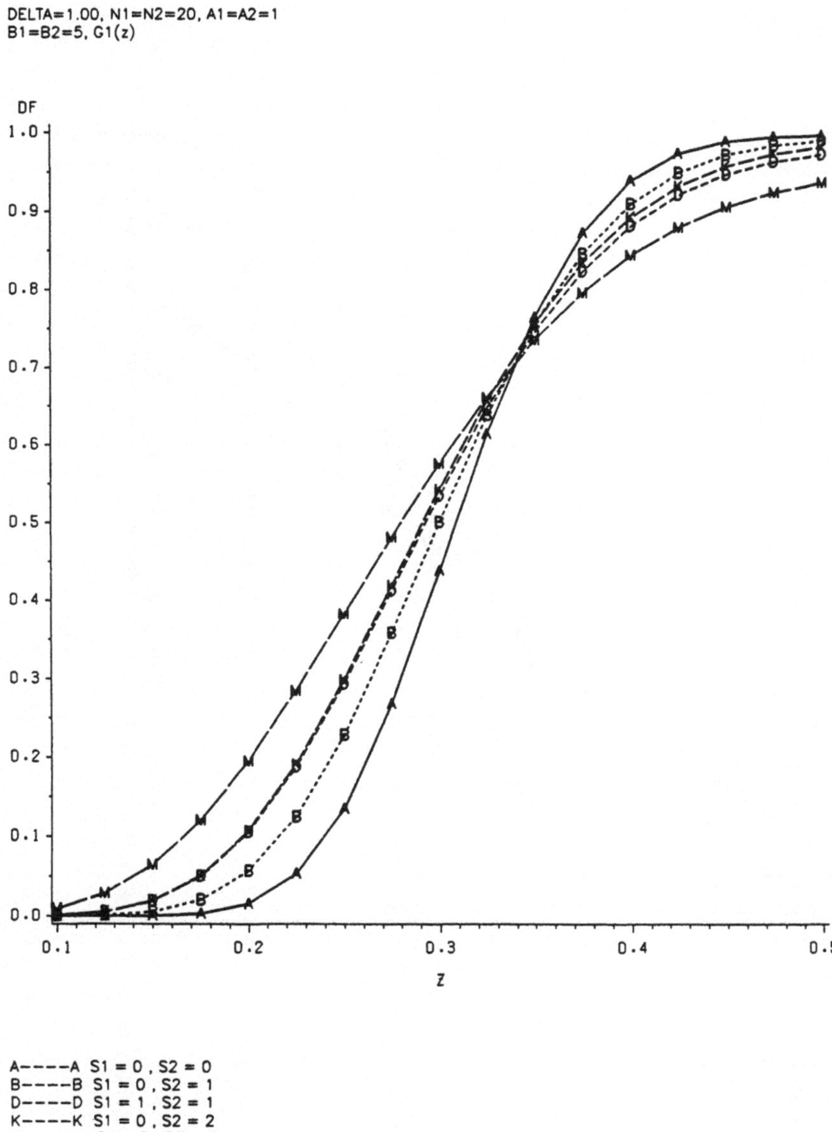
DELTA=1.00, N1=N2=20, A1=A2=1
B1=B2=5, G1(z)
DF
1.0
0.9
0.8
0.7
0.6
0.5
0.4
0.3
0.2
0.1
0.0
0.1
0.2
0.3
0.4
0.5
Z
A----A S1 = 0 , S2 = 0
B----B S1 = 0 , S2 = 1
D----D S1 = 1 , S2 = 1
K----K S1 = 0 , S2 = 2
M----M S1 = 2 , S2 = 2

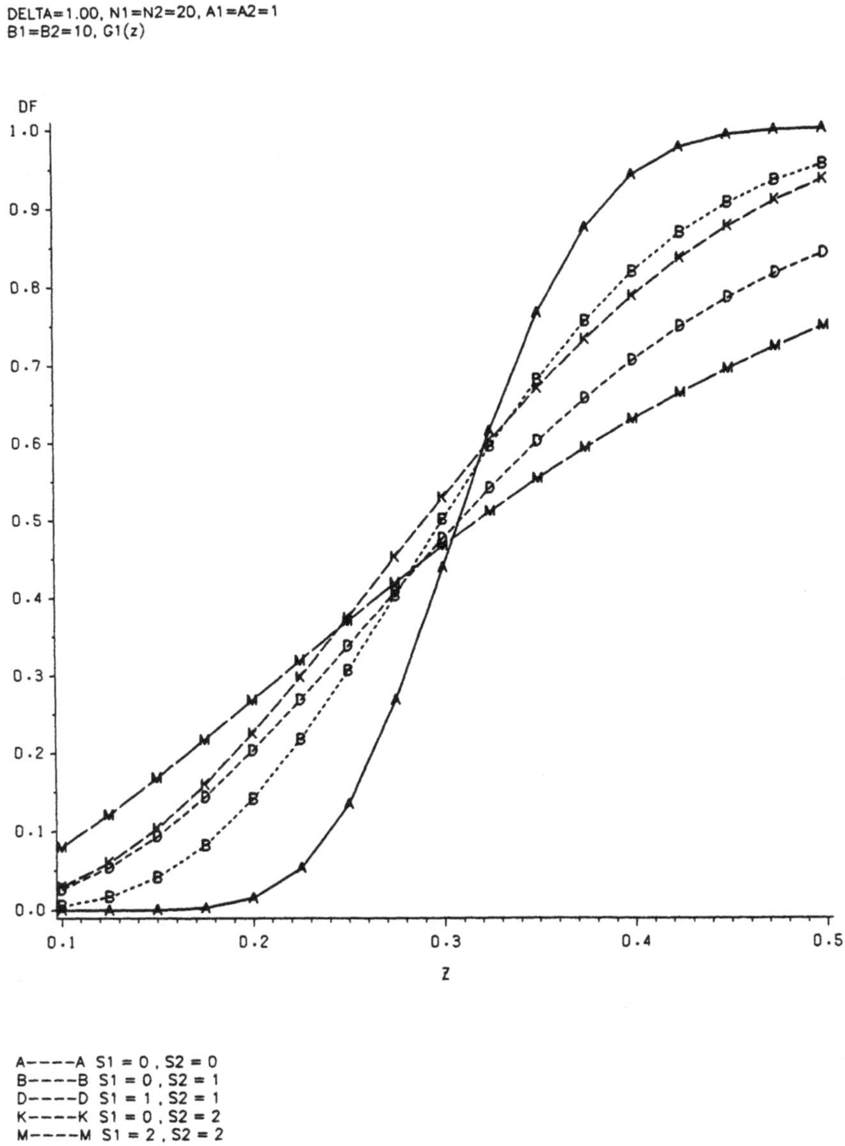
DELTA=1.00, N1=N2=20, A1=A2=1
B1=B2=10, G1(z)
DF
1.0
0.9
0.8
0.7
0.6
0.5
0.4
0.3
0.2
0.1
0.0
0.1
0.2
0.3
0.4
0.5
Z
A----A S1 = 0 , S2 = 0
B----B S1 = 0 , S2 = 1
D----D S1 = 1 , S2 = 1
K----K S1 = 0 , S2 = 2
M----M S1 = 2 , S2 = 2

Table 1

Values of $E(e_{12})$ for $a_1 = a_2 = 1$ and $n_1 = n_2 = 20$

(s_1, s_2)	$b_1 = b_2$	$\delta = 0.25$	$\delta = 0.50$	$\delta = 1.00$
(0,0)	Normal	.4720	.4136	.3110
(0,1)	1	.4613	.4017	.3022
	5	.4505	.4028	.3044
	10	.4213	.3914	.3092
(0,2)	1	.4498	.3902	.2936
	5	.4333	.3911	.2977
	10	.3905	.3675	.3009
(1,1)	1	.4608	.3968	.2938
	5	.4725	.4180	.3029
	10	.4835	.4467	.3421
(2,2)	1	.4498	.3802	.2771
	5	.4716	.4199	.3011
	10	.4850	.4552	.3659

3.5 Conclusions

The presence of the outliers in one sample in general does not affect the behaviour of the error rates. If both samples have outliers the error rates tend to increase. As the variance in the outlier samples increase the error rates tend to get inflated. This stands to reason. If, on the other hand, the variance of the outliers is small we can proceed to use the normal rule for classification. Similar conclusions are true when no outliers are present only in one of the samples.

ACKNOWLEDGEMENTS

The NSERC of Canada is to be thanked for its support of this research.

S. Kocherlakota and K. Kocherlakota
University of Maitoba
Winnipeg, Manitoba, Canada
R3T 2N2

N. Balakrishnan
McMaster University
Hamilton, Ontario, Canada
L8S 4K1

REFERENCES

Amoh, R.K. and Kocherlakota, K. (1986). Errors of misclassification associated with the inverse Gaussian distribution. Commun. Statist. - Theor. Meth. 15, 589-612.

Balakrishnan, N. and Kocherlakota, S. (1985). Robustness to nonnormality of the linear discriminant function: Mixtures of normal distributions. Commun. Statist. - Theor. Meth. 14, 465-478.

Balakrishnan, N., Kocherlakota, S. and Kocherlakota, K. (1986a). On the errors of misclassification based on dichotomous and normal variables. Ann. Inst. Statist. Math. (to appear).

Balakrishnan, N., Kocherlakota, S. and Kocherlakota, K. (1986b). Errors of misclassification using dichotomous and nonnormal continuous variables - I (mixtures of normals). Technical Report #175, Department of Statistics, University of Manitoba.

Balakrishnan, N., Kocherlakota, S. and Kocherlakota, K. (1986c). Errors of misclassification using dichotomous and nonnormal continuous variables - II (truncated normal distribution). Technical Report #176, Department of Statistics, University of Manitoba.

Chinganda, E.F. (1976). Misclassification Errors and their Distributions, MS.c. Thesis, University of Manitoba, Canada.

Chinganda, E.F. and Subrahmaniam, K. (1979). Robustness of the linear discriminant function to nonnormality: Johnson's system. J. Statist. Planning and Inference 3, 69-77.

John, S. (1961). Errors in discrimination. Ann. Math. Statist. 32, 1125-1144.

Kocherlakota, S., Balakrishnan, N. and Kocherlakota, K. (1986). The linear discriminant function: Sampling from the truncated normal distribution. Biometrical J. (to appear).

Subrahmaniam, K. and Chinganda, E.F. (1978). Robustness of the linear discriminant function to nonnormality: Edgeworth series distribution. J. Statist. Planning and Inference 2, 79-91.

Sadanori Konishi

TRANSFORMATIONS OF STATISTICS IN MULTIVARIATE ANALYSIS

ABSTRACT

There have been a lot of suggestions on how to transform variables in order to get some desirable properties. This paper is mainly concerned with the problem of normalizing and variance stabilizing transformations of statistics in multivariate analysis. A method for finding normalizing transformations is constructed based on an Edgeworth expansion. Investigation is made in connection with the problem which arises in deriving higher order Edgeworth expansions.

1. INTRODUCTION

A classical and practical example of transformations is Fisher's (1921) z-transformation for the correlation coefficient, r, in a bivariate normal sample. Transformations of this kind are made for two different purposes: one is to improve a normal approximation, and the other is to stabilize an asymptotic variance. Fisher's z-transformation for r is of particular interest in practical applications, since these two requirements can be simultaneously achieved by the same transformation $z(r) = (1/2)\log\{(1+r)/(1-r)\}$.

Theoretical approach to Fisher's z-transformation has been made by Hotelling (1953), Borges (1971), Konishi (1978, 1981) and Efron (1982). Konishi (1981) has given a general procedure for finding normalizing transformations in order to obtain simple and accurate approximations to the distributions of multivariate statistics. This procedure was used by Hayakawa (1986) to obtain normalizing transformations for ordinary, multiple and canonical correlation coefficients in an elliptical sample. Taniguchi, Krishnaiah and Chao (1986) also applied his procedure to some maximum likelihood estimators in Gaussian ARMA processes.

A. K. Gupta (ed.), Advances in Multivariate Statistical Analysis, 213–231.

Another example of normalizing transformations is the cube root transformation of Wilson and Hilferty (1931) for the chi-square distribution. The approach used by Wilson-Hilferty was adapted to develop normal approximations to the distributions of a definite quadratic form in normal variables by Jensen and Solomon (1972), and of the likelihood ratio statistic for testing the multivariate general linear hypothesis by Mudholkar and Trivedi (1980). Mudholkar and Trivedi (1981) also used this approach to obtain a normal approximation to the distribution of a sample variance for nonnormal populations.

It may be noted here that the normalizing property appears to be ambiguous, while the variance stabilizing property is clear and succinct: the problem can be reduced to find a function which renders an asymptotic variance independent of unknown parameters. The most common approach to normalization is to choose a transformation in such a way that the skewness of the distribution vanishes or becomes smaller. The correction of the skewness may be usefull to make the distribution of a transformed variable nearly symmetrical, but this idea gives no guarantee that a normal approximation for the transformed variable becomes accurate over the whole domain. In fact $z(r)$ approximated by a normal variate with mean $z(\rho)$ and variance $1/(N-3)$, where ρ and N are, respectively, a population correlation coefficient and a sample size, achieved normality only in the tails of the distribution of r.

Recently Niki and Konishi (1986) pointed out that the correction of the asymptotic skewness is essential for the problem of deriving higher order Edgeworth expansions for the distributions of multivariate statistics.

The purpose of this paper is to reconsider various transformations from a common standard point. To this end a general principle of normalization is constructed based on the rate of convergence to the normal distribution in an Edgeworth expansion. The asymptotic unbiasedness property and the relation between normalization and variance stabilization are also discussed. The basic idea and some examples have already been given in Konishi (1981, 1984), *etc.*, but more thorough investigation is made in this paper in connection with the problem which arises in deriving higher order Edgeworth expansions.

2. TRANSFORMATION THEORY

2. 1. *Normalization*

Suppose $X_1, X_2, \cdots, X_n$ be a random sample of size n from a p-variate distribution $F(x;\theta)$ with finite sixth moments, where θ denotes the parameter vector. Let T_n be an estimator of a parameter $\mu = \mu(\theta)$ where $\mu(\cdot)$ is a real-valued function. We consider a class of statistics that $\sqrt{n}(T_n - \mu)$ has a limiting normal distribution with zero mean as n tends to infinity. Define the asymptotic bias b, variance σ^2 and skewness κ_3 of T_n by

$$E[\sqrt{n}(T_n - \mu)] = \frac{1}{\sqrt{n}} b + O(n^{-3/2}),$$

$$E[\{\sqrt{n}(T_n - E[T_n])\}^2] = \sigma^2 + O(n^{-1}), \qquad (2.1)$$

$$E[\{\sqrt{n}(T_n - E[T_n])\}^3] = \frac{1}{\sqrt{n}} \kappa_3 + O(n^{-3/2}),$$

respectively, where the expectation is taken over the distribution F.

Let $f(T_n)$ be a one-to-one and twice continuously differentiable function in a neighborhood of $T_n = \mu$. Then the asymptotic bias, variance and skewness of the transformed statistic $f(T_n)$ is, respectively, given by

$$E[\sqrt{n}\{f(T_n) - f(\mu)\}] = \frac{1}{\sqrt{n}}\{bf'(\mu) + \frac{1}{2}\sigma^2 f''(\mu)\} + O(n^{-3/2}),$$

$$E[n\{f(T_n) - E[f(T_n)]\}^2] = \sigma^2 f'(\mu)^2 + O(n^{-1}) \qquad (2.2)$$

$$E[n\sqrt{n}\{f(T_n) - E[f(T_n)]\}^3] = \frac{1}{\sqrt{n}}\{\kappa_3 f'(\mu)^3 + 3\sigma^4 f''(\mu) f'(\mu)^2\} + O(n^{-3/2}),$$

where f', f'' are derivatives at $T_n = \mu$. Then an Edgeworth expansion for the distribution of $f(T_n)$ is, up to order $n^{-1/2}$,

$$\Pr[\sqrt{n}\{f(T_n) - f(\mu) - \frac{1}{n}(bf'(\mu) + \frac{1}{2}\sigma^2 f''(\mu))\}/\{\sigma f'(\mu)\} < x]$$
$$= \Phi(x) - \frac{1}{\sqrt{n}}\left\{\frac{\kappa_3}{6\sigma^3} + \frac{1}{2}\sigma f''(\mu)f'(\mu)^{-1}\right\}H_2(x)\varphi(x) + O(n^{-1}), \tag{2.3}$$

where $\Phi(x)$ and $\varphi(x)$ are, respectively, the standard normal distribution function and density, and $H_2(x) = x^2-1$.

The problem is what function should be chosen to obtain an accurate normal approximation over the whole domain of T_n. An accuracy of the normal approximation mainly depends on the values of the term of $O(n^{-1/2})$: large values of this term may give a poor fit. Hence it is natural to search for a function which reduces the coefficient of $H_2(x)\varphi(x)$ in (2.3) to zero for any value of x, so that the term of $O(n^{-1/2})$ can be made to vanish. This can be realized by finding a function which satisfies the second order differential equation

$$\frac{\kappa_3}{6\sigma^3} + \frac{1}{2}\sigma f''(\mu)f'(\mu)^{-1} = 0\ . \tag{2.4}$$

Let f_0 be a solution of this differential equation: note that κ_3 and σ may be obtained as a function of μ. Then, taking $f(T_n) = f_0(T_n)$ in (2.3), we have

$$\Pr[\sqrt{n}\{f_0(T_n) - f_0(\mu) - \frac{1}{n}(bf_0'(\mu) + \frac{1}{2}\sigma^2 f_0''(\mu))\}/\{\sigma f_0'(\mu)\} < x]$$
$$= \Phi(x) + O(n^{-1})\ . \tag{2.5}$$

This implies that by making a suitable transformation with an asymptotic bias correction, the term of $O(n^{-1/2})$ in (2.3) can be made to vanish, so the error involved is of order n^{-1}.

The standardized quantity $\sqrt{n}\{f_0(T_n) - f_0(\mu)\}/\{\sigma f_0'(\mu)\}$ has an Edgeworth expansion

$$\Pr[\sqrt{n}\{f_0(T_n) - f_0(\mu)\}/\{\sigma f_0'(\mu)\} < x]$$
$$= \Phi(x) - \frac{1}{\sqrt{n}}\left\{\frac{b}{\sigma} + \frac{1}{2}\sigma f_0''(\mu) f_0'(\mu)^{-1}\right\}\varphi(x) + O(n^{-1}) . \tag{2.6}$$

This type of the normal approximation is relatively good in the tail areas, but still poor in the mid-range. It is worth noting that approximate errors are mainly caused by $\varphi(x)$ in the mid-range and by $H_2(x)\varphi(x)$ in parts of the tail areas, since the absolute values of x are small in the mid-range and become larger as x moves into the tail areas. This will be illustrated through examples of the sample correlation coefficient and a chi-square variate in Section 3.

Another type of normalization has been developed on the basis of Cornish-Fisher or Cornish-Fisher type expansion by Blom (1954), Bol'shev (1959), Takeuchi (1975), Hall (1983a, 1983b), Shimizu and Yuasa (1984) and so on. The use of transformations to parametrize models has been discussed by Hougaard (1982).

2.2 *Asymptotic Variance Stabilization*

The asymptotic variance is independent of unknown parameters if and only if $\sigma f'(\mu)$ in (2.2) is constant. Hence the problem of finding a variance stabilizing transformation is reduced to solving the first order differential equation

$$\sigma f'(\mu) = 1 . \tag{2.7}$$

The transformation of this type has been used on data with particular reference to analysis of variance (See Bartlett (1947)).

If the two differential equations (2.4) and (2.7) have the same solution, or equivalently if

$$\frac{\kappa_3}{6\sigma^3} + \frac{1}{2}\sigma\frac{d}{d\mu}\left\{\log\frac{1}{\sigma}\right\} = 0, \tag{2.8}$$

then the normalization and variance stabilization for T_n are simultaneously achieved by the same function. It is of interest to note that the left-side of (2.8) can be calculated directly from the asymptotic variance and skewness of T_n, without any knowledge of a function f.

As a generalization of the variance stabilization to p-dimensional case, Holland (1973) considered a covariance stabilizing transformation given as a solution of certain matrix differential equation. He conjectured that no unique solution exists for p higher than 2.

2. 3. *Asymptotic Bias Reduction*

It is desirable that the transformed statistic, having the properties of normalization and/or variance stabilization, also satisfies the asymptotic unbiasedness property. The reduction of the asymptotic bias can be accomplished by choosing a function f which removes the first order term in the bias of $f(T_n)$ given in (2. 2), that is,

$$b f'(\mu) + \frac{1}{2}\sigma^2 f''(\mu) = 0 . \tag{2. 9}$$

However, it seems to be rare that the normalizing and/or variance stabilizing transformations satisfy this condition simultaneously.

2. 4. *Higher Order Edgeworth Expansion*

For small or moderate sample size, the normal approximation for T_n itself is, in general, considerably less accurate. One refinement is to use the normalizing transformation technique discussed in Subsection 2. 1.

Another work of improvement is concerned with the derivation of an asymptotic expansion for the distribution of the statistic T_n: the distribution function $G_n(x)$ of the standardized quantity $Z_n = \sqrt{n}(T_n - \mu)/\sigma$ is approximated by a finite sum

$$G_n(x) \approx \Phi(x) - \varphi(x)\left\{\frac{1}{\sqrt{n}}a_1(x) + \frac{1}{n}a_2(x) + \cdots + \frac{1}{n^{r/2}}a_r(x)\right\}, \tag{2. 10}$$

where a_j's are polynomials in x whose coefficients depend on the cumulants of the population distribution $F(x;\theta)$. This type of expansion is called the Edgeworth expansion of order $n^{-r/2}$.

An asymptotic expansion is said to be valid uniformly in x up to order $n^{-r/2}$ if

$$\operatorname*{Sup}_x \left| G_n(x) - \left\{ \Phi(x) - \varphi(x) \sum_{j=1}^{r} n^{-j/2} a_j(x) \right\} \right| = o(n^{-r/2}) . \qquad (2.11)$$

where the supremum is over the whole domain of Z_n. For the theory of Edgeworth expansions, see Wallace (1958), Petrov (1975), Bhattacharya and Ghosh (1978), Pfanzagl (1980), Siotani, Hayakawa and Fujikoshi (1985, Chapter 4).

To obtain highly accurate values even for small n using (2.10), higher order terms may be required. However, it must be noted that the Edgeworth expansion is not generally convergent infinite series for any prescribed n, and that the addition of the next term does not always improve the approximation. Supremums in (2.11) may increase rapidly with r to be over a certain r_0 and are always attained in the tails of the distribution. This fault of the Edgeworth expansion was pointed out by Wallace (1958). Niki and Konishi (1986) have shown that an Edgeworth expansion for the distribution of the transformed statistic $f_0(T_n)$ given as a solution of (2.4) may overcome this weakness and produce highly accurate values over the whole domain of T_n.

To be more specific, consider an Edgeworth expansion for the distribution of the transformed statistic $f(T_n)$. It is assumed that the derivatives of $f(T_n)$ of requisite order are continuous in a neighborhood of $T_n = \mu$, and that F has finite moments of arbitrary order.

Suppose the standardized quantity $W_n = \sqrt{n}\{f(T_n) - f(\mu)\}/\{\sigma f'(\mu)\}$ admits an Edgeworth expansion

$$\Pr[W_n < x] = \Phi(x) - \varphi(x)\left\{\frac{1}{\sqrt{n}}b_1(x) + \frac{1}{n}b_2(x) + \cdots + \frac{1}{n^{r/2}}b_r(x)\right\} + o(n^{-r/2}),$$

where $b_j(x)$ is a polynomial in x whose coefficients depend upon the cumulants of F and the derivatives of $f(T_n)$ at $T_n = \mu$. Arranging terms in each $b_j(x)$ according to the order of the Hermite polynomials $H_i(x)$, we have

$$b_j(x) = \frac{(\kappa_1^{(3)})^j}{(3!)^j j!} H_{3j-1}(x)$$

$$+ \text{ terms involving lower degrees of } H_i(x),$$

where $\kappa_1^{(3)}$ denotes the asymptotic skewness of W_n and in general $\kappa_i^{(j)}$ are the coefficient of the terms of $O(n^{-i/2})$ in the j th cumulant of W_n.

For large r, $(\kappa_1^{(3)})^j H_{3r-1}(x)/\{(3!)^r r!\}$, the term of the highest degree of $H_i(x)$ in $b_r(x)$, may cause errors which show up in parts of the tails of the distribution of T_n, since $H_{3r-1}(x)$ is a polynomial of degree r in x and the values of x are fairly large in the tail areas (for details, see Niki and Konishi (1986)).

It is of interest to note that the coefficients of $H_{3j-1}(x)$ include $\kappa_1^{(3)}$ for $j=1, 2, \ldots r$. Hence a transformation which reduces $\kappa_1^{(3)}$ to zero may be effective to improve the approximation based on the Edgeworth expansion up to order $n^{-r/2}$. As discussed in Subsection 2.1, the condition $\kappa_1^{(3)} = 0$ can be realized by finding a function which satisfies the second order differential equation (2.4). Then the coefficients $b_j(x)$ are given as follows :
when j is even

$$b_j(x) = \frac{(\kappa_2^{(4)})^{j/2}}{(4!)^{j/2}(\frac{j}{2})!} H_{2j-1}(x)$$

$$+ \text{ terms involving lower degrees of } H_i(x),$$

and when j is odd

$$b_j(x) = (\kappa_2^{(4)})^{(j-3)/2} \left\{\frac{2\kappa_1^{(1)}\kappa_2^{(4)}}{j-1} + \frac{\kappa_3^{(5)}}{5}\right\} \times \frac{H_{2(j-1)}(x)}{(4!)^{(j-1)/2}\left(\frac{j-3}{2}\right)!}$$

$$+ \text{ terms involving lower degrees of } H_i(x) .$$

That is, the Hermite polynomial of the highest degree in each $b_j(x)$ can be reduced to $H_{2j-1}(x)$ when j is even, and to $H_{2(j-1)}(x)$ when j is odd.

The efficiency of the transformation applied this

method was illustrated by Niki and Konishi (1986) through examples of the sample correlation coefficient and a chi-square variate.

Nonparametric bootstrap methods introduced by Efron (1979) offer an alternative approach to the distribution problem. Transformation theory for the nonparametric approach will be discussed in another paper.

3. EXAMPLES

3.1. *Sample Correlation Coefficient*

The transformation theory discussed in the last section is illustrated through an example of the sample correlation coefficient. Let r be the correlation coefficient based on a sample of size $n+1$ from a bivariate normal distribution with population correlation coefficient ρ. It is known (see Hotelling (1953, p. 212)) that the asymptotic bias, variance and skewness of r defined by (2.1) are, respectively,

$$b = -\frac{1}{2}\rho(1-\rho^2), \quad \sigma^2 = (1-\rho^2)^2 \text{ and } \kappa_3 = -6\rho(1-\rho^2)^3. \tag{3.1}$$

Hence an Edgeworth expansion for the distribution of $f(r)$ which corresponds to (2.3) is given by

$$\Pr\left[\frac{\sqrt{n}[f(r)-f(\rho)-\frac{1}{n}\{-\frac{1}{2}\rho(1-\rho^2)f'(\rho)+\frac{1}{2}(1-\rho^2)^2f''(\rho)\}]}{(1-\rho^2)f'(\rho)} < x\right] \tag{3.2}$$

$$= \Phi(x) - \frac{1}{\sqrt{n}}\{-\rho + \frac{1}{2}(1-\rho^2)f''(\rho)f'(\rho)^{-1}\}H_2(x)\varphi(x) + O(n^{-1}).$$

Following a general transformation theory, we search for a function which makes the coefficient of $H_2(x)\varphi(x)$ vanish, that is,

$$-\rho + \frac{1}{2}(1-\rho^2)f''(\rho)f'(\rho)^{-1} = 0 .$$

A solution of this differential equation is $f_0(r) = (1/2)\log\{(1+r)/(1-r)\}$ $(= z(r))$, Fisher's z-transformation. Taking $f(r) = z(r)$ in (3.2), we have

$$\Pr[\sqrt{n}\{z(r) - z(\rho) - \frac{\rho}{2n}\} < x] = \Phi(x) + O(n^{-1}). \tag{3.3}$$

The most commonly used approximation is to regard $z(r)$ as a normal variate with mean $z(\rho)$ and variance $(n-2)^{-1}$. It follows from (2.6) that

$$\Pr[\sqrt{n-2}\{z(r) - z(\rho)\} < x] = \Phi(x) + O(n^{-1/2}), \tag{3.4}$$

and that for the normal approximation of r

$$\Pr[\sqrt{n}(r - \rho)/(1 - \rho^2) < x] = \Phi(x) + O(n^{-1/2}). \tag{3.5}$$

We use the notations Z_1, Z_2 and R standing for the normal approximations (3.3), (3.4) and (3.5), respectively.

It can be shown on the basis of the rate of convergence to normality that the normal approximation Z_1 compares favorably with Z_2 and R, but of course this does not imply that Z_1 is superior to Z_2 and R uniformly in x.

Figure 1 compares errors in approximating the values of the probability $\Pr(r<r_0)$ $(|r_0|<1)$ by using the three normal approximations: error = (approximate value - exact value). Here the notations Z_1, Z_2 and R refer to error curves of (3.3), (3.4) and (3.5), respectively. It may be seen that Z_2 achieves normality only in the tails of the distribution, while Z_1 over the whole domain of r.

The terms of order $n^{-1/2}$ of Z_2 and R are, respectively, given by

$$-\frac{\rho}{2}\varphi(x) \qquad \textit{and} \qquad \rho\left\{\frac{1}{2} + H_2(x)\right\}\varphi(x).$$

Figure 2 presents a graph of $-H_2(x)\varphi(x)$, which is similar in form to the error curve R. It may be noted that approximation errors are caused by $\varphi(x)$ in the mid-range and by $H_2(x)\varphi(x)$ in parts of tail areas, since the absolute values of x are small in the mid-range and become larger as x moves into the tails. We may confirm that the transformation given as a solution of the differential equation (2.4) is effective to improve the approximation in the tail areas.

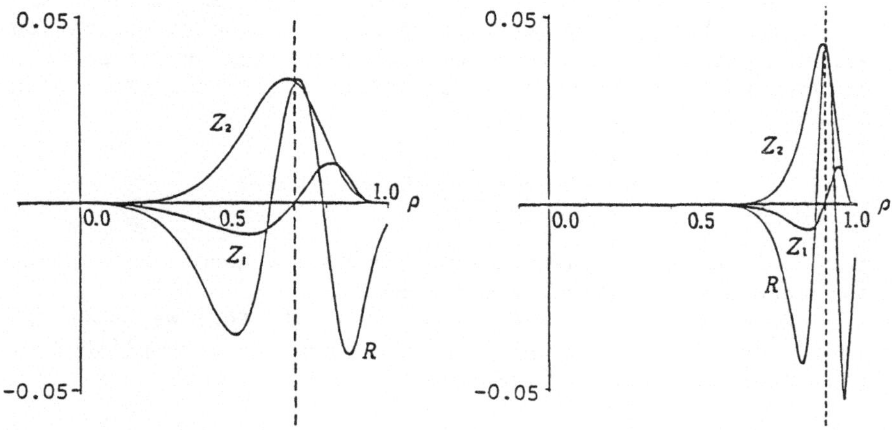

(a) n = 19, ρ = 0.7 (b) n = 19, ρ = 0.9

Figure 1. Errors in approximating the values of $\Pr(r < r_0)$ by using Z_1 [approximation (3.3)], Z_2[(3.4)] and R [(3.5)] : Error = (Approximate value - Exact value).

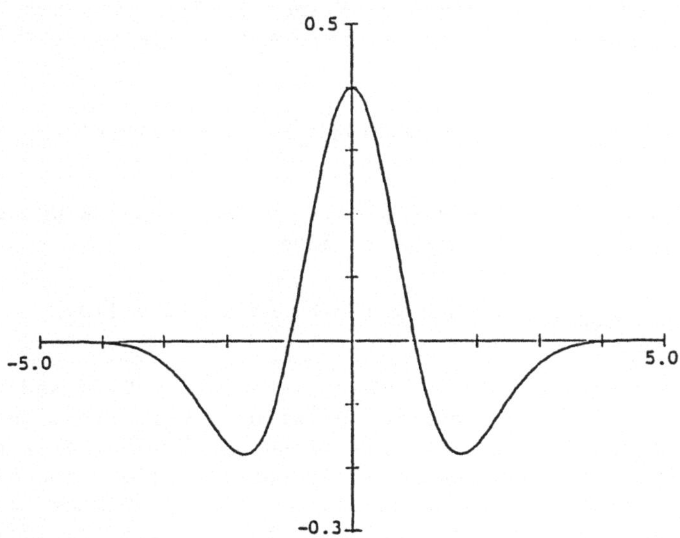

Figure 2. Graph of $h(x) = -H_2(x)\varphi(x)$

The equation (2.8) in Subsection 2.2 gives the condition that the normalization and variance stabilization are simultaneously achieved by the same transformation. It is easy to check that by substituting (3.1) into (2.8) with $\mu = \rho$

$$-\rho + \frac{1}{2}(1-\rho^2)\frac{d}{d\rho}\{\log(1-\rho^2)^{-1}\} = 0 .$$

Hence Fisher's z-transformation yields the normalization and variance stabilization, simultaneously.

From (2.9) in Subsection 2.3 it follows that the reduction of the asymptotic bias can be accomplished by solving the differential equation

$$-\frac{1}{2}\rho(1-\rho^2)f'(\rho) + \frac{1}{2}(1-\rho^2)^2 f''(\rho) = 0 .$$

A solution of this differential equation is found to be $f(r) = \sin^{-1}r$. This implies that $E[\sin^{-1}r] = \sin^{-1}\rho + O(n^{-1})$, but it is known (see Harley (1956)) that $\sin^{-1}r$ is exactly an unbiased estimator for $\sin^{-1}\rho$.

Now let r_I be the maximum likelihood estimator of an intraclass correlation coefficient, ρ_I, in a p-variate normal sample. It is known (Fisher (1958), Chapter 7) that the variance stabilizing transformation for r_I is, for any dimension p,

$$Z_p(r_I) = \{(p-1)/(2p)\}^{1/2}\log[\{1+(p-1)r_I\}/(1-r_I)] .$$

Substituting the asymptotic variance and skewness of r_I in the left-side of (2.8), we have

$$\frac{2}{3kp(p-1)}\{p-2-3(p-1)\rho_I\} - \frac{1}{2}k\{p-2-2(p-1)\rho_I\}$$

where $k = \{2/p(p-1)\}^{1/2}$ This is equal to 0 if and only if $p = 2$. It is worth pointing out that Fisher's z-transformation $Z_p(r_I)$ satisfies, only in the case where $p=2$, the two requirements of normalization and variance stabilization, simultaneously. For p higher than 2, normalization can be achieved by a power transformation (see, for details, Konishi (1985)).

3. 2. *Definite Quadratic Form in Normal Variables*

Suppose $X_1, X_2, \cdots, X_k$ be independent standard normal variables. Let $\lambda = (\lambda_1, \lambda_2, \cdots, \lambda_k)$ with $\lambda_j > 0$ and $\mu = (\mu_1, \mu_2, \cdots, \mu_k)$, where the elements λ_j and μ_j are all constants. Then the distribution of a definite quadratic form in normal variables is the same as that of

$$Q_k = \sum_{j=1}^{k} \lambda_j (X_j - \mu_j)^2 .$$

It is known (e. g., Johnson and Kotz (1970, p. 152)) that the r-th cumulant of Q_k is

$$\kappa_r = 2^{r-1}(r-1)!\, m_r, \qquad r=1, 2, \cdots$$

with $m_r = \Sigma_{j=1} \lambda_j (1+r\mu_j)$. It is assumed that for the mean $\kappa_1 = m_1$, $w_j = m_j/m_1 = O(1)$, j=2, 3, ... as m_1 tends to infinity. Following a general procedure discussed in Subsection 2. 1, we have

$$\Pr\left[\sqrt{m_1}\left\{\left(\frac{Q_k}{m_1}\right)^h - 1 - \frac{1}{m_1}h(h-1)w_2\right\}\Big/\sqrt{2h^2 w_2} < x\right] = \Phi(x) + O(m_1^{-1}) \qquad (3.\ 6)$$

where $h = 1-2m_1 m_3/(3m_2^2)$ and $w_2 = m_2/m_1$.

Jensen and Solomon (1972) used the Wilson-Hilferty method to obtain the normalizing transformation, in which the form of transformation was restricted to a class of power transforms. Konishi, Niki and Gupta (1986) derived a higher order Edgeworth expansion for the distribution of the transformed variable in (3. 6).

In the special case when $\lambda = (1, 1, \ldots, 1)$ and $\mu = 0$, Q_k has a central chi-square (χ^2) distribution with k degrees of freedom, for which m_r, w_j and h in (3. 6) are, respectively, reduced to k, 1 and 1/3. Then it follows that

$$\Pr\left[\sqrt{\frac{9k}{2}}\left\{\left(\frac{\chi^2}{k}\right)^{1/3} - 1 + \frac{2}{9k}\right\} < x\right] = \Phi(x) + O(k^{-1}), \qquad (3.\ 7)$$

the Wilson-Hilferty approximation for the central χ^2 distribution.

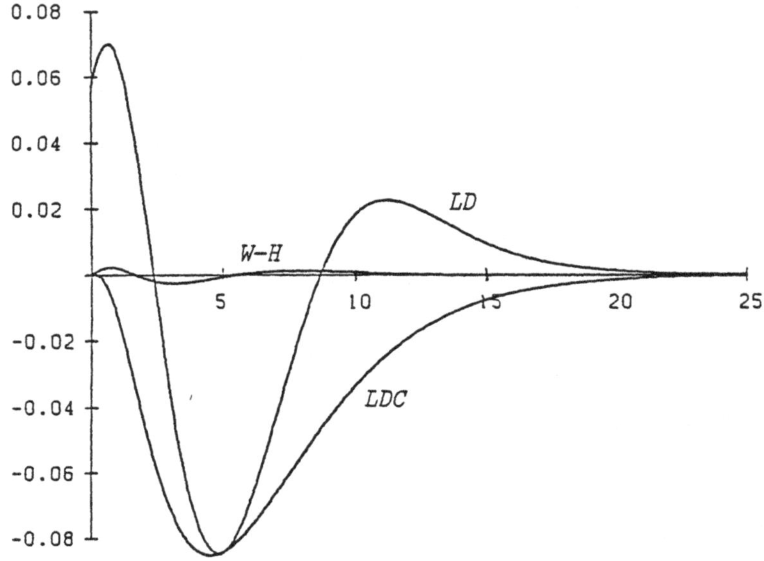

Figure 3. Errors in approximating the values of $\Pr(\chi^2 < x_0)$ by using *W-H* [approximation (3.7)], *LD* [(3.8)] and *LDC* [(3.9)] for k = 5: Error = (Approximate value - Exact value).

Figure 3 compares errors in approximating the values of the probability $\Pr(\chi^2 < x_0)$ by using the Wilson-Hilferty approximation (*W-H*) and the limiting distributions of χ^2 (*LD*) and of $(\chi^2)^{1/3}$(*LDC*) given as follows :

$$\Pr\left[\frac{\chi^2 - k}{\sqrt{2k}} < x\right] = \Phi(x) + O(k^{-1/2}), \tag{3.8}$$

and

$$\Pr\left[\sqrt{\frac{9k}{2}}\left\{\left(\frac{\chi^2}{k}\right)^{1/3} - 1\right\} < x\right] = \Phi(x) + O(k^{-1/2}) . \tag{3.9}$$

We can see in Figure 3 (also in Figure 1) that the asymptotic bias correction is essential to achieve normality.

3.3. *Canonical Correlation Coefficients*

Let $1>r_1>\cdots>r_p>0$ be the sample canonical correlation coefficients between variates $x_1, \cdots, x_p$ and $x_{p+1}, \cdots, x_{p+q}$ ($p\leq q$) based on a sample of size $n+1$ from a $(p+q)$-variate normal distribution, and let $1>\rho_1\geq\cdots\geq\rho_p>0$ be the population canonical correlations.

If ρ_i is distinct from other $p-1$ canonical correlations, then a normalizing transformation is

$$\Pr\left[\sqrt{n}\left\{\frac{1}{2}\log\frac{1+r_i}{1-r_i} - \frac{1}{2}\log\frac{1+\rho_i}{1-\rho_i} - \frac{c_i}{n}\right\} < x\right] = \Phi(x) + O(n^{-1}),$$

with $c_i = (1/2)\{p+q-2+\rho_i^2+2(1-\rho_i^2) \times \Sigma_{\alpha\neq i}\rho_\alpha^2/(\rho_i^2-\rho_\alpha^2)\}/\rho_i$ (Konishi (1981)). It is easy to see that Fisher's z-transformation produces stable asymptotic variance as well as normality.

The results for multiple and partial correlation coefficients are straightforward.

3.4. *Latent Roots of a Sample Covariance Matrix*

Let S be a sample covariace matrix based on a sample of size $n+1$ from a p-variate normal distribution with positive definite covariance matrix Σ. Let $l_1>\cdots>l_p>0$ and $\lambda_1\geq\cdots\geq\lambda_p>0$ be the latent roots of S and Σ, respectively. If λ_i is a simple root, then a normalizing transformation for l_i is given by

$$\Pr\left[\sqrt{\frac{9n}{2}}\left\{\left(\frac{l_i}{\lambda_i}\right)^{1/3} - 1 - \frac{1}{3n}\left(d_i-\frac{2}{3}\right)\right\} < x\right] = \Phi(x) + O(n^{-1}), \qquad (3.10)$$

with $d_i = \Sigma_{\alpha\neq i}^{p}\lambda_\alpha/(\lambda_i-\lambda_\alpha)$ (Konishi (1981)).

The asymptotic variance stabilizing transformation is $(1/\sqrt{2})\log l_i$, which can be obtained by solving the differential equation (2.7) with $\sigma = \sqrt{2}\lambda_i$ and $\mu = \lambda_i$. We can also see that the left-side of (2.8) is

$$\frac{\sqrt{2}}{3} + \frac{\sqrt{2}}{2}\lambda_i\frac{d}{d\lambda_i}\left\{\log\frac{1}{\sqrt{2\lambda_i}}\right\} \neq 0 .$$

We now consider transformations for the ratio of two distinct latent roots. It follows from Konishi (1977) that the asymptotic variance, bias and skewness of l_i/l_j are, respectively, given by $\sigma^2 = 4(\lambda_i/\lambda_j)^2$, $b = (\lambda_i/\lambda_j)(d_i-d_j+2)$ and $\kappa_3 = 48(\lambda_i/\lambda_j)^3$, where d_i are defined in (3.10). Substituting these in (2.4) and solving the differential equation yields

$$\Pr\left[\frac{\sqrt{n}}{2}\left\{\log(l_i/l_j) - \log(\lambda_i/\lambda_j) - \frac{1}{n}(d_i-d_j)\right\} < x\right] = \Phi(x) + O(n^{-1}) .$$

Writing $\lambda_{ij} = \lambda_i/\lambda_j$, it then follows from (2.8) that

$$1 + \lambda_{ij}\frac{d}{d\lambda_{ij}}\left\{\log\frac{1}{2\lambda_{ij}}\right\} = 0 .$$

Hence the transformation is stabilizing as well as normalizing.

ACKNOWLEDGEMENTS

The author would like to thank Ms. Motoko Yuasa for her help with programming. This work was supported in part by the Ministry of Education Grant 61530018.

The Institute of Statistical Mathematics
4-6-7, Minami-Azabu, Minato-ku
Tokyo 106, Japan

REFERENCES

Bartlett, M. S. (1947). 'The use of transformations'. *Biometrics* 3, 39-52.

Bhattacharya, R. N. and Ghosh, J. K. (1978). 'On the validity of the formal Edgeworth expansion'. *Annals of Statistics* 6, 434-451.

Blom, G. (1954). 'Transformations of the binomial, negative binomial, Poisson and χ^2 distributions'. *Biometrika* 41, 302-316.

Bol'shev, L. N. (1959). 'On transformations of random variables'. *Theory of Probability and its Applications* 4, 129-141.

Borges, R. (1971). 'Derivation of normalizing transformations with an error of order $1/n$'. *Sankhya Ser.* A 33, 441-460.

Efron, B. (1979). 'Bootstrap methods: another look at the jackknife'. *Annals of Statistics*, 7, 1-26

Efron, B. (1982). 'Transformation theory: How normal is a family of distributions?'. *Annals of Statistics* 10, 323-339.

Fisher, R. A. (1921). 'On the "probable error" of a coefficient of correlation deduced from a small sample'. *Metron* 1, 1-32.

Fisher, R. A. (1958). *'Statistical Methods for Research Workers'*. Oliver & Boyd, Edinburgh.

Hall, P. (1983a). 'A unified approach to the correction of normal approximations'. *SIAM Journal on Applied Mathematics* 43, 1187-1193.

Hall, P. (1983b). 'Inverting an Edgeworth expansion'. *Annals of Statistics* 11, 569-576.

Harley, B. I. (1956). 'Some properties of an angular transformation for the correlation coefficient'. *Biometrika* 43, 219-224.

Hayakawa, T. (1986). 'Normalizing and variance stabilizing transformation of multivariate statistics under an elliptical population'. To appear in *Annals of the Institute of Statistical Mathematics*.

Holland, P. W. (1973). 'Covariance stabilizing transformations'. *Annals of Statistics* 1, 84-92.

Hotelling, H. (1953). 'New light on the correlation coefficient and its transforms'. *Journal of the Royal Statistical Society* B 15, 193-232.

Hougaard, P. (1982). 'Parametrizations of non-linear models'. *Journal of the Royal Statistical Society* B 44, 244-252.

Jensen, D. R. & Solomon, H. (1972). 'A Gaussian approximation to the distribution of a definite quadratic form'. *Journal of the American Statistical Association* 67, 898-902.

Johnson, N. L. & Kotz, S. (1970). *Distributions in Statistics : Continuous Univariate Distributions* -2, Boston : Houghton Mifflin.

Konishi, S. (1977). 'Asymptotic expansion for the distribution of a function of latent roots of the covariance matrix'. *Annals of the Institute of Statistical Mathematics* 29, 389-396.

Konishi, S. (1978). 'An approximation to the distribution

of the sample correlation coefficient'. *Biometrika* 65, 654-656.

Konishi, S. (1981). 'Normalizing transformations of some statistics in multivariate analysis'. *Biometrika* 68, 647-651.

Konishi, S. (1984). 'Normalizing and variance stabilizing transformations of multivariate statistics'. *Proceedings of the Institute of Statistical Mathematics* 32, 159-171, (in Japanese).

Konishi, S. (1985). 'Normalizing and variance stabilizing transformations for intraclass correlations'. *Annals of the Institute of Statistical Mathematics* 37, 87-94.

Konishi, S., Niki, N. & Gupta, A. K. (1986). 'Asymptotic expansions for the distribution of quadratic forms in normal variables'. To appear in *the Annals of the Institute of Statistical Mathematics*.

Mudholkar, G. S. & Trivedi, M. C. (1980). 'A normal approximation for the distribution of the likelihood ratio statistic in multivariate analysis of variance'. *Biometrika* 67, 485-488.

Mudholkar, G. S. & Trivedi, M. C. (1981). 'A Gaussian approximation to the distribution of the sample variance for nonnormal populations'. *Journal of the American Statistical Association* 76, 479-485.

Niki, N. & Konishi, S. (1986). 'Effects of transformations in higher order asymptotic expansions'. To appear in *Annals of the Institute of Statistical Mathematics*. 38-3.

Petrov, V. V. (1975). *Sums of Independent Random Variables*. Springer - Verlag, Berlin.

Pfanzagl, J. (1980). 'Asymptotic expansions in parametric statistical theory'. *Developments in Statistics*, Vol. 3 (P. R. Krishnaiah, ed.), 1-97, Academic Press, New York.

Shimizu, R. & Yuasa, M. (1984). 'Normal approximation for asymptotic distributions'. *Proceedings of the Institute of Statistical Mathematics* 32, 141-157 (in Japanese).

Siotani, M., Hayakawa, T. & Fujikoshi, Y. (1985). *Modern Multivariate Statistical Analysis*. American Sciences Press, Ohio.

Takeuchi, K. (1975). *Kakuritsu - Bunpu no Kinji* (Approximation to probability distributions). Tokyo Kyoiku - Shuppan (in Japanese).

Taniguchi, M., Krishnaiah, P. R. and Chao, R. (1986). 'Normalizing transformations of some statistics of Gaussian ARMA processes'. Technical Report No. 86-5,

Center for Multivariate Analysis, University of Pittsburgh.

Wallace, D. L. (1958). 'Asymptotic approximations to distributions'. *Annals of Mathematical Statistics* 29, 635-654.

Wilson, E. B. & Hilferty, M. M. (1931). 'The distribution of chi-square'. *Proceedings of the National Academy of Sciences* 17, 684-688.

G. J. McLACHLAN

ERROR RATE ESTIMATION IN DISCRIMINANT ANALYSIS: RECENT ADVANCES

1. Introduction

An important problem in discriminant analysis is the estimation of the error rates associated with a given discriminant rule for allocating an object of unknown origin to one of a finite number, say g, of distinct classes or populations. The rule is based on the observed value of a random vector X of p measurements on the object. Over the years there have been many investigations on this problem; see, for example, Hills (1966), Lachenbruch and Mickey (1968), and McLachlan (1974a,b,c), and the references therein. Toussaint (1974) has compiled an extensive bibliography, which has been updated recently by Hand (1986b). An overview of error rate estimation has been given by McLachlan (1986), while recent work on robust error rate estimation has been summarized by Knoke (1986).

Following the major studies done in the 1960's, there has always been a high level of interest in this topic, promoted more recently by the appearance of Efron's (1979) paper on the bootstrap, with its important applications to all aspects of error rate estimation. In the last year or so, there has been a considerable number of papers produced in this area. Indeed, in the time since the article by McLachlan (1986) was written in late 1984, there have been over fifteen papers published on error rate estimation. Much of the recent work has arisen from consideration of the new ideas presented on the subject in the seminal paper of Efron (1983). The main thrust of the recent studies has been to highlight the usefulness of resampling technqiues in producing improved estimators of the error rates by appropriate bias correction of the apparent error rate. Attention has also been given to ways of smoothing the apparent error rate in an attempt to reduce its variance.

In this paper we wish to focus on these more recent advances on this problem, and to provide an up to date account of results in the area. Also, in presenting these

A. K. Gupta (ed.), Advances in Multivariate Statistical Analysis, 233–252.

results, we wish to provide further clarification of some issues on which there has been some confusion in the literature.

As most of the recently proposed error rate estimators are essentially nonparametric, parametric estimators are not reviewed here. For an account of the latter type of estimators, the reader is referred to McLachlan (1974a,1986), Page (1985), and Snapinn and Knoke (1984). Also, Vlachonikolis (1986) has considered recently a parametric approach to the estimation of the expected error rate of the location model of Krzanowski (1975) for discrimination with mixed binary and continuous variables. It is worth noting that although one may be willing to use a particular parametric rule due to its known robustness to mild departures from the adopted model, the parametric estimators of the error rates of the rule may not be robust. This is another reason why much attention has been given to the development of nonparametric error rate estimators.

2. Allocation Rules

The g possible classes are denoted by $C_1,\ldots,C_g$ and their prior probabilities by $\pi_1,\ldots,\pi_g$ respectively. The elements of $\pi = (\pi_1,\ldots,\pi_g)$ are nonnegative and sum to 1. We let Z denote the random vector (X,Y) associated with an object, where Y is a random variable taking on values 1 to g so that the value of Y specifies the class to which the object belongs. For an object of unknown origin, we let r(x) denote a rule for predicting the unknown value of Y, y, having observed X = x, where r(x) = i implies that the object is allocated to $C_i (i=1,\ldots,g)$.

In order to construct a suitable allocation rule it is assumed that there is a training set

$$t = \{z_1=(x_1,y_1),\ldots,z_n = (x_n,y_n)\}$$

for which $y_1,\ldots,y_n$ are known and $x_1,\ldots,x_n$ are independently and identically distributed. We let n_i denote those realizations of X from $C_i (i=1,\ldots,g;\ n_1+n_2+\ldots+n_g=n)$. It is assumed here that the observations have been sampled from a mixture of the classes in proportions $\pi_1,\ldots,\pi_g$, so that

$$n_i \sim \text{bin}(n,\pi_i),$$

providing $\hat{\pi}_i = n_i/n$ as an estimate of $\pi_i (i=1,\ldots,g)$. The error rates associated with $r(x;t)$ are useful in summarizing its global performance as an allocation rule. Of course for a specific case with $X = x$, it is more appropriate to concentrate on the estimation of the posterior probabilities of class membership $\xi_i(x)$, where

$$\xi_i(x) = \pi_i f_i(x) / \left\{ \sum_{k=1}^{g} \pi_k f_k(x) \right\}$$

is the posterior probability that an object with $X = x$ belongs to C_i and $f_i(x)$ is the probability density function of X in $C_i (i=1,\ldots,g)$. The density functions are with respect to arbitrary measure, so $f_i(x)$ can be a mass function by the adoption of counting measure. We let $f(z)$ be the density function of Z.

The estimated posterior probabilities lead to the formation of an allocation rule, where $r(x;t) = i$ if

$$\hat{\xi}_i(x) > \hat{\xi}_j(x) \qquad (j=1,\ldots,g;j\neq i). \tag{2.1}$$

For known posterior probabilities, (2.1) defines the optimal or Bayes rule of allocation which minimizes the error rate averaged with respect to the prior probabilities; see Anderson (1984, Chapter 6).

In this article we are concerned with the estimation of the error rates of a given rule rather than with the design of an appropriate rule. For an account of the latter topic the reader is referred to the books devoted to discriminant analysis, which have been supplemented recently by the special issue of *Computers and Mathematics with Applications* on *Statistical Methods of Discrimination and Classification* (1986, 12A, 173-308), edited by S.C. Choi. But briefly, one way of proceeding is to adopt some parametric form for each class density $f_i(x)$, or at least for each $\xi_i(x)$ as in logistic regression, leading to a parametric form $r(x;\theta)$ for the allocation rule. The vector of unknown parameters θ under the adopted model can be estimated using, say, maximum likelihood or a Bayesian approach. For the commonly studied model of $g = 2$ normal classes with the same covariance matrix,

$$X \sim N(\mu_i, \Sigma) \quad \text{in} \quad C_i (i=1,2), \tag{2.2}$$

the former approach leads to $r(x;t)$ being defined to be 1 or 2 according as to whether the sample linear discriminant function,

$$W(x) = \{x - \tfrac{1}{2}(\bar{x}_1 + \bar{x}_2)\} \, S^{-1} (\bar{x}_1 - \bar{x}_2)',$$

is greater or less than c, where $\bar{x}_i$ denotes the sample mean of those observations from C_i and S the pooled sample covariance matrix. Apart from the cut-off point,

$$c = \log(\pi_2/\pi_1),$$

which is often taken to be zero, $W(x)$ is Fisher's linear discriminant function as modified by Anderson (1951). For this model, it can be seen that $r(x;t)$ is symmetrically defined in $z_1, \ldots, z_n$. This will generally be the case in practice, and it is implicitly assumed to be so in the subsequent work.

One nonparametric approach to the formation of a sample rule is to use the kernel method to form estimates of the class densities for use in (2.1); see Hand (1982). Note that Hall (1986) has shown that a kernel with thin tails, such as the standard normal or the double exponential, is often not a good choice when using likelihood cross-validation to choose the smoothing parameter. A suggested kernel is

$$q(x) = 0.1438 \exp[-\tfrac{1}{2}\{\log(1+|x|)\}^2] \tag{2.3}$$

in the univariate case. For multivariate data the kernel can be taken to be the product of (2.3) over the p components of x. The choice of error rate estimator is not so crucial if the rule is based on nonparametric density estimation since the sample size has to be large for the latter to be a practical exercise. And for large sample sizes, there is little difference between the apparent error rate and its nonparametric competitors.

3. Definition of Error Rates

For a given training set t, it is the actual or conditional error rates $ec_i(f_i;t)$ which are of prime concern, where

$$ec_i(f_i;t) = \mathrm{pr}\{r(X;t) \neq i \mid X \in C_i, t\} \qquad (i=1,\ldots,g)$$

is the probability that a randomly chosen member of C_i is misallocated. The expectation of $ec_i(f_i;t)$ over the sampling distribution of the training data t gives the expected or unconditional error rate for the ith class,

$$\begin{aligned} eu_i(f) &= E\{ec_i(f_i;T)\} \\ &= \mathrm{pr}\{r(X;T) \neq i \mid X \in C_i\} \qquad (i=1,\ldots,g), \end{aligned}$$

where T is the random quantity with t as a realization.

We let

$$eo_i(f) = ec_i(f_i;t_\infty),$$

be the error rate associated with $r(x;t_\infty)$, the rule that would be obtained if the size of the training set were increased to infinity. We shall refer to the $eo_i(f)$ as the optimal error rates, although $r(x;t_\infty)$ may not be optimal in the sense of being the Bayes rule.

The overall actual error rate is given by

$$ec(f;t) = \sum_{i=1}^{g} \pi_i ec_i(f_i;t)$$

and, similarly, eu(f) and eo(f) denote the overall expected and optimal error rates, respectively. Unless we specifically refer to the individual class rates in the subsequent work, we shall be implicitly referring to the overall rate in considering an error of either actual, expected, or optimal type. In practice interest in the optimal rate is limited to the extent that it represents the error of the best obtainable version of the given rule, while the expected rate provides a guide to the performance of the rule before it is actually formed.

4. Apparent Error Rate

In defining the various error rate estimators in this an the subsequent sections, we shall focus on the estimation of the actual error rate. An obvious and easily computed nonparametric estimator of the actual error rate is the apparent error rate A of r(x;t) when it is applied to the training observations in t; that is,

$$A = \sum_{j=1}^{n} Q[y_j, r(x_j;t)]/n$$

where, for any i and j, Q[i,j] = 0 for i = j and 1 for i ≠ j The apparent error rate, or the resubstitution estimator as it is often called, provides too optimistic an assessment of the actual rate, since it is obtained by applying r(x;t) to the same data used in its construction; see McLachlan (1976) for the bias of A under the normal model (2.2). More recently, Efron (1986) has investigated the bias of A under logistic regression model.

5. Cross-Validation

One way of nearly eliminating the bias of the apparent error rate is to use cross-validation, as employed by Lachenbruch and Mickey (1968). The cross-validated estimate so obtained is given by

$$A(CV) = \sum_{j=1}^{n} Q[y_j, r(x_j;t_{(j)})]/n,$$

where $t_{(j)}$ denotes t with the point $z_j = (x_j, y_j)$ deleted.

Recently, for g = 2 classes, Hand (1986a) considered shrinking A(CV) in an attempt to reduce its variance at the expense of increasing its bias. For shrinking towards the origin in the case where there is no prior knowledge of the true error rate, Hand (1986a) studied two shrunken estimators, $A(SCV) = \gamma A(CV)$, where $\gamma = n/(n+1)$ or $n/(n+3)$. In terms of mean squared error (MSE),

$$MSE\{A(SCV)\} = E\{A(SCV)-ec(f;T)\}^2,$$

where E refers to expectation over the distribution of T, Hand (1986a) compared these two estimators with A(CV) and some other estimators, using simulations for the sample linear discriminant function W(x) applied under (2.2) with the common covariance matrix taken to be known. For the combinations of the parameters considered, these two shrunken estimators were found to be superior to A(CV) although, as anticipated, they were more biased. Sometimes, however, they were more biased than A.

6. The Jackknife

It follows from Efron (1982, Chapter 7) that the jackknifed version of A appropriate for the estimation of the actual error rate is given by

$$A(J) = A + (n-1)(A^{\dagger}-A_{(\cdot)}),$$

where

$$A^{\dagger} = \frac{1}{n}\sum_{j=1}^{n}\sum_{k=1}^{n}\frac{1}{n}Q[y_k, r(x_k;t_{(j)}],$$

$$A_{(\cdot)} = \sum_{j=1}^{n} A_{(j)}/n,$$

and $A_{(j)}$ denotes the apparent error rate of $r(x;t_{(j)})$ when applied to the members of $t_{(j)}$; that is

$$A_{(j)} = \sum_{k\neq j}^{n} Q[y_k, r(x_k;t_{(j)})]/(n-1).$$

In exhibiting the close relationship between cross-validation and the jackknife methods of bias correction of the apparent error rate in its estimation of the actual rate, Efron (1982) showed that the jackknifed version of A can be expressed as

$$A(J) = A(CV) + (A-A^{\dagger}).$$

Under smoothness conditions on Q, G. Gong in an unpublished Ph.D. thesis from Stanford University showed that A(J) - A and A(CV) - A have asymptotic correlation 1. Also, Efron (1982) showed how A - A(J) can be considered as a quadratic approximation to the nonparametric bootstrap estimate of the bias of A to be defined in the next section.

As discussed by McLachlan (1986), a frequent mistake in practice is to use

$$A + (n-1)(A-A_{(\cdot)}) \tag{6.1}$$

as the jackknifed version of A for the estimation of the actual error rate ec(f,t). This version, however, is appropriate for eliminating the first order bias of A in its estimation of the optimal error rate eo(f). This misunderstanding would appear to still exist in the literature. For example, for g = 2 classes, Rao and Dorvlo (1985) used the formula corresponding to (6.1) to jackknife $G(-\tfrac{1}{2}D)$, $G(-\tfrac{1}{2}DS)$, and D for use in $G(-\tfrac{1}{2}D)$, for the purpose of estimating the optimal error rate of an individual class, but then considered these jackknifed versions as estimators also of the expected individual error rates. The estimator $G(-\tfrac{1}{2}D)$, where G denotes the standard normal distribution function and

$$D = \{(\bar{x}_1-\bar{x}_2)S^{-1}(\bar{x}_1-\bar{x}_2)'\}^{\frac{1}{2}}$$

is the sample Mahalanobis distance, is the so-called plug-in estimator under the normal model (2.2). The version $G(-\tfrac{1}{2}DS)$, where

$$DS = \{(n-p-3)/(n-2)\}^{\frac{1}{2}}D,$$

partially reduces its optimistic bias, which was derived by McLachlan (1973).

Wang (1986) used the formula (6.1) to jackknife the apparent error rate for the estimation of the expected error rate in the case of g = 2 multinomial distributions. The inappropriate choice of the jackknifed version of the apparent error rate was reflected in his simulation results, where it was generally well below that of the cross-validated estimate

7. The Bootstrap

In addition to cross-validation and the jackknife as

methods of bias correction, there is the general bootstrap methodology of Efron (1979) which can be applied as follows to correct the apparent error rate for bias in its estimation of the actual error rate. Of course the bootstrap is a very powerful methodology and it can be used to assess the variability of all aspects of the allocation rule; see McLachlan (1980, 1986).

Step 1. A new training set,

$$t^* = \{z_1^* = (x_1^*, y_1^*), \ldots, z_n^* = (x_n^*, y_n^*)\},$$

called the bootstrap sample, is generated according to $\hat{f}(z)$, an estimate of the density formed from the original training data t. That is, t* consists of the observed values of an independent and identically (i.i.d.) sample $Z_1^*, \ldots, Z_n^*$ from $\hat{f}(z)$.

Step 2. The rule r(x;t*) is formed from the bootstrap training data t* in precisely the same manner as r(x;t) was from the original set t.

Step 3. The apparent error rate of r(x;t*) is computed by noting the proportion of the members in t* misallocated by r(x;t*). Also, the difference

$$u^* = A^* - ec(\hat{f};t^*) \tag{7.1}$$

is computed.

Step 4. Let U* be the random variable defined according to (7.1). Then its expectation, the bootstrap bias of the apparent error rate, can be approximated by averaging U* over over M repeated independent realizations (say, M = 50 or 100) of bootstrap samples $t_m^*(m=1,\ldots,M)$. Thus

$$E^*(U^*) \approx \bar{u}^*,$$

where

$$\bar{u}^* = \sum_{m=1}^{M} u_m^*/M$$

and where E* refers to expectation with respect to the distribution of the bootstrap data T*, and u_m^* denotes the value of U* for the mth bootstrap realization t_m^*. This bootstrap estimate of the bias of A, b, is

$$b(B) = \bar{u}^*,$$

and so the apparent error rate corrected for bias according to the bootstrap is

$$A(B) = A - b(B).$$

In Step 1 of the above algorithm, the nonparametric version of the bootstrap would take $\hat{f}(z)$ to be the empirical probability function with mass $1/n$ at each original data point $z_j = (x_j, y_j)$ in $t(j=1,\ldots,n)$. In this case $ec(\hat{f};t^*)$ is given by

$$ec(\hat{f};t^*) = \sum_{j=1}^{n} Q[y_j, r(x_j;t^*)]/n.$$

The bootstrap estimator $b(B)$ of the bias of A has much less variability than the cross-validated estimator, but unfortunately $b(B)$ is negatively correlated with $A - ec(f;T)$. As a consequence, the MSE of $A(B)$ is inflated, although it is still, in general, less than that of $A(CV)$.

8. The 0.632 Estimator

For $g = 2$ classes, Efron (1983) developed more sophisticated variants of his ordinary bootstrap estimator of the actual error rate, including the double bootstrap which corrects its downward bias without an increase in its MSE, and the randomized bootstrap and 0.632 estimators which appreciably lower its MSE. The latter estimator, which is the most promising of the variants, is a weighted sum of the apparent error rate and ε, the bootstrap error rate at an original data point not in the training set. It is defined by

$$A(.632) = 0.368A + 0.632\varepsilon,$$

where

$$\varepsilon = \sum_{m=1}^{M} \sum_{j=1}^{n} \delta_{mj} Q[y_j, r(x_j;t_m^*)] / \sum_{m=1}^{M} \sum_{j=1}^{n} \delta_{mj}$$

and $\delta_{mj} = 1$, if x_j is not present in the bootstrap training set t_m^*, and zero otherwise. Efron (1983) developed the 0.632 estimator by consideration of the distribution of the distance δ between the point at which the rule is applied and the nearest point in the training set. The distribution of δ is quite different in the bootstrap context than in the actual situation, with the bootstrap distance having a high probability of being zero. This probability is equal to the probability that the point at which the rule is applied is included in the bootstrap sample, which is $1 - (1-1/n)^n$ and which tends to 0.632, as $n \to \infty$. In his ingenious argument, Efron (1983) showed that the points which contribute to ε (that is, those with $\delta > 0$ in the bootstrap context) are about 1/0.632 too far away from the training set than in the actual situation. This led to

$$b(.632) = 0.632(A-\varepsilon)$$

as the estimator of the bias of A, and hence

$$A(.632) = A - b(.632)$$

$$= 0.368A + 0.632\varepsilon,$$

as the bias corrected version of A.

Efron (1983) showed that ε is almost the same as A(HCV), the estimated rate obtained after a cross-validation that leaves out half of the observations at a time. Hence A(.632) is almost the same as

$$0.368A + 0.632A(HCV).$$

Estimators of this type were considered previously by Toussaint and Sharpe (1975) and McLachlan (1977) in the context of choosing the weight τ so that

$$A(\tau) = (1-\tau)A + \tau A(CVL) \qquad (8.1)$$

is an unbiased estimator of the actual error rate; A(CVL) denotes the estimated rate after cross-validation removing n/L observations at a time. For the rule based on the sample linear discriminant function W(x) with zero cut-off point, McLachlan (1977) calculated the value of τ, τ_0, for which $A(\tau)$ has zero first order bias under the normal model (2.2). For L = 2, so that A(CVL) = A(HCV), it was found that τ_0

ranged from 0.6 to 0.7 for the combinations of the parameters taken. Hence under (2.2), $A(\tau_0)$ is about the same as Efron's 0.632 estimator; see McLachlan (1986) for further discussion of this. The estimator $A(\tau)$ has been considered too by Wernecke, Kalb, and Stürzebecher (1980) and Wernecke and Kalb (1983).

Concerning the use of ε as an estimator in its own right, it has been suggested in the literature (Chatterjee and Chatterjee, 1983, page 650) that ε will be an unbiased estimator, since it is computed by applying the rule to those original data points not in the bootstrap sample. Actually, Chatterjee and Chatterjee (1983) proposed a slightly different version where ε was computed for each single bootstrap sample and then averaged over the M replications. However, it follows from the work of Efron (1983) briefly outlined above, that ε is biased upward, which has been confirmed empirically in the simulation experiments of Chernick, Murthy, and Nealy (1985, 1986a). These experiments did suggest, though, that in some instances where the error rate is high, ε is superior to A(.632).

9. Smoothing of the Apparent Error Rate

Various attempts have been made to smooth the apparent error rate with a view to reducing its variance in estimating the actual error rate. For convenience of expression we relabel the training data $x_1,\ldots,x_n$ so that $x_{ij}(j=1,\ldots,n_i)$ denote those n_i observations from $C_i (i=1,\ldots,g)$. Let h(v) denote a function ranging from zero to one, which is continuous, monotonic increasing, and satisfies

$$h(v) = 1 - h(-v)$$

over the real line. For g = 2, suppose that r(x;t) is 1 or 2 according as some statistic V(x;t) is less or greater than zero. Then a smoothed version of the apparent error rate A is given by

$$A(S) = \frac{1}{n} \sum_{i=1}^{2} \sum_{j=1}^{n_i} K_i(x_{ij}),$$

where $K_1(x) = 1 - K_2(x)$ and $K_1(x) = h(V(x;t))$.

For univariate normal and uniform distributions, Glick (1978) performed some simulations to study A(S) with h(v) having straight-line slope. Tutz (1985) proposed the logistic form for h(v),

$$h(v;\gamma) = \exp(\gamma v)/\{1+\exp(\gamma v)\},$$

where γ determines the slope of the function, and hence, the properties of the smoothed estimator A(S). For example, for $\gamma = 0$, A(S) is ½, while in the other direction, it approaches the apparent error rate as $\gamma \to \infty$. Tutz (1985) suggested that γ be based on the observed sample, by letting $\gamma = \alpha D^2$, where α is chosen to depend on the sample sizes. Tutz (1985) subsequently showed how using the single function $h(v;\gamma)$, a smoothed version of the apparent error rate for $g > 2$ classes can be defined recursively. Some simulations were performed to demonstrate the usefulness of this smoothing.

Snapinn and Knoke (1985) extended the ideas of Glick (1978) to the multivariate case. In particular, they considered a smoothed estimator for $g = 2$ by taking

$$K_1(x;\gamma) = G[\{c-W(x)\}/(\gamma D)], \tag{9.1}$$

where the smoothing constant

$$\gamma = [\{(p+2)(n_1-1)+(n_2-1)\}/\{n(n-p-3)\}]^{\frac{1}{2}}$$

was chosen so that the bias of the smoothed estimator of the actual error rate for the first class is approximately equal to that of the modified plug-in estimator $G(-\frac{1}{2}DS)$ under (2.2) for $c = 0$.

10. Posterior Probability Estimator

In many situations, the allocation rule r(x;t) is defined by (2.1) in terms of the estimated posterior probabilities. In which case a natural way of smoothing the apparent error rate is to define the smoothing function in terms of the available estimates of the posterior probabilities. The smoothed estimator so obtained is sometimes referred to as the posterior probability estimator, and is given by

$$A(PP) = \sum_{j=1}^{n} \min_i \hat{\xi}_i(x_j)/n.$$

For known posterior probabilities, A(PP) provides an unbiased estimator of the optimal error rate eo(f) of the Bayes rule, with smaller variance than the apparent error rate A (Fukunaga and Kessell, 1973); see also Schwemer and Dunn (1980) for class specific results. For $g = 2$, Matloff and Pruitt (1984) have given sufficient conditions on the form of the posterior probabilities for $\sqrt{n}\{A(PP)-ec(f,T)\}$ to have a distribution converging to the normal with mean zero and finite variance.

Empirical evidence (for example, Glick, 1978 and Hora and Wilcox, 1982) suggests that A(PP) has smaller mean squared error than A in the estimation of the actual error rate, but it is biased in the same direction as A. The posterior probability estimator is closely related to the plug-in estimator (Ganesalingam and McLachlan, 1980). The component of the bias of A(PP) due to using the estimated posterior probabilities of the training observations rather than independent data in its formation can be reduced to the second order by cross-validation. Let

$$A(CVPP) = \sum_{j=1}^{n} \min_i \hat{\xi}_i(x_j;t_{(j)})/n.$$

Ignoring terms of the second order, A(CVPP) will have the same bias as A(PP) formed using the estimated posterior probabilities of n observations independent of the training data. The coefficient of 1/n in the leading term of the asymptotic bias of the latter estimator was derived by Ganesalingam and McLachlan (1980) for the posterior probabilities and consequent rule formed parametrically for $g = 2$ under the normal model (2.2). Their tabulated results for a variety of the combinations of π_1, p, and the Mahalanobis Δ, suggest that the bias of A(CVPP) is still of practical concern for small Δ or large p relative to n. Further, their results suggest that A(CVPP) will provide a more optimistic assessment if a logistic model were adopted rather than directly using the appropriate normal model in obtaining estimates of the posterior probabilities. An obvious way of correcting A(PP) for bias is to use the bootstrap. Another consideration is the weighting of the posterior probability estimator A(PP) with an estimator biased in the opposite direction. Hand (1986a) reported some encouraging results for A(PP) weighted with A(CV) with arbitrarily chosen weights 0.1 and 0.9 respectively.

An advantage of the posterior probability estimator is that it does not require knowledge of the origin of the data in its formation, and so is particularly useful when data of known origin are in short supply. Recently, Basford and McLachlan (1985) made use of this to propose an estimator of the same form as A(PP) for assessing the performance of a clustering rule in the case where all the available observations are unclassified with respect to the underlying classes. The bootstrap was found to be quite effective in reducing the optimistic bias of the assessment so obtained.

11. Further Conclusions of Simulation Experiments

Most of the work in comparing the error rate estimators has concentrated on classes with normal distributions and then essentially for Fisher's linear discriminant function applied in the case of $g = 2$ classes with the same covariance matrix. In this context the simulation results of Efron (1983) and Chernick et al. (1985, 1986a) on nonparametric error rate estimators clearly suggest in terms of MSE that the 0.632 estimator is superior to its competitors in those situations where bias correction of the apparent error rate is warranted. In the latter study, which also considered $g = 3$ classes, some new variants of the bootstrap were proposed, including the convex bootstrap which takes convex combinations of neighbouring training observations in the resampling. Preliminary results by Chernick et al. (1986b) for $g = 2$ and 3 nonnormal classes led to similar conclusions as in the case of normality. However, they caution about the use of the 0.632 estimator for heavy-tailed distributions such as the Cauchy, as their simulations suggest that ε does not have its usual positive bias then to compensate for the downward bias of the apparent error rate.

For complicated allocation rules overfitting is a real danger, resulting in a grossly optimistic apparent error. Gong (1986) reported some results for simulations and real data for a moderately complicated rule formed using forward logistic regression. The bootstrap was found to offer a significant improvement over cross-validation and the jack-knife for estimation of the bias of the apparent error and for providing a bias corrected estimate. For the same purposes Wang (1986) also found the bootstrap to be the best of these three methods for some simulations performed for an allocation rule in the case of two multinomial classes.

The above simulation experiments did not include any smoothed estimators. For $g = 2$ classes with nonnormal distributions as well as the normal, Snapinn and Knoke (1985) compared their normally smoothed estimator (9.1) and one with uniform smoothing with the apparent error rate and its cross-validated version A(CV), and the ideal constant estimator, $A(IC) = A - b$. They concluded that the normally smoothed estimator has smaller MSE than A(IC) if the sample size is sufficiently large relative to p. An estimator worthy of further investigation is the posterior probability estimator corrected for bias according to the bootstrap. If it were formed using the appropriate parametric forms for the class densities or the posterior probabilities, then it should be superior or at least comparable to its competitors in terms of MSE.

Department of Mathematics,
University of Queensland,
St. Lucia, Q. 4067, Australia.

References

Anderson, T.W. (1951). 'Classification by multivariate analysis'. Psychometrika 16, 31-52.

Anderson, T.W. (1984). 'An Introduction to Multivariate Statistical Analysis'. New York: Wiley.

Basford, K.E. and McLachlan, G.J. (1985). 'Estimation of allocation rates in a cluster analysis context'. J. Amer. Statist. Assoc. 80, 286-293.

Chatterjee, S. and Chatterjee, S. (1983). 'Estimation of misclassification probabilities by bootstrap methods'. Commun. Statist.-Simula. Computa. 12, 645-656.

Chernick, M.R., Murthy, V.K. and Nealy, C.D. (1985). 'Application of bootstrap and other resampling techniques: evaluation of classifier performance'. Pattern Recognition Letters 3, 167-178.

Chernick, M.R., Murthy, V.K., and Nealy, C.D. (1986a). 'Correction note to Application of bootstrap and other resampling techniques: evaluation of classifier performance'. Pattern Recognition Letters 4, 133-142.

Chernick, M.R., Murthy, V.K., and Nealy, C.D. (1986b). 'Estimation of error rate for linear discriminant functions by resampling: Non-Gaussian populations'. Unpublished manuscript.

Efron, B. (1979). 'Bootstrap methods: another look at the jackknife'. Ann. Statist. 7, 1-26.

Efron, B. (1982). 'The Jackknife, the Bootstrap, and Other Resampling Plans'. Philadelphia: SIAM.

Efron, B. (1983). 'Estimating the error rate of a prediction rule: improvement on cross-validation'. J. Amer. Statist. Assoc. 78, 316-331.

Efron, B. (1986). 'How biased is the apparent error rate of a logistic regression?' J. Amer. Statist. Assoc. 81, 461-470.

Fukunaga, K. and Kessell, D.L. (1973). 'Nonparametric Bayes error estimation using unclassified samples'. IEEE Trans. Inf. Theory IT-19, 434-440.

Ganesalingam, S. and McLachlan, G.J. (1980). 'Error rate estimation on the basis of posterior probabilities'. Pattern Recognition 12, 405-413.

Glick, N. (1978). 'Additive estimators for probabilities of correct classification'. Pattern Recognition 10, 211-222.

Gong, G. (1986). 'Cross-validation, the jackknife, and the bootstrap: excess error estimation in forward logistic regression'. J. Amer. Statist. Assoc. 81, 108-113.

Hall, P. (1986). 'Cross-validation in nonparametric density estimation'. Proc. XIIIth Int. Biometric Conference, 15pp. Seattle: Biometric Society.

Hand, D.J. (1982). 'Kernel Discriminant Analysis'. Chichester: Wiley.

Hand, D.J. (1986a). 'Cross-validation in error rate estimation'. Proc. XIIIth Int.Biometric Conference, 15pp. Seattle: Biometric Society.

Hand, D.J. (1986b). 'Recent advances in error rate estimation'. Pattern Recognition Letters (to appear).

Hills, M. (1966). 'Allocation rules and their error rates'. J. R. Statist. Soc. B 28, 1-31.

Hora, S.C. and Wilcox, J.B. (1982). 'Estimation of error rates in several-population discriminant analysis'. J. Marketing Res. 19, 57-61.

Knoke, J.D. (1986). 'The robust estimation of classification error rates'. Comp. & Math. with Appls. 12A, 253-260.

Krzanowski, W.J. (1975). 'Discrimination and classification using both binary and continuous variables'. J. Amer. Statist. Assoc. 70, 782-792.

Lachenbruch, P.A. and Mickey, M.R. (1968). 'Estimation of error rates in discriminant analysis'. Technometrics 10, 1-11.

Matloff, N. and Pruitt, R. (1984). 'The asymptotic distribution of an estimator of the Bayes error rate'. Pattern Recognition Letters 2, 271-274.

McLachlan, G.J. (1973). 'An asymptotic expansion of the expectation of the estimated error rate in discriminant analysis'. Austral. J. Statist. 15, 210-214.

McLachlan, G.J. (1974a). 'Estimation of the errors of misclassification on the criterion of asymptotic mean square error'. Technometrics 16, 255-260.

McLachlan, G.J. (1974b). 'The relationship in terms of asymptotic mean square error between the separate problems of estimating each of the three types of error rate of the linear discriminant function'. Technometrics 16, 569-575.

McLachlan, G.J. (1974c). 'An asymptotic unbiased technique for estimating the error rates in discriminant analysis'. Biometrics 30, 239-249.

McLachlan, G.J. (1976). 'The bias of the apparent error rate in discriminant analysis'. Biometrika 63, 239-244.

McLachlan, G.J. (1977). 'A note on the choice of a weighting function to give an efficient method for estimating the probability of misclassification'. Pattern Recognition 9, 147-149.

McLachlan, G.J. (1980). 'The efficiency of Efron's bootstrap approach applied to error rate estimation in discriminant analysis'. J. Statist. Comput. Simul. 11, 273-279.

McLachlan, G.J. (1986). 'Assessing the performance of an allocation rule'. Comp. & Maths. with Appls. 12A, 261-272.

Page, J.T. (1985). 'Error-rate estimation in discriminant analysis'. Technometrics 27, 189-198.

Rao, P.S.R.S. and Dorvlo, A.S. (1985). The jackknife procedure for the probabilities of misclassification'. Commun. Statist.-Simula. Computa. 14, 779-790.

Schwemer, G.T. and Dunn, O.J. (1980). 'Posterior probability estimators in classification simulations'. Commun. Statist.-Simula. Computa. B9, 133-140.

Snapinn, S.M. and Knoke, J.D. (1984). 'Classification error rate estimators evaluated by unconditional mean squared error'. Technometrics 26, 371-378.

Snapinn, S.M. and Knoke, J.D. (1985). 'An evaluation of smoothed classification error-rate estimators'. Technometrics 27, 199-206.

Toussaint, G.T. (1974). 'Bibliography on estimation of misclassification'. IEEE Trans. Inf. Theory IT-20, 472-479.

Toussaint, G.T. and Sharpe, P.M. (1975), 'An efficient method for estimating the probability of misclassification applied to a problem in medical diagnosis'. Comput. Biol. Med. 4, 269-278.

Tutz, G.E. (1985). 'Smoothed additive estimators for non-error rates in multiple discriminant analysis'. Pattern Recognition 18, 151-159.

Vlachonikolis, I.G. (1986). 'Estimation of the expected probability of misclassification'. Comp & Maths. with Appls. 12A, 187-195.

Wang, M-C. (1986). 'Re-sampling procedures for reducing bias of error rate estimation in multinomial classification'. Comput. Statist. & Data Analysis 4, 15-39.

Wernecke, K-D and Kalb, G. (1983). 'Further results in estimating the classification error in discriminance analysis'. Biom. J. 25, 247-258.

Wernecke, K-D, Kalb, G., and Stürzebecher, E. (1980). 'Comparison of various procedures for estimation of the classification error in discriminance analysis'. Biom. J. 22, 639-649.

Govind S. Mudholkar and Perla Subbaiah

SOME SIMPLE OPTIMAL TESTS IN MULTIVARIATE ANALYSIS

ABSTRACT

The class of tests reviewed in this paper pertains mainly to standard hypothesis testing problems in multivariate analysis. The tests are simple in that they are combinations of familiar univariate tests. In the multivariate context they are asymptotically equivalent to the B-optimal likelihood ratio tests; and even in small to moderate size samples they compare well with the B-optimal tests in terms of the power. Their construction and optimality are reviewed using common problems such as MANOVA, testing complete independence with a special emphasis on the sphericity hypothesis. The applicability of the method of test construction for developing solutions to less standard problems, for example those involving missing data or restricted parametric spaces, is outlined. Its possible role in constructing robust tests is indicated.

1. INTRODUCTION

As in the history of univariate statistics the development of methods for analyzing multiresponse data started with convenient multivariate normal assumptions for the response vectors. Now, because of its orderliness, simplicity and mathematical elegance the multivariate normal theory dominates this sector of the education of statisticians, and availability of ready tools, for example, software packages, have made the associated methods dominant in the statistical practice. The bulk of nonnormal theory portion of multivariate analysis consists of nonparametric rank methods, and a small residual part of the literature is on the robust methods. For a thoughtful perspective on the subject see Sen (1986).

The reasons underlying the central role played by the normal assumption in statistics are many and well known. One is the analytical simplicity and computational ease of the maximum likelihood estimates and the likelihood ratio tests in the numerous problems encountered at the introductory level. However, these fortuitous advantages disappear quickly if the parametric space is restricted in any way. A more basic and obvious reason is the central limit effect which is used in the large sample justification of simple normal theory solutions and implementation of nonparametric

A. K. Gupta (ed.), Advances in Multivariate Statistical Analysis, 253–275.

methods in non-normal case. In a similar manner the univariate normal theory has served as a take off point in the development of many univariate robust methods, e.g., those based upon the trimmed means and Winsorized standard deviations. However this course of growth towards robust methods has not yet significantly materialized in the multivariate analysis. The possible reasons for this include the natural fascination for invariance, desire to be coordinate free and the intrinsic indeterminacies and difficulties inherent in multidimensional mathematics.

The purpose of this paper is to review an approach for constructing multivariate tests which works well for most commonly considered hypothesis testing problems in the normal theory and can also be applied to many less common problems, e.g., those involving restricted parameter spaces and missing data, and holds promise for developing robust methods. This approach which leads to simple and powerful tests without posing additional null distribution problems has its roots in the stepwise methods initiated by C. R. Rao (1948), S. N. Roy and Bargmann (1958), J. Roy (1958), and in the techniques for combining independent tests initiated by Fisher (1932) and studied further by many including Oosterhoff (1969), Littell and Folks (1973), Mudholkar and George (1979). The status of likelihood ratio tests in the normal theory is summarized in Section 2. The stepdown methods are outlined in Section 3. The theory of combining independent tests is discussed in Section 4. In Section 5 the modification of stepwise procedure based on combining tests of significance and their optimality are outlined. In Section 6, a special case, that of the sphericity hypothesis, is examined in some detail and the results of a simulation study for estimating the power function and mean significance probabilities are given. The conclusions and some directions of work in progress and future possibilities are indicated in the final section.

2. LIKELIHOOD RATIO TESTS

A large body of early work in multivariate analysis, after the introduction of the D^2, T^2 tests by Mahalanobis and Hotelling, was devoted to a systematic construction of the likelihood ratio tests for a broad family of hypotheses regarding multivariate normal populations by S. S. Wilks and development of their null distrubitons. It was followed by investigations involving the power functions and optimality properties such as unbiasedness and invariance of these tests. In the fifties S. N. Roy and his associates, using a method of test construction now known as the Union-Intersection principle, introduced competitors to the likelihood ratio tests. These alternatives had an advantage over the original tests in that they provided simultaneous confidence regions which could be used for post-hoc analyses. See Mudholkar, Davidson, and

Subbaiah (1974 a,b), Krishnaiah, Mudholkar, and Subbaiah (1980), Subbaiah and Mudholkar (1981). Now it is recognized that these tests are among invariant tests based on multidemensional maximal invariants resulting from the invariance reductions (Anderson (1984), Lehmann (1959)) of the testing problems. Actually many other tests including those proposed by Pillai (1955) which use symmetric functions of the maximal invariants as the test statistic are still current in practice. Many of these tests have desirable properties such as monotone power functions (see Das Gupta, Anderson, and Mudholkar (1964), Mudholkar (1965)); but none dominates the others in the Neyman Pearson framework. For a decision theoretic optimality properties see Kiefer (1966), Schwartz (1966) and references there in.

It is now well known that in terms of a method of test comparison introduced by Bahadur (1960, 1967) the likelihood ratio tests are in general optimal. Clearly, for any specified alternative, the significance probability P-value of any reasonable test should converge to zero as the sample size increases to infinity. The faster exponential rate of decay of the P-value indicates a better test. Specifically, let $T^{(n)}$ be a statistic based on n observations for testing $H_0: \theta \varepsilon \omega$ against $H_1 : \theta \varepsilon \Omega - \omega$, ω being a subset of parametric space Ω, such that $T^{(n)}$ has a continuous c.d.f. $F_\theta^{(n)}$, and large values of $T^{(n)}$ indicate significance. Also assume that the null distribution of $T^{(n)}$ does not depend on $\theta \varepsilon \omega$, i.e., for $\theta \varepsilon \omega$, $P_\theta (T^{(n)} < t) = F_0^{(n)}(t)$ and P-value is $P^{(n)} = 1 - F_0^{(n)}(t)$.

DEFINITION 2.1. *The exact slope* $c(\theta)$ *of* $\{ T^{(n)} \}$ *with P-values* $\{ P^{(n)}\}$ *is defined as*

$$c(\theta) = \lim_{n \to \infty} (- 2/n) \log P^{(n)}, \tag{2.1}$$

for $\theta \varepsilon \Omega - \omega$.

The computation of exact slope is simplified by the following result due to Bahadur (1971, p.27).

THEOREM 2.2. *Suppose that*

$$\lim_{n \to \infty} T^{(n)}/\sqrt{n} = b(\theta) \qquad (2.2)$$

a.s. for each $\theta \in \Omega - \omega$, *where* $-\infty < b(\theta) < \infty$, *and that*

$$\lim_{n \to \infty} n^{-1} \log(1 - F_0^{(n)}(\sqrt{n}\; t)) = -f(t) \qquad (2.3)$$

for each t *in an open interval* I, *where* f *is a continuous function on* I *and* $\{ b(\theta) : \theta \in \Omega - \omega \} \subset I$. *Then (2.1) holds with* $c(\theta) = 2 f(b(\theta))$ *for each* $\theta \in \Omega - \omega$.

It is well known (Bahadur (1967)) that the maximum exact slope can be expressed as a function of Kullback-Leibler information number $I(\theta, \theta_0) = E_\theta \{ \log [f(\mathbf{X};\theta) / f(\mathbf{X};\theta_0)] \}$, for $\theta \in \Omega$ and $\theta_0 \in \omega$, where $f(\mathbf{X};\theta)$ is the p.d.f. of $\mathbf{X}$. Specifically

$$c(\theta) \le 2 J(\theta) \qquad (2.4)$$

where $J(\theta) = \inf \{ I(\theta, \theta_0) : \theta_0 \in \omega \}$. Bahadur (1967), Bahadur and Raghavachari (1972) discuss the general conditions for the likelihood ratio to attain the optimal slope $c(\theta) = 2 J(\theta)$ which can be essentially stated as follows:

Suppoose $L(\hat{\omega})$ and $L(\hat{\Omega})$ denote the maximum likelihood functions under ω and Ω respectively, and $T^{(n)} = -n^{-1} \log (L(\hat{\omega}) / L(\hat{\Omega}))$. Then $\{ T^{(n)} \}$ has optimal slope if (i) for $\theta \in \Omega - \omega$, $\lim\inf T^{(n)} \ge J(\theta)$ a.e. P_θ, and (ii) $\lim\sup n^{-1} \log P_{\theta 0} [T^{(n)} \ge t] \le -t$ holds for any $t > 0$ and $\theta_0 \in \omega$. Hsieh (1979a) refines these conditions and shows that the optimal slope is attained under conditions (i) and (ii)' where (ii)' is as follows: (ii)'. The null distribution of $L(\hat{\omega}) / L(\hat{\Omega})$ is same as $W^{(n)} = [K^{(n)}]^{-1} [\prod Z_{1i}^{a_i(n)} (1 - Z_{1i})^{b_i(n)}] \prod Z_{2j}^{c_j(n)}$, Z_{ij}'s being distributed as independent beta variables such that p.d.f. of Z_{ij} has parameters depending on n as on p.594 of Hsieh (1979a). Furthermore Hsieh shows that under the multivariate normality assumption, the likelihood ratio statistic for a number of problems such as MANOVA, equality of covariance matrices, independence of sets of variates, sphericity of a covariance matrix attains the optimal slope. Moreover, for the MANOVA problem, Hsieh (1979b) provides a detailed comparison of several invariant procedures in terms of the exact slopes.

3. STEPWISE PROCEDURES

The stepwise procedures in multivariate analysis intrinsically present in Rao (1948, 1956) and formally inunciated by Roy and Bargmann (1958), were extended to various problems by J. Roy (1958), and generalized further by Dempster (1963). A version of this which appears in Anderson (1958) is further developed in Anderson (1984). For a review of the stepwise tests for MANOVA problem see Mudholkar and Subbaiah (1980). In the stepwise approach a nested decomposition $H_{01} \supset H_{02} \supset \ldots \supset H_{0k}$ of the null hypothesis H_0 is identified, so that $H_0 = \cap H_{0i}$. The hypotheses H_{0i} in the nested hierarchy are then tested sequentially using statistics T_i with independent null distributions. The nested decomposition and the independence of the test statistics can result variously.

For example, an a priori ordering of the p responese variates $Y' = (Y_1, \ldots, Y_p)$ in the MANOVA probem into k subsets Y_i of p_i dimension, $p_1 + \ldots + p_k = p$, induces a hierarchical decomposition of the MANOVA hypothesis H_0 into k subhypotheses H_{0i}, each a multivariate general linear hypothesis regarding the conditional mean of Y_i given the variables in the earlier sets Y_j, $j = 1, 2, \ldots, (i-1)$. The subhypothesis H_{0i} may be tested using any MANOVA test statistic such as Roy's largest root, Hotelling-Lawley's trace, Pillai's trace criterion. However, the use of likelihood ratio at every stage in the hierarchy would be the optimal choice in the sense of Section 2. In view of the nature of the conditional distributions of Y_i, given Y_j, $j = 1, 2, \ldots, (i-1)$, the k statistics for the subhypotheses H_{0i}, are independent under H_0. For details see Mudholkar and Subbaiah (1975), Subbaiah and Mudholkar (1982). Note that the ordering of the variables as mentioned above can occur if p_1 of the respones are of primary importance in the investigation, but p_2 more respones are measured as add-ons.

Another example of an a priori ordering occurring naturally among the response variables, which induces a suitable decomposition of H_0, is the missing data situation considered by Rao (1956) or Kleinbaum (1973). The problem discussed by Rao may be described in terms of a p-variate general linear model in which all the p responses are measured on n_1 of the n observations; but only first p_1 of the p measurements exist in the remaining $n_2 = n - n_1$ data points, the other $p_2 = p - p_1$, being missing. A MANOVA hypothesis H_0 in this case can be decomposed into H_{01}, H_{02} where H_{01} is the restriction of H_0 to the first p_1 variables, and H_{02} is in terms of the analysis of covariance model with the first n_1 observations, the p_1 variables being the covariates to the remaining p_2 which are considered to be the responses. Rao's solution then amounts to computing the likelihood ratio statistics for the two hypotheses, showing them independent and taking their product as the test statistic for H_0; Kleinbaum (1973), on the other hand, considers the missing data, which might happen by design or disaster in a growth curve model. He proposes a BAN estimator together with an iteractive computational scheme for the parameters of this model labelled GGCM, and then develops the Wald statistic for testing a general linear hypothesis in this situation. It is well known that a growth curve model testing problem can be reduced to a MANOVA problem. Hence with missing data, the above approach can be adapted for obtaining the hierarchical decomposition and the independent test statistics.

As a third example consider the class of problems involving covariance matrices discussed by Anderson (1984). In a typical problem, for example that of sphericity hypothesis, H_0: $\Sigma = \sigma^2 \mathbf{I}$, he decomposes the null hypothesis as $H_{01} \cap H_{02}$, where $H_{01} : \sigma_{ij} = 0, \ i \neq j$, the hypothesis of multiple independence and $H_{02} : \sigma_{ij} = 0, \ i \neq j$; and $\sigma_{11} = \ldots = \sigma_{pp}$. Anderson shows that in this situation, the likelihood ratio λ for H_0 is the product of the likelihood ratio statistics λ_1, λ_2 for simpler hypotheses H_{01}, H_{02} and that λ_1 and λ_2 are independent. He (1958) uses the result to obtain λ and its distribution; and in 1984 incorporates that into a stepwise procedure. In a later section we shall discuss this problem in some detail.

In the traditional stepwise procedures, having obtained the independent tests for the component hypotheses, the type I error probability α of the test for the overall null hypothesis H_0 is controlled by conducting the individual tests at level α_i such that $\alpha = 1 - \prod (1 - \alpha_i)$, and rejecting H_0 if at least one of the components is rejected. In Rao's (1956) and Anderson's (1958) solution the two independent statistics are multiplied to obtain the likelihood ratio statistic and its distribution under the overall null hypothesis . In Section 5 we discuss some alternative procedures.

4. COMBINATION OF INDEPENDENT TESTS

The problem of combining independent tests arises at the conclusion of an enquiry into a scientific hypothesis which consists of several independent investigations differing in time, space and quantitative or qualitative aspects of design. It becomes crucial if the individual studies are only marginally conclusive and pooling of the diverse pieces of evidence becomes imperative. In this section we state the combination problem and summarize the optimality results related to the various combination methods discussed in the literature. Refer Oosterhoff (1969), Littell and Folks (1973), Mudholkar and George (1979), Berk and Cohen (1979). The B-optimality results discussed in these papers assume that the test statistics are independently distributed. However these results are valid even in the present framework. i.e., when the test statistics are independent only under H_0.

Let T_i denote the test statistic for testing H_{0i}, and P_i the associated P-value, $i = 1, \ldots, k$. Some commonly used combination methods for testing $H_0 = \cap H_{0i}$ are (i) $\psi_T = -\min \{P_i\}$ due to Tippett, (ii) $\psi_F = \Sigma\ -2 \log P_i$ due to Fisher, (iii) $\psi_N = \Sigma\, \Phi^{-1}(1 - P_i)$ considered by Liptak (1958), where Φ denotes the standard normal c.d.f., and (iv) $\psi_L = \Sigma - \log (P_i/(1-P_i))$ introduced in George (1977), Mudholkar and George (1979). Under H_0, ψ_F is a χ^2_{2k} variable, ψ_N is the N(0,k) - variable, ψ_T is distributed as negative of the smallest uniform order statistic, and ψ_L is distributed as a sum of k i.i.d. logistic variables. The null distribution of ψ_L can be computed exactly or well approximated by a multiple $a \cdot t_\upsilon$ of the student's t with υ d.f., where $a = \pi \{ k (5k + 2) / (3 (5k + 4)) \}^{1/2}$ and $\upsilon = 5k + 4$.

Many combination methods considered for testing H_0 are of the form

$$\psi (P_1 \ldots, P_k) = \Sigma\ H^{-1} (1 - P_i) \qquad \textbf{(4.1)}$$

where H is a convenient distribution function. When H_0 is true P_i's are i.i.d. uniform (0,1) variables, and ψ is distributed as k-fold convolution G^{k*} of the distribution function G. Another class of combination methods introduced by Lancaster (1961) is generated by functions of the form

$$\psi\,(P_1, \ldots, P_k) = \Sigma\, G_i^{-1}\,(1 - P_i) \tag{4.2}$$

where G_i is a χ^2 c.d.f. with ν_i d.f. , which allows different transformations of P-values, reflecting varying levels of importance. The null distribution of ψ is χ^2 with $\Sigma\upsilon_i$ d.f.

For the problem of combining independent tests Littell and Folks (1973), discuss the optimality of the Fisher's method in a class of monotone combination methods M $(P_1, \ldots, P_k)$. A combination method ψ belongs to M if rejection of H_0 for $(P_1, \ldots, P_k)$ also implies its rejection for $(P_1', \ldots, P_k')$ provided $P_i' \le P_i$, $i = 1, \ldots, k$. Littell and Folks (1973) show that the exact slope c_ψ of ψ for any $\psi \,\varepsilon\, M$, cannot exceed the exact slope c_F of the Fishers method. i.e.,

$$c_\psi\,(\,\theta\,) \;\le\; c_F\,(\,\theta\,). \tag{4.3}$$

Mudholkar and George (1979) show that the exact slopes of the logit method ψ_L and the Fisher's method ψ_F are same. Berk and Cohen (1979) discuss the class of methods with the same exact slope as ψ_F. In particular, they show that the combination methods belonging to the class considered by Lancaster (1961) given in (4.2) are B-optimal.

Let $B(T_1, \ldots, T_k)$ denote the class of combination methods that are B-optimal. i.e., $\psi \,\varepsilon\, B$ if $c_\psi = c_F$. If we consider $T_i = -2 \log \lambda_i$, λ_i being the likelihood ratio for H_{0i}, then it follows from (4.3), that $c_\lambda \le c_F$, where c_λ is the exact slope of $\psi_\lambda = -2 \log \lambda = \Sigma\, -2 \log \lambda_i$. Moreover, if the likelihood ratio λ satisifes the conditions discussed in Section 2, and attains the optimal slope, then $c_\lambda = c_F$. This can be summarized as follows:

THEOREM 4.1. *Suppose $T_i = -2 \log \lambda_i$, $i = 1, \ldots, k$, are independent under H_0. If the likelihood ratio test ψ_λ is B-optimal, then any combination method $\psi \,\varepsilon\, B$ is asymptotocially equivalent to the likelihood ratio test. i.e, $c_\psi = c_\lambda = c_F$.*

In the next section we discuss the modified stepwise procedures and their optimality, as a consequence of the above theorem, and their application for several common problems related to multivariate normal populations.

5. MODIFIED PROCEDURES - B-OPTIMALITY

In the outline of the stepwise procedures in Section 3, we had a hierarchical decomposition of the null hypothesis H_0 such that the components H_{0i} could be tested at level α_i using independent test statistics, subject to $\alpha = 1 - \prod (1 - \alpha_i)$, the overall type I error. The consequences of distributing overall type I error in this manner were examined by Mudholkar and Subbaiah (1976) in the context of the simple Hotelling's T^2 problem. It was found that in such a scheme, which in case of equal α_i amounts to the combining independent tests according to Tippett's method, high power is obtained in one direction at a substantial sacrifice of the power in the remaining directions. This unsatisfactory method was improved by combining the tests using some B-optimal procedures (Mudholkar and Subbaiah (1980)). The resulting procedures are shown to be optimal and their finite sample properties are seen to be comparable to those of the likelihood ratio test. This is also true for the types of solutions in Section 3 attributed to Anderson and Rao. We may combine the independent P-values of the component tests instead of combining the tests by taking the product of the likelihood ratio statistics of the component tests as the overall test statistic. In the general case, the above observation holds if the component hypotheses are tested using likelihood ratio statistics which are independent under H_0, and satisfy the conditions mentioned in Section 2. In this section we summarize some previous investigations and present a detailed evidence in the context of sphericity in the next section.

5.1. **Hotellings T^2 and MANOVA Problems.** Consider the MANOVA model for the data matrix **Y** (nxp). The rows of **Y** are independently distributed as p-variate normal with dispersion matrix Σ and means given by $E(Y) = \mathbf{A\Theta}$ where **A**(nxm) is a known matrix of rank m. The MANOVA problem is that of testing H_0 $\mathbf{\Phi} = \mathbf{B}\ \mathbf{\Theta} = 0$ where **B**(txm) is a given matrix of rank t. The invariant tests for this problem are based on the eigenvalues of $\mathbf{HE}^{-1}$, where $\mathbf{H}$ and $\mathbf{E}$ are "hypothesis" and "error" sum of squares and products matrices. The stepdown procedure of J. Roy (1958) involves in decomposing H_0 as $H_0 = \cap \{H_{0i} : \eta_i = 0\}$, where η_i are the parameters in the conditional distribution of $\mathbf{Y}_i$ given $\mathbf{Y}_{(i-1)} = (\mathbf{Y}_1, \ldots, \mathbf{Y}_{i-1})$, $i = 1,2, \ldots, p$. The hypotheses H_{0i} are tested sequentially using the likelihood ratio statistics or equivalently the F statistics F_i, $i = 1, \ldots, p$, which are independently distributed under H_0. The modified stepwise procedure then involves combining P-values associated with these tests. Refer Mudholkar and Subbaiah (1986) for details.

The Hotelling's T^2 problem, a particular case of MANOVA, has been studied by Mudholkar and Subbaiah (1980). It is shown that the Fisher's combination method ψ_F is B-optimal, and the Tippett's method is suboptimal. The power functions for Hotelling's T^2, ψ_F and ψ_T estimated from a simulation study confirmed these results even with $n = 20$. The MANOVA problem has been discussed by P. K. Sen (1983), and Mudholkar and Subbaiah (1986). In order to describe the exact slopes, some additional notation is needed. Let $\mathbf{H}/n \to \mathbf{K}$ and $\mathbf{E}/n \to \Sigma$ as $n \to \infty$. Let $l_1 \geq \ldots \geq l_p$ denote the eigenvalues of $\mathbf{HE}^{-1}$. Then the exact slope of the likelihood ratio test is $c_\lambda = \prod(1 + \lambda_i) = \log(|\mathbf{K} + \Sigma| / |\Sigma|)$. The exact slopes of Roy's largest root $R = l_1$, Hotelling- Lawley's trace $T = \sum l_i$ are, respectively, $c_R = \log(1 + \lambda_1)$ and $c_T = \log(1 + \sum \lambda_i)$. It is easy to see that $c_R \leq c_T \leq c_\lambda$. If $\mathbf{K}\Sigma^{-1}$ is of rank 1, then $c_R = c_T = c_\lambda$. If all eigenvalues are equal, then $c_R < c_T < c_\lambda$. Mudholkar and Subbaiah (1986) showed that $c_\lambda = c_F = c_L$.

5.2. **Complete Independence.** Let $X_1, \ldots, X_N$ be a random sample from $N_p(\mu, \Sigma)$. The problem of interest is that of testing $H_0 : \Sigma$ is diagonal. i.e., $\sigma_{ij} = 0$, for $i \neq j$. The stepdown procedure of Roy and Bargmann (1958) involves decomposing H_0 as $H_0 = \cap \{H_{0i}: \rho^2_{i.12\ldots(i-1)} = 0\}$, where $\rho^2_{i.12\ldots(i-1)}$ is the multiple correlation between X_i and $(X_1, \ldots, X_{i-1})$, $i = 2, \ldots p$. The likelihood ratio test for H_{0i} is equivalent to $F_i = (N-i)R^2_{i.12\ldots(i-1)} / (i-1)(1-R^2_{i.12\ldots(i-1)})$, $i = 2, \ldots, p$ which are independently distributed under H_0 as F with d.f. (i-1, N-i). It is shown by Mudholkar and Subbaiah (1981) that $c_\lambda = c_F = -\log|\boldsymbol{P}|$, where $|\boldsymbol{P}|$ denotes the determinant of population correlation matrix. A simulation study comparing the power functions for the likelihood ratio test, Nagao's test, Fisher and logit combination methods is also conducted.

5.3. **Testing equality of k multivariate normal populations.** Let X_{ij}, $j = 1, \ldots, N_i$ denote a random sample from $N_p(\mu_i, \Sigma_i)$, $i = 1, \ldots, k$. Also let $n_i = N_i - 1$, $N = N_1 + \ldots + N_k$, $n = N - k$, and $N_i/N \to \gamma_i$ as $N \to \infty$. Thus $0 < \gamma_i < 1$ and $\gamma_1 + \ldots + \gamma_k = 1$. The hypothesis of interest $H_0 : \mu_1 = \ldots = \mu_k$ and $\Sigma_1 = \ldots = \Sigma_k$ can be expressed as $H_0 = H_{01} \cap H_{02}$, where $H_{01} : \Sigma_1 = \ldots \Sigma_k$ and $H_{02} : \mu_1 = \ldots = \mu_k$ given $\Sigma_1 = \ldots = \Sigma_k$. Then it is well known (Anderson (1984)) that the likelihood ratios λ, λ_1, λ_2 for testing H_0, H_{01}, H_{02}, respectively, can be expressed as $\lambda = \lambda_1 \lambda_2$. The exact slopes of these tests are $c_\lambda = \log \{ | \Sigma \gamma_i (\Sigma_i + (\mu_i - \mu)(\mu_i - \mu)' | / \Pi | \Sigma_i |^{\gamma_i} \}$, $c_1 = \log \{ | \Sigma \gamma_i \Sigma_i | / \Pi | \Sigma_i |^{\gamma_i} \}$ and $c_2 = \log \{ | I + \Sigma \gamma_i (\mu_i - \mu)(\mu_i - \mu)' \Sigma^{-1} | \}$, where $\mu = \Sigma \gamma_i \mu_i$ and $\Sigma = \Sigma \gamma_i \Sigma_i$. Then it is easy to see that $c_\lambda = c_1 + c_2$. Moreover, $c_\lambda = c_F$.

6. TESTING SPHERICITY

The hypothesis of sphericity in multivariate analysis amounts to postulating the covariance structure of the response variables to be known up to a constant. i.e., $\Sigma = \sigma^2 \psi_0$, which ψ_0 is known. In the canonical form the problem is that of testing $H_0 : \Sigma = \sigma^2 I$ on the basis of a random sample X_i, $i = 1, \ldots, N$ from $N_p(\mu, \Sigma)$. The likelihood ratio test for the problem was obtained by Mauchley (1940) who also provided approximations to the null distribution of the test statistic. The work on the exact null distribution was done by many including Mathai and Rathie (1970), Pillai and Nagarsenker (1971), Nagarsenker and Pillai (1973). Using invaraince considerations it can be shown that the invariant tests of H_0 are functions of the eigenvalues $l_1, \ldots, l_p$ of $\mathbf{S} = (1/n)\mathbf{A}$, where $\mathbf{A} = \Sigma (\mathbf{X}_i - \bar{\mathbf{X}})(\mathbf{X}_i - \bar{\mathbf{X}})'$, and $n = N-1$. In this notation the likelihood ratio statistic is

$$\lambda = | \mathbf{A} |^{N/2} / \{ \mathrm{tr}(\mathbf{A}) / p \}^{Np/2}$$
$$= \{ \Pi\, l_i / (\Sigma\, l_i/p)^p \}^{N/2}. \quad \textbf{(6.1)}$$

Venables (1976) gave a union-intersection interpretation for the likelihood ratio test. Actually he obtained a class of union-intersection statistics including λ and l_1/l_p, the test statistic proposed by Krishnaiah and Waikar (1971, 1972). Another prominent alternative test statistic for the problem is

$$T = (n\, p^2/2) \quad \mathrm{tr}\ \{ \mathbf{A}\, (\mathrm{tr}\, \mathbf{A})^{-1} - p^{-1}\, \mathbf{I} \}^2 \quad \textbf{(6.2)}$$

proposed by Nagao (1973). Nagao argues that the variance of the limiting distribution of $-2 \log \lambda$ which is proportional to $\tau^2 = \mathrm{tr}\, \{ \Sigma\, (\mathrm{tr}\, \Sigma\ )^{-1} - p^{-1}\, \mathbf{I}\, \}^2$ may be treated as a measure of departure from the null hypothesis and estimated by replacing Σ by $\mathbf{S}$, leading to (6.2). Using the likelihood ratio statistics for the components in Anderson's decomposition of H_0 given in Section 3 and the combination strategy of Section 4, we can also construct asymptotic equivalents of the likelihood ratio. Specifically the likelihood ratios λ_1 and λ_2 for H_{01} and H_{02} are

$$\lambda_1 = | \mathbf{R} |^{N/2}, \quad \lambda_2 = (\Pi\, a_{ii})^{N/2} / (\mathrm{tr}\, \mathbf{A}/p)^{Np/2}. \quad \textbf{(6.3)}$$

It is easy to see that $\lambda = \lambda_1\, \lambda_2$. In order to compute the P-valeus P_1, P_2 associated with λ_1 and λ_2, we may use Bartlett's approximations for the null distribution of these statistics. Thus P_1 can be computed using the fact that $-\ [N - 1 - ((2p + 5) / 6)]\ \log | \mathbf{R} |$ is distributed as χ^2 with d.f. $p(p - 1)/2$. (Refer Morrison (1976) p. 118). Alternatively the normal approximation for the null distribution of λ_1 given by Mudholkar, Trivedi and Lin (1982) may be used. P_2 can be computed using the null distribution of $M' = M / [1 + ((p + 1) / (3np))]$ which is χ^2 with (p-1) d.f., where $M = np \log (\Sigma\ a_{ii}/p) - \Sigma\ n \log (a_{ii}/n)$. (Refer Rao (1952), p. 226-228)). The null distributions of $\psi_F = -2 \log P_1 P_2$ and $\psi_L = \Sigma\ - 2 \log (P_i / (1 - P_i))$ are mentioned in Section 4. In the following Theorem the relationship between the exact slopes of λ_1, λ_2 and λ is given.

THEOREM 6.1. *If c_1, c_2 and c_λ are the exact slopes of the likelihood ratios λ_1, λ_2 and λ respectively, then*

$$c_\lambda = c_1 + p\, c_2.$$

Proof. It is easy to show that $c_1 = \lim_{N \to \infty} (-2/N) \log \lambda_1 = \log [\prod \sigma_{ii} / |\Sigma|]$, $c_2 = \lim_{N \to \infty} (-2/Np) \log \lambda_2 = \log [(\mathrm{tr}\,\Sigma / p) / (\prod \sigma_{ii})^{1/p}]$, and $c_\lambda = \lim_{N \to \infty} (-2/N) \log \lambda = \log [(\mathrm{tr}\,\Sigma / p)^p / |\Sigma|]$. Hence the result.

It may be noted that as stated in Theorem 4.1. $c_\lambda = c_F = c_L$. i.e., the three tests are equivalent in terms of the Bahadur efficiency. Their finite sample comparisons involve the following empirical investigation.

6.1. A Monte Carlo Experiment: The simulation study was conducted on Honeywell Multics system at the Oakland University using the IMSL random number generator routine GGNSM. 3000 samples of size N=21 were obtained from the p-variate normal distribution with mean 0 and covariance matrix Σ. The sphericity test statistics λ, T, ψ_F, ψ_L given above were computed for each sample. The significance of each test at 5% level was determined by comparing the computed statistics with the corresponding 5% points, and also by computing the associated P-values. The formulae in Anderson (1984,(18) p.431 for p=2 and (23), (24) p.432 for p>2) were used in conducting these calculations for the likelihood ratio test. The corresponding formulae for Nagao's test were taken from Nagao (1973, (8.2) p.708 and (5.3) p.706).

The powers of the different tests at the alternative Σ were obtained from the proportion of times the null hypothesis was rejected in the 3000 iterations. The distributions of the P-values of various tests at the alternative Σ were examined using the descriptive statistics including the means and the standard deviations of the 3000 values of each test. The process was repeated for various alternatives Σ, and also for the sample size N=35. A selection of the results is presented in Tables 1 and 2.

One of the striking features of the Table 1 is the poor quality of Nagao's approximation for the null distribution of his test statistic, especially for small values of p. This under estimation of the P-value, seen in Table 2, equivalently the over estimation of the critical constants as evident in Table 1 also stood out in results not included in the tables. Obviously, because of this reason, the empirical power function of Nagao's test is not comparable with those of the other tests.

A general feature of all the power functions is however their sensitivity to the heterogeneity of the variances. The powers of all tests seem relatively less affected by the changes in the correlations. In view of the standard error ($\leq (1/12000)^{1/2} = .009$) of the estimated power and the standard error ($(1/36000)^{1/2} = .005$) of the mean P-value, the null distributions of other statistics are satisfactory. In terms of the relative operating characteristics in finite samples the two combination tests and the likelihood ratio test seem comparable. However in view of the point estimates of the powers and the mean P-values the likelihood ratio appears slightly superior to the combination tests when variances are not too different. On the other hand the combination tests based upon the Fisher's method and Logit method are preferable to the likelihood ratio for detecting substantial differences among the variances.

7. MISCELLANEOUS REMARKS AND FUTURE DIRECTIONS

1. Development. The logic underlying the tests reviewed in this paper is simple and its applicability is quite general. Yet the specifics and details in the development of a good test are many. Thus the construction of test for a problem involves identification of different decompositions of the overall null hypohtesis and a choice among them with respect to the availability of reasonable tests for the compnents. It may involve construction of approximations for computing the P-values if only select percentiles of the null distribution are available. The choice among competing solutions involves computation of power functions or simulations. The construction and evaluation of the tests for several problems is currently in progress.

2. Confidence Regions. A desirable adjunct of a test in multivariate analysis is the associated simultaneous confidence intervals used in the post-hoc analysis. (Refer Mudholkar, Davidson, and Subbaiah (1974b), Subbaiah and Mudhokar (1981)). For most likelihood ratio tests such results are unavailable. For the tests in this paper some such results exist in unpublished form. The general probelm appears feasible but difficult.

3. General Properties. The literature on the classical multivariate tests is varied and substantial. The theoretical papers pertain to the traditional properties such as unbiasedness, invariance and monotonicity of power functions, to the decision theoretic aspects such as admissibility, mimimaxity and Bayes character, and to the asymptotic relative efficiencies. The empirical studies pertain to the simulation of power functions of competing tests. Some properties of the modified tests of this paper are reported in Mudholkar and Subbaiah (1979, 1981, 1986 a,b), several aspects are under investigation, but much of the work lies ahead.

4. Known Covariance Structure. In many realistic multivariate normal problems the underlying covariance structure is assumed either to be reducible to some known symmetric pattern or to be restricted in some manner. In several of these forms the maximum likelihood estimates lack a simple, explicit and exact form. Hence the likelihood ratio tests for these problems are often approximate with respect to both the statistics and their null distributions. As an example consider a generalization of the Pitman-Morgan test, the test for homogeneity of variances in a multivariate normal population with a symmetric correlation structure. This large sample approximate likelihood ratio test is constructed and studied by Han (1968). For the same problem Mietlowski (1974) has constructed some stepdown tests and the related confidence bounds. The null hypothesis of homogeneity admits various decompositions. One of these takes into account the restricted correlation structure and in combination with the reviewed method appears to yield an exact test with excellent operating characteristics.

5. Multivariate One-sided Tests. For the practically important problem of testing H_0: $\mu = \mathbf{0}$ against the orthant alternative $H_1 : \mu \geq \mathbf{0}$, the partial order $>$ being coordinate wise, only intricate large sample approximate tests are available, e.g., see Barlow, Bartholomew, Bremner and Brunk (1972). However the problem can be solved along the lines of Mudholkar and Subbaiah (1980) using the techniques considered in this paper. A comparative study of various simple exact tests so obtained is in progress. This problem is a particular case of the more general problem of inference under order restrictions.

6. Elliptically Contoured Distributions. It is well known that many normal theory methods can be easily modified when elliptically contoured vector or matrix distributions replace the normal distribution in the model. This is particlarly so in case of the stepwise tests. Hence it is possible to construct simple exact tests for many restricted and unrestricted testing of hypothesis problems in these general models.

7. Robust Analogues. It is well known that the trimmed means can be studentized using Winsorized standard deviations. Small sample t-

approximations for the studentized trimmed means, proposed by Tukey and McLaughlin (1963) and Huber (1970), are examined with respect to robustness and efficiency by Patel (1981), Patel and Mudholkar (1985). They recommend the 15% studentized trimmed mean as a reasonably efficient robust solution for the problem of testing of hypotheses or confidence interval estimation of the location parameter of a possibly nonnormal population. The robust equivalents alluded to in Section 1 for the standard solutions of the multiparameter of multivariate location problem can be constructed by using the studentized trimmed means or their minor variations in place of the student's t statistics appearing in the stepwise solutions for the problem.

Govind S. Mudholkar, Department of Statistics, University of Rochester, Rochester, NY 14627 and Perla Subbaiah, Department of Mathematical Sciences, Oakland University, Rochester, MI 48063

REFERENCES

Anderson, T.W. (1958, 1984). *An Introduction to Multivariate Statistical Analysis*, first and second editions. John Wiley and Sons, Inc., New York.

Bahadur, R. R. (1960). Stochastic comparison of tests. *Ann. Math. Statist.*, **31**, 276-295.

Bahadur, R. R. (1967). An optimal property of the likelihood ratio statistic. *Proc. Fifth Berkeley Symp. Math. Statist. Probability*, **1**, 13-26.

Bahadur, R. R. (1971). *Some Limit Theorems in Statistics*. Regional Conference Series in Applied Mathematics. No. 4, SIAM, Philadelphia.

Bahadur, R. R., and Raghavachari, M. (1972). Some asymptotic properties of likelihood ratio on general sample spaces. *Proc. Sixth Berkeley Symp. Math. Statist. Probability*, **1**, 129-152.

Barlow, R. E. , Bartholomew, D. J., Bremner, J. M., and Brunk, H. D. (1972). *Statistical Inference Under Order Restrictions*. John Wiley and Sons, Inc., New York.

Berk, R. H. , and Cohen, A. (1979): Asympotically optimal methods of combining tests. *J. Amer. Statist. Assoc.*, **74**, 812-814.

Das Gupta, S., Anderson, T. W., and Mudholkar, G. S. (1964). Monotonicity of the power functions of some tests of the multivariate linear hypothesis. *Ann. Math. Statist.,* **35**, 200-205.

Dempster, A. P. (1963). Multivariate theory for general step-wise methods. *Ann. Math. Statist.,* **36**, 873-883.

Fisher, R. A. (1932). *Statistical Methods for Research Workers.* 4th ed., Oliver and Boyd, Edinburgh.

George, E. O. (1977. *Combining Independent One-sided and Two-sided Statistical Tests - Some Theory and Applications.* Unpublished Ph.D. Dissertation, University of Rochester, Rochester, New York.

Han, C. P. (1968). Testing the homogeneity of a set of correlated variances. *Biometrika,* **55**, 317-326.

Hsieh, H. K. (1979a). On asymptotic optimality of likelihood ratio tests for multivariate normal distributions. *Ann. Statist.,* **7**, 592-598.

Hsieh, H. K. (1979b). Exact Bahdur efficiencies for tets of the multivariate linear hypothesis. *Ann. Statist.,* **7**, 1231-1245.

Huber, P.J. (1970). *Studentizing Robust Estimates. In Nonparametric Techniques in Statiscal Inference,* 453-463. (M. L. Puri, Editor, Cambridge University Press).

Kiefer, J. (1966). Multivariate optimality results. *Multivariate Analysis,* 255-274. (P. R. Krishnaiah, Editor), Academic Press, New York.

Kleinbaum, D. G. (1973). A generalization of the growth curve model which allows missing data. *J. Mult. Analys.,* **3**, 117-124.

Krishnaiah, P. R., Mudholkar, G. S., and Subbaiah, P. (1980) Simultaneous test procedures for mean vectors. *Handbook of Statistics,* **1**, *Analysis of Variance* (P. R. Krishnaiah, Editor), North Holland Publishing Company, Amsterdam.

Krishnaiah, P. R., and Waikar, V. B. (1971). Simultaneous tests for equality of latent roots against certain alternatives - I. *Ann. Inst. Statist. Math.,* **23**, 451-468.

Krishnaiah, P. R., and Waikar, V. B. (1972). Simultaneous tests for equality of latent roots against certain alternatives - II., *Ann. Inst. Statist. Math.,* **24**, 81-85.

Lancaster, H.O. (1961). The combination of probabilities: An application of orthonormal functions. *Australian Journal of Statistics*, **3**, 20-33.

Lehmann, E. L. (1959). *Testing Statistical Hypotheses*, John Wiley and Sons, Inc., New York.

Liptak, T. (1958). On the combination of independent tests. *Akad. Mat. Kutato. Int. Kozl.* **3**, 171-197.

Littell, R. C., and Folks, J. L. (1973). Asymptotic optimality of Fisher's method of combining independent tests II. *J. Amer. Statist. Assoc.*, **68**, 193-194.

Mathai, A. M. and Rathie, P. N. (1970). The exact distribution for the sphericity test. *J. Statist. Res.*, **4**, 140-159.

Mauchly, J. W. (1940). Significence test for sphericity of a normal n-variate distribution. *Ann. Math. Statist.*, **11**, 204-209.

Mietlowski, W. L. (1974). On the comparison of the dispersions of correlated vecotrs. Unpublished Ph.D. Dissertation, University of Rochester, Rochester, New York.

Mudholkar, G. S. (1965). A class of tests with monotone power functions for two problems in multivariate statistical analysis. *Ann. Math. Statist.*, **36**, 1794-1801.

Mudholkar, G. S., Davidson, M. L., and Subbaiah, P. (1974a). Extended linear hypotheses and simultaneous tests in multivariate analysis of variance. *Biometrika*, **61**, 467-477.

Mudholkar, G. S., Davidson, M. L., and Subbaiah, P. (1974b). A note on the union-intersection character of some MANOVA procedures. *J. Multivariate Analysis*, **4**, 486-493.

Mudholkar, G. S., and George E. O. (1979). The logit statistic for combining probabilities - An Overview. *Proceedings of International Symposium on Optimization and Statistics*, p. 345-365, (J. S. Rustagi, Editor, Academic Press, New York).

Mudholkar, G. S., and Subbaiah, P. (1975). A note on MANOVA multiple comparisons based upon step-down procedure. *Sankhya ,Series B*, **37**, 300-307.

Mudholkar, G. S., and Subbaiah, P. (1980). Testing significance of a mean vector - a possible alternative to Hotellings T^2. *Ann. Inst. Statist. Math.* **32**, 43-52.

Mudholkar, G. S., and Subbaiah, P. (1980). A review of step-down procedures for multivariate analysis of variance. *Multivariate Statistical Analysis*, 161-178. (R. P. Gupta, Editor), North-Holland Publishing Company, Amsterdam.

Mudholkar, G. S., and Subbaiah, P. (1981). Complete independence in the multivariate normal distribution. *Statistical Distributions in Scientific Work*, **5**, 157-168. (C. Tallie et. al., Editors). D. Reidel Publishing Company, Boston.

Mudholkar, G. S., and Subbaiah, P. (1986a). On a Fisherian detour of the step-down procedure for MANOVA. Submitted for publication.

Mudholkar, G. S., and Subbaiah, P. (1986b). Some asymptotic (Bahadur) equivalents of the likelihood ratio tests. Submitted for publication.

Mudholkar, G. S., Trivedi, M C., and Lin, C. C. (1982). An approximation to the distribution of the likelihood ratio statistic for testing complete independence. *Technometrics*, **24**, 139-143.

Nagarsenker, B. N., and Pillai, K. C. S.. (1973). The distribution of the sphericity test criterion. *J. Mult. Analysis*, **3**, 226-235.

Oosterhoff, J. (1969). *Combination of One-sided Statistical Tests.* Mathematics Centre Tract 28, The Mathematical Centre, Amsterdam.

Patel, K.R. (1981). A study of some estimators, their studentizations and applications. Unpublished Ph.D. Dissertation, University of Rochester, Rochester, New York.

Patel, K.R. and Mudholkar, G. S., and Fernando, J.L.I. (1986). An evaluation of Student's t-approximations for studentized trimmed means, trimean and Gastwirth's estimator. To appear in J. Amer. Statist.Assoc.

Pillai, K. C. S. (1955). Some new test criteria in multivariate analysis. *Ann. Math. Statist.*, **26**, 117-121.

Pillai, K. C. S., and Nagarsenker, B. N. (1971). On the distribution of the sphericity test criterion in classical and complex normal populations having unknown covariance matrices. *Ann. Math. Statist.*, **42**, 764-767.

Rao, C. R. (1948). Tests of significance in multivariate analysis. *Biometrika*, **35**, 58-79.

Rao, C. R. (1952). *Advanced Statistical Methods in Biometric Research.* John Wiley and Sons, Inc., New York.

Rao, C. R. (1956). Analysis of dispersion with incomplete observations on one of the characters. *J. Roy. Statist. Soc., Series B., 18*, 259 -264.

Roy, J. (1958). Step-down procedure in multivariate analysis. *Ann. Math. Statist.*, **29**, 1177-1187.

Roy, S. N., and Bargmann, R. E. (1958). Tests of multiple independence and the associated confidence bounds. *Ann. Math. Statist.*, **29**, 491-503.

Schwartz, R. (1966). Fully invariant proper Bayes tests. *Multivariate Analysis*, 275-284. (P. R. Krishaiah, Editor, Academic Press, New York).

Sen, P. K. (1986). Contemporary text books on multivariate statistical analysis: A panoramic appraisal and critique. *J. Amer. Statist. Assoc.* **81**, 560-564.

Subbaiah, P., and Mudholkar, G. S. (1981). The union-intersection principle and the simultaneous test procedures for means. *Topics in Applied Statistics* , 641-657. (Y. P. Chaubey and T. D. Dwivedi, Editors). Concordia University, Montreal.

Subbaiah, P., and Mudholkar, G. S. (1982). MANOVA multiple comparisons using the generalized step-down procedure. *Biom. J.*, **24**, 17-26.

Tukey, J. W., and McLaughlin, D.H. (1963). Less vulnerable confidence and significance procedures for location based on a single sample: Trimming/Winsorization I, *Sankhya, Series A*, **25**, 331-352.

Venables, W. (1976). Some implications of the union-intersection principle for tests of sphericity. *J. Mult. Analysis*, **6**, 185-190.

TABLE 1. *The empirical power functions based on 3000 samples of size N=21 and α=.05*

p	Population Parameters Variances			Correlations			LR Test	Nagao Test	Fisher's Comb.	Logit Comb.
2	σ_{11}	σ_{22}		ρ_{12}						
	1	1		0			.0413	.0177	.0420	.0440
				.2			.1053	.0513	.0990	.0937
				.4			.3457	.2100	.3227	.2840
				.6			.7570	.6243	.7277	.6373
				.8			.9913	.9773	.9897	.9653
	1	2		0			.2363	.1360	.2400	.2213
				.2			.2983	.1783	.3047	.2963
				.4			.5513	.4003	.5563	.5623
				.6			.8710	.7667	.8703	.8647
				.8			.9963	.9887	.9963	.9933
	1	3		0			.5480	.4017	.5460	.4997
				.2			.6123	.4563	.6187	.5897
				.4			.7777	.6503	.7893	.7943
				.6			.9470	.8990	.9517	.9590
				.8			.9993	.9983	.9993	.9993
	2	3		0			.1033	.0507	.1027	.0993
				.2			.1750	.0920	.1793	.1870
				.4			.4150	.2727	.4080	.4007
				.6			.8160	.6957	.8050	.7670
				.8			.9933	.9830	.9913	.9793
3	σ_{11}	σ_{22}	σ_{33}	ρ_{12}	ρ_{13}	ρ_{23}				
	1	1	1	0	0	0	.0430	.0393	.0447	.0460
				.2			.0770	.0783	.0733	.0790
				.4			.2193	.2013	.1953	.1830
				.6			.5857	.4990	.5417	.4723
				.8			.9733	.8913	.9640	.9223
				.2	.2		.1263	.1143	.1177	.1113
				.4	.4		.5190	.4383	.4710	.4213
				.6	.6		.9940	.9537	.9923	.9743

(Table 1 continued)

σ_{11}	σ_{22}	σ_{33}	ρ_{12}	ρ_{13}	ρ_{23}				
			.2	.2	.2	.1467	.1633	.1347	.1313
			.4	.4.	.4	.5640	.6153	.5263	.4563
			.6	.6	.6	.9350	.9563	.9250	.8783
			.8	.8	.8	.9997	1.0000	.9997	.9977
1	2	3	0	0	0	.3673	.3173	.4253	.3810
			.2			.4267	.3523	.4723	.4480
			.4			.6277	.4767	.6493	.6547
			.6			.8913	.6987	.8883	.8963
			.8			.9990	.9660	.9987	.9977
			.2	.2		.4963	.4007	.5413	.5297
			.4	.4		.8310	.6487	.8340	.8470
			.6	.6		1.0000	.9923	1.0000	1.0000
			.2	.2	.2	.5083	.4633	.5513	.5337
			.4	.4	.4	.8243	.8157	.8240	.8390
			.6	.6	.6	.9807	.9813	.9783	.9800
			.8	.8	.8	1.0000	1.0000	1.0000	1.0000

TABLE 2. *Estimated Means of P-values based on 3000 samples of size $N = 21$ and $\alpha = .05$*

p	Population Parameters Variances		Correlations	LR Test	Nagao Test	Fisher's Comb.	Logit Comb.
2	σ_{11}	σ_{22}	ρ_{12}				
	1	1	0	.5010	.4729	.5006	.5017
			.2	.4101	.3935	.4118	.4210
			.4	.2175	.2230	.2238	.2504
			.6	.0522	.0650	.0579	.0880
			.8	.0024	.0035	.0031	.0085
	1	2	0	.2807	.2795	.2803	.2995
			.2	.2432	.2458	.2412	.2516
			.4	.1239	.1363	.1227	.1283
			.6	.0282	.0373	.0282	.0313
			.8	.0010	.0016	.0011	.0017

(Table2 continued)

	σ_{11}	σ_{22}		ρ_{12}						
	1	3		0			.1194	.1322	.1202	.1448
				.2			.0948	.1087	.0941	.1094
				.4			.0511	.0631	.0490	.0500
				.6			.0106	.0154	.0099	.0095
				.8			.0003	.0005	.0003	.0003
	2	3		0			.4081	.3913	.4075	.4161
				.2			.3311	.3241	.3305	.3361
				.4			.1708	.1811	.1730	.1878
				.6			.0406	.0519	.0430	.0556
				.8			.0016	.0025	.0020	.0039
3	σ_{11}	σ_{22}	σ_{33}	ρ_{12}	ρ_{13}	ρ_{23}				
	1	1	1	0	0	0	.5114	.5095	.5108	.5130
				.2			.4365	.4343	.4441	.4436
				.4			.2781	.2854	.2946	.3128
				.6			.0945	.1170	.1084	.1421
				.8			.0064	.0194	.0081	.0171
				.2	.2		.3826	.3858	.3945	.4072
				.4	.4		.1213	.1403	.1361	.1662
				.6	.6		.0015	.0093	.0021	.0060
				.2	.2	.2	.3592	.3514	.3713	.3828
				.4	.4	.4	.1220	.1095	.1353	.1617
				.6	.6	.6	.0131	.0089	.0162	.0278
				.8	.8	.8	.0001	.0000	.0002	.0005
	1	2	3	0	0	0	.1845	.2003	.1606	.1896
				.2			.1575	.1782	.1409	.1563
				.4			.0813	.1123	.0765	.0777
				.6			.0208	.0497	.0211	.0205
				.8			.0007	.0104	.0008	.0008
				.2	.2		.1303	.1553	.1175	.1263
				.4	.4		.0331	.0600	.0330	.0312
				.6	.6		.0002	.0049	.0002	.0002
				.2	.2	.2	.1226	.1346	.1117	.1179
				.4	.4	.4	.0398	.0408	.0389	.0371
				.6	.6	.6	.0040	.0033	.0043	.0044
				.8	.8	.8	.0000	.0000	.0000	.0000

Robb J. Muirhead

DEVELOPMENTS IN EIGENVALUE ESTIMATION

1. INTRODUCTION

The area of decision-theoretic multiparameter estimation in multivariate statistics has been one of intense activity and wide interest over the past few years. Many classical procedures revolve around the eigen structures of random and parameter matrices. Invariance and other considerations tend to focus a great deal of attention on the eigenvalues. The purpose of this paper is to review some of the work relating to eigenvalue estimation.

The eigenvalue estimation problem may be stated thus: In a classical multivariate procedure the parameters of interest are eigenvalues $\omega_1, \ldots, \omega_m$ $(\omega_1 \geq \ldots \geq \omega_m \geq 0)$ of some parameter matrix. We have at our disposal a natural set of sample eigenvalues $\ell_1, \ldots, \ell_m$ $(\ell_1 > \ldots > \ell_m > 0)$, and we wish to use these to construct estimates of $\omega_1, \ldots, \omega_m$. In most situations the usual estimate of ω_i is either ℓ_i or a simple linear function of ℓ_i; such an estimate ignores information about ω_i in ℓ_j, for $j \neq i$. This paper is concerned with possible decision-theoretic approaches to this problem. Ideally, such an approach would specify a loss function in terms of the parameters $\omega_1, \ldots, \omega_m$, and risk calculations would involve expectations of this loss taken with respect to the joint distribution of $\ell_1, \ldots, \ell_m$. This, however, poses a major obstacle and seems infeasible at the moment, due to the complexity of the distributions of sample eigenvalues.

One way of overcoming this problem, and the one on which this paper concentrates, is to construct a random matrix F whose eigenvalues are $\ell_1, \ldots, \ell_m$, and a parameter matrix Δ whose eigenvalues are $\omega_1, \ldots, \omega_m$. Sometimes the choice of these matrices is obvious (as in the Wishart and noncentral Wishart settings). In MANOVA and canonical correlations however, the choice is not as clear-cut and it turns out that much may be gained, at least in terms of simplicity of risk calculations, by the choice of a non-observable F with observable eigenvalues $\ell_1, \ldots, \ell_m$. The approach may then be summarized: Act as if F is observable and use it to construct an "estimate" $\hat{\Delta}(F)$ of Δ, using a particular loss function. The eigenvalues of $\hat{\Delta}(F)$ may then be regarded

A. K. Gupta (ed.), Advances in Multivariate Statistical Analysis, 277–288.

as estimates of $\omega_1, \ldots, \omega_m$, the eigenvalues of Δ . At this point we must insist that these be proper estimates in that they depend on F only through $\ell_1, \ldots, \ell_m$, and hence are observable. To this end we consider only orthogonally invariant estimates of Δ which have the same eigenvectors as F and whose eigenvalues are functions only of $\ell_1, \ldots, \ell_m$, i.e. estimates of the form

$$\hat{\Delta}(F) = H\phi(L)H' \tag{1.1}$$

where $F = HLH'$ with H the matrix of normalized eigenvectors ($HH' = H'H = I$), $L = \mathrm{diag}(\ell_1, \ldots, \ell_m)$, and $\phi(L) = \mathrm{diag}(\phi_1(L), \ldots, \phi_m(L))$. This class of estimates was introduced by Stein (1975, 1977a, 1977b) in connection with the problem of estimating a covariance matrix Σ , and it has attracted wide attention since then. For $i = 1, \ldots, m$, the observable random variable $\phi_i(L)$ may then be regarded as an estimate of ω_i . In the five situations reviewed in this paper, Δ has an orthogonally invariant unbiased estimate taking the form $\hat{\Delta}_U \equiv \alpha F + \beta I$ for certain constants α and β . Of particular interest are orthogonally invariant estimates which dominate $\hat{\Delta}_U$.

2. EIGENVALUES OF A COVARIANCE MATRIX

Suppose that the $m \times m$ matrix S has the Wishart distribution with n degrees of freedom and covariance matrix Δ , written $S \sim W_m(n, \Delta)$. The usual (unbiased) estimate of Δ is $F = n^{-1}S$; its eigenvalues $\ell_1, \ldots, \ell_m$, however, tend to be much more dispersed than the eigenvalues $\omega_1, \ldots, \omega_m$ of Δ , especially when $\Delta \not\approx cI$. In general, a simple convexity argument shows that ℓ_1 overestimates ω_1 and that ℓ_m underestimates ω_m , so that an intuitively appealing approach is to shrink the sample eigenvalues towards some central value. A great deal of work has been done on estimating Δ . It is the subset of this that is particularly relevant to the eigenvalue estimation problem that we wish to concentrate on here. This includes work by Stein, Haff, and Dey and Srinivasan. The loss function which has come under most scrutiny so far (because it is reasonably tractable) is the convex function

$$L(\hat{\Delta}, \Delta) = \mathrm{tr}(\hat{\Delta}\Delta^{-1}) - \ln\det(\hat{\Delta}\Delta^{-1}) - m \quad , \tag{2.1}$$

introduced by James and Stein (1961) and commonly referred

to as "Stein's loss". This discussion here will concentrate on results relating to this loss.

In a major development Stein (1975, 1977a) used a fundamental Wishart identity to derive an unbiased estimate of the risk $R(\hat{\Delta}, \Delta) = E_{\Sigma}[L(\hat{\Delta}, \Delta)]$ of an orthogonally invariant estimate $\hat{\Delta}$ of the form (1.1). This unbiased estimate is

$$\hat{R}(\hat{\Sigma}) = \frac{n - m + 1}{m} \sum_{i=1}^{m} \psi_i + \frac{2}{n} \sum_{i \neq j} \sum \frac{\psi_j \ell_j}{\ell_j - \ell_i} - \sum_{i=1}^{m} \ln \psi_j + 2 \sum_{i=1}^{m} \ell_i \frac{\partial \psi_i}{\partial \ell_i} - k_{m,n} \qquad (2.2)$$

where

$$k_{m,n} = E\left[\sum_{i=1}^{m} \ln \chi^2_{n-i+1}\right] - m \ln n + m$$

and $\psi_i = \phi_i / \ell_i$, $i = 1, \ldots, m$. Stein derived an estimate which approximately minimizes the risk by ignoring the derivatives $\partial\psi_i/\partial\ell_i$ and choosing ψ_i to minimize the remaining terms in (2.2). This procedure leads to the estimate of Δ determined by $\phi_i \equiv \phi_i^{(s)}$, where

$$\phi_i^{(s)} = n\ell_i \Big/ \left[n - m + 1 + 2\ell_i \sum_{j \neq i} \frac{1}{\ell_i - \ell_j}\right], \quad i = 1, \ldots, m .$$

Since it is not generally true that the reasonable condition $\phi_1^{(s)} \geq \ldots \geq \phi_m^{(s)} \geq 0$ holds, Stein modified his eigenvalue estimates using isotonic regression. This modification is described in detail in Lin and Perlman (1985). These authors also note that $\phi_i^{(s)}$ "differs most from the usual estimates ℓ_i when some or all of the ℓ_i are nearly equal and m/n is not small".

Haff (1982) investigated Bayes estimates of the covariance matrix Δ in the class restricted by orthogonal invariance, and showed that an approximation to the Bayes rule leads to the estimate determined by $\phi_i \equiv \phi_i^{(H)}$, where

$$\phi_i^{(H)} = n\ell_i \Big/ \left[n - m - 1 + 2\ell_i \sum_{j \neq i} \frac{1}{\ell_i - \ell_j}\right], \quad i = 1, \ldots, m .$$

The only difference between $\phi_i^{(S)}$ and $\phi_i^{(H)}$ is that $m - 1$ appears in the denominator of $\phi_i^{(H)}$ rather than $m + 1$. Again, isotonic regression is necessary to ensure that $\phi_1^{(H)} \geq \ldots \geq \phi_m^{(H)} \geq 0$.

Considered as an estimate of ω_i, the ith largest eigenvalue of Δ, both $\phi_i^{(S)}$ and $\phi_i^{(H)}$ utilize information, not only from ℓ_i, but also from the other sample eigenvalues ℓ_j for $j \neq i$, with adjacent eigenvalues tending to have the most influence. It is not known whether the corresponding covariance matrix estimates dominate the unbiased estimate $n^{-1}S$. A major Monte Carlo study by Lin and Perlman (1985) comparing four estimates of Δ indicated that Stein's modified estimate provides substantial improvement over $n^{-1}S$ for a wide range of covariance structures. Haff (1982) noted that the same is true for his modified estimate. It is interesting to note that the eigenvalue estimates of Stein and Haff are close to estimates obtained by Anderson (1965) which incorporate a bias correction (see Muirhead (1982), page 405).

Using Stein's unbiased risk estimate (2.2), Dey and Srinivasan (1984, 1985) derived eigenvalue estimates which are somewhat different in character from those of Stein and Haff, and some of these lead to minimax estimates of Δ. An estimate of Δ which they included in a Monte Carlo study is determined by $\phi_i = \phi_i^{(DS)}$, where

$$\phi_i^{(DS)} = \frac{n\ell_i}{n + m + 1 - 2i} - \frac{nc\ell_i \ln(n\ell_i)}{b + \sum_{i=1}^{m} [\ln(n\ell_i)]^2}$$

with

$$c = 6(m - 2)/[5(n + m - 1)^2]$$

and

$$b = 5\cdot 8(m - 2)^2/(n + m - 1)^2 .$$

This particular estimate (one of a whole class of minimax estimates) was chosen by Dey and Srinivasan because their numerical studies indicated that it is close to being optimal in terms of minimum risk. Again, isotonic regression is needed to ensure that $\phi_1^{(DS)} \geq \ldots \geq \phi_m^{(DS)} \geq 0$. The sign of the correction term in $\phi_i^{(DS)}$ depends on the magnitude of ℓ_i, being positive if $\ell_i < 1$ and negative if $\ell_i > 1$. The magnitude of the correction depends on all the sample eigenvalues. Dey and Srinivasan report that this covariance matrix estimate performs substantially better than the unbiased estimate $n^{-1}S$ over a wide range of eigenstructures, but that the estimates of Stein and Haff described earlier do significantly better (in terms of risk) if $\Delta \approx cI$.

The discussion has been concentrated on estimates which have a tendency to move the sample eigenvalues together in an intuitively appealing way, and has been restricted to the particular loss function (2.1). Another loss function which has been considered is $\mathrm{tr}(\hat{\Delta}\Delta^{-1} - I)^2$, but with this, theoretical results of the nature described above are somewhat more difficult to come by. In addition to the papers referenced already there is a host of other work on decision-theoretic covariance matrix estimation and we conclude by referencing some of these. They include Eaton (1970), Haff (1977, 1979a, 1979b, 1980), Efron and Morris (1976), Olkin and Selliah (1977), Takemura (1984), and Verathaworn (1983).

3. EIGENVALUES IN A TWO-SAMPLE PROBLEM

Let S_1 and S_2 be independent Wishart matrices, with $S_i \sim W_m(n_i, \Sigma_i)$, $i = 1, 2$. The problem of estimating the eigenvalues $\omega_1, \ldots, \omega_m$ ($\omega_1 \geq \ldots \geq \omega_m > 0$) of $\Sigma_1\Sigma_2^{-1}$ has been considered by Muirhead and Verathaworn (1985). These eigenvalues are important in the problem of testing $\Sigma_1 = \Sigma_2$, as they form maximal invariants under a natural group of transformations leaving the testing problem invariant.

It is convenient here to transform the eigenvalue estimation problem and follow the approach outlined in Section 1. Let L be an $m \times m$ nonsingular matrix such that $L\Sigma_2L' = I$ and define an $m \times m$ parameter matrix Δ as $\Delta = L\Sigma_1L'$. Put $A = LS_1L'$, $B = LS_2L'$, so that $A \sim W_m(n_1, \Delta)$, $B \sim W_m(n_2, I)$, and define the random matrix F as $F = A^{1/2}B^{-1}A^{1/2}$. Note that the eigenvalues of Δ are $\omega_1, \ldots, \omega_m$, the eigenvalues of $\Sigma_1\Sigma_2^{-1}$ and, although F is not observable, its eigenvalues $\ell_1, \ldots, \ell_m$ ($\ell_1 > \ldots > \ell_m > 0$) are the same as those of $S_1S_2^{-1}$. We may overcome the nonobservability problem by restricting attention to orthogonally invariant estimates of the form (1.1), so that the eigenvalues of $\hat{\Delta}(F)$, which are observable, may be regarded as estimates of $\omega_1, \ldots, \omega_m$.

Given F, Muirhead and Verathaworn (1985) considered the problem of estimating Δ using Stein's loss function (2.1). An unbiased estimate of Δ is

$$\hat{\Delta}_U = \frac{n_2 - m - 1}{n} F$$

which leads to the eigenvalue estimates $\hat{\omega}_i = [(n_2 - m - 1)/n] \ell_i$, $i = 1, \ldots, m$. Following an approach similar to that used by Haff (1982) in the problem of estimating a covariance matrix, Muirhead and Verathaworn showed that a rather ad-hoc approximation to the Bayes rule leads to the estimate determined by $\phi_i \equiv \hat{\phi}_i$, where

$$\hat{\phi}_i = \ell_i \Bigg/ \left[\frac{n_1 - m - 1}{n_2} + \frac{2(n_1 + n_2 - m - 1)}{n_2(n_2 - m - 1)} \cdot \ell_i \sum_{j \neq i} \frac{1}{\ell_i - \ell_j} \right]$$

for $i = 1, \ldots, m$. An isotonized modification of these estimates ensures that $\hat{\phi}_1 \geq \ldots \geq \hat{\phi}_m \geq 0$. A Monte Carlo study indicated that these modified estimates perform substantially better than the eigenvalues $\hat{\omega}_i$ of $\hat{\Delta}_U$, particularly when $\Delta \approx cI$ or when Δ has groups of equal eigenvalues. As far as the author is aware, no other loss functions or other decision-theoretic properties of other estimates have been studied in connection with this problem.

4. THE NONCENTRAL WISHART NONCENTRALITY MATRIX

Suppose that $F \sim W_m(n, I, \Delta)$, i.e. F has the noncentral Wishart distribution with n degrees of freedom, identity covariance matrix, and noncentrality matrix Δ . Here we consider the problem of estimating Δ . This problem is of interest in the usual multivariate linear model where the covariance structure of the observations is known. The power of tests of interest here are functions of Δ or, more accurately, of its eigenvalues. An unbiased estimate of Δ is $\hat{\Delta}_U = F - nI$.

Muirhead and Leung (1986) considered the problem of estimating Δ by $\hat{\Delta}(F)$ using the squared error loss function

$$L(\hat{\Delta}, \Delta) = \mathrm{tr}(\hat{\Delta} - \Delta)^2 \quad . \qquad (4.1)$$

They showed that non-linear estimates of the form

$$\hat{\Delta}_\alpha = \hat{\Delta}_U + \frac{\alpha}{\mathrm{tr}F} I$$

dominate the unbiased estimate $\hat{\Delta}_U$ provided $mn > 4$ and $0 < \alpha < 4(mn - 4)/m$, with an optimal choice for α being $\alpha^* = 2(mn - 4)/m$. This result is a multivariate extension of one due to Perlman and Rasmussen (1975) obtained when considering the problem of estimating the noncentrality parameter in a noncentral χ^2 distribution. From the point of view of eigenvalue estimation it seems clear that one should be able to do better than the eigenvalues of $\hat{\Delta}_\alpha$. No decision-theoretic results are yet available on estimates which move the eigenvalues of $\hat{\Delta}_U$ closer together.

5. EIGENVALUES IN MANOVA

Consider a typical MANOVA setting where independent matrices S_1 and S_2 are observed, with $S_1 \sim W_m(n_1, \Sigma, \Omega)$ and $S_2 \sim W_m(n_2, \Sigma)$. In this setting the eigenvalues $\omega_1, \ldots, \omega_m$ $(\omega_1 \geq \ldots \geq \omega_m \geq 0)$ of Ω are of great interest; they form maximal invariants under a natural group of transformations leaving the problem of testing $\Omega = 0$ invariant. The problem of estimating them in a decision-theoretic framework has been considered by Muirhead and Leung (1986).

The distribution theory and risk calculations are greatly simplified by making the following transformations: Put $A = \Sigma^{-1/2}S_1\Sigma^{-1/2}$, $B = \Sigma^{-1/2}S_2\Sigma^{-1/2}$, so that $A \sim W_m(n_1, I, \Delta)$, with $\Delta = \Sigma^{1/2}\Omega\Sigma^{-1/2}$, and $B \sim W_m(n_2, I)$. The eigenvalues of Δ are $\omega_1, \ldots, \omega_m$ and the eigenvalues $\ell_1, \ldots, \ell_m$ $(\ell_1 > \ldots > \ell_m > 0)$ of $S_1S_2^{-1}$ are also the eigenvalues of the nonobservable random matrix F defined as $F = A^{1/2}B^{-1}A^{1/2}$. The nonobservability problem is overcome by restricting attention to orthogonally invariant estimates of the form (1.1) which have observable eigenvalues. An unbiased estimate of Δ is provided by the orthogonally invariant estimate $\hat{\Delta}_U = (n_2 - m - 1)F - n_1I$.

Muirhead and Leung (1986) considered the problem of estimating Δ by $\hat{\Delta}(F)$ using the squared error loss function (4.1). They showed that, for certain values of α, the estimate $\alpha\hat{\Delta}_U$ dominates $\hat{\Delta}_U$, with an optimal value of α being

$$\alpha^* = (n_2 - m - 3)/(n_2 - m - 1) \quad .$$

They showed also that, in turn, $\alpha\hat{\Delta}_U$ is itself dominated by estimates of the form

$$\hat{\Delta}_{\alpha,\beta} = \alpha\hat{\Delta}_U + \frac{\beta}{\mathrm{tr}F} I$$

for certain choices of α and β . If α is taken to be α^* , an optimal value for β is

$$\beta^* = \frac{2(n_2 - m - 3)(n_1 + n_2 - m - 1)(mn_1 - 4)}{m(n_2 - m - 1)(n_2 - m + 3)(n_2 - m + 1)} \quad .$$

When $m = 1$ these results reduce to ones of Perlman and Rasmussen (1975) obtained in connection with the problem of estimating the noncentrality parameter in a noncentral F distribution.

Of great interest in this MANOVA setting (and also in the canonical correlations situation described in the next section) would be the development of eigenvalue estimates similar in character to those obtained by Stein and Haff and discussed in Section 2. The derivation and properties of such estimates remain open problems.

6. EIGENVALUES IN CANONICAL CORRELATIONS

Suppose that $S \sim W_{p+q}(n, \Sigma)$, and partition S and Σ as

$$S = \begin{bmatrix} S_{11} & S_{12} \\ S_{21} & {}_{22} \end{bmatrix} \quad , \quad \Sigma = \begin{bmatrix} \Sigma_{11} & \Sigma_{12} \\ \Sigma_{21} & \Sigma_{22} \end{bmatrix}$$

where S_{11} and Σ_{11} are $p \times p$, S_{22} and Σ_{22} are $q \times q$, with $p \leq q$. The population canonical correlation coefficients are $\rho_1, \ldots, \rho_p$ $(1 \geq \rho_1 \geq \ldots \geq \rho_p \geq 0)$, where ρ_1^2 , $\ldots, \rho_p^2$ are the eigenvalues of $\Sigma_{11}^{-1}\Sigma_{12}\Sigma_{22}^{-1}\Sigma_{21}$. The eigenvalues $r_1^2, \ldots, r_p^2$ of $S_{11}^{-1}S_{12}S_{22}^{-1}S_{21}$ are the squares of the sample canonical correlation coefficients $r_1, \ldots, r_p$ $(1 > r_1 > \ldots > r_p > 0)$. The problem of estimating the parameters $\omega_i = \rho_i^2/(1 - \rho_i^2)$, $i = 1, \ldots, p$, in a decision-theoretic framework has been considered by Muirhead and Leung (1985, 1986). The estimates they consider are certain linear and non-linear functions of $\ell_i = r_i^2/(1 - r_i^2)$, $i = 1, \ldots, p$.

The random variables $\ell_1, \ldots, \ell_p$ are the eigenvalues of the (nonobservable) random matrix

$$F = A^{1/2}B^{-1}A^{1/2}$$

where $A = \Sigma_{11.2}^{-1/2}S_{12}S_{22}^{-1}S_{21}\Sigma_{11.2}^{-1/2}$, $B = \Sigma_{11.2}^{-1/2}S_{11.2}\Sigma_{11.2}^{-1/2}$, with $\Sigma_{11.2} = \Sigma_{11} - \Sigma_{12}\Sigma_{22}^{-1}\Sigma_{21}$, $S_{11.2} = S_{11} - S_{12}S_{22}^{-1}S_{21}$, and F has a distribution which depends on Σ only through the parameter matrix Δ given by

$$\Delta = \Sigma_{11.2}^{-1/2}\Sigma_{12}\Sigma_{22}^{-1}\Sigma_{21}\Sigma_{11.2}^{-1/2} ,$$

which has as its eigenvalues the parameters we wish to estimate, namely $\omega_i = \rho_i^2/(1 - \rho_i^2)$, $i = 1, \ldots, p$. Attention is restricted to orthogonally invariant estimates of the form (1.1) with observable eigenvalues. An unbiased estimate of Δ is the orthogonally invariant estimate

$$\hat{\Delta}_U = \frac{n - p - q - 1}{n} F - \frac{q}{n} I .$$

Muirhead and Leung (1985, 1986) considered the problem of estimating Δ by $\hat{\Delta}(F)$ using the squared error loss function (4.1). They showed that, for certain α , $\alpha\hat{\Delta}_U$ dominates $\hat{\Delta}_U$, with an optimal value of α being

$$\alpha^* = \frac{n(n - p - q - 3)}{(n + 2)(n - p - q - 1)} .$$

They showed also that $\alpha\hat{\Delta}_U$ is dominated by estimates of the form

$$\hat{\Delta}_{\alpha,\beta} = \alpha\hat{\Delta}_U + \frac{\beta}{\text{tr}F} I$$

for certain choices of α and β . If α is taken to be α^* , an optimal choice for β is

$$\beta^* = \frac{2(n-p-q-3)(n-p-1)(pq-4)}{p(n+2)(n-p-q-1)(n-p-q+3)(n-p-q+1)} .$$

These results are multivariate generalizations of ones obtained by Muirhead (1985) in connection with the problem of estimating $\bar{R}^2/(1 - \bar{R}^2)$, where $\bar{R}$ is a multiple correlation coefficient. The comments made at the end of Section 5 are again applicable here.

ACKNOWLEDGEMENT

This work was supported by the National Science Foundation.

AUTHOR AFFILIATION

The Department of Statistics
The University of Michigan
1440 Mason Hall
Ann Arbor, Michigan 48109-1027

REFERENCES

Anderson, G.A. (1965). 'An asymptotic expansion for the distribution of the latent roots of the estimated covariance matrix'. Ann. Math. Statist., 36, 1153-1173.

Dey, D.K. and Srinivasan, C. (1984). 'Estimation of a covariance matrix under Stein's loss'. Technical Report, Texas Tech. University.

Dey, D.K. and Srinivasan, C. (1985). 'Estimation of a covariance matrix under Stein's loss'. Ann. Statist., 13, 1581-1591.

Eaton, M.L. (1970). 'Some problems in covariance matrix estimation'. Technical Report No. 49, Department of Statistics, Stanford University.

Efron, B. and Morris, C. (1976). 'Multivariate empirical Bayes estimation of covariance matrices'. Ann. Statist. 4, 22-32.

Haff, L.R. (1977). 'Minimax estimators for a multinormal precision matrix'. J. Multivariate Anal., 7, 374-385.

Haff, L.R. (1979a). 'Estimation of the inverse covariance matrix: Random mixtures of the inverse Wishart matrix and the identity'. Ann. Statist., 7, 1264-1276.

Haff, L.R. (1979b). 'An identity for the Wishart matrix with applications'. J. Multivariate Anal., 9, 531-5421

Haff, L.R. (1980). 'Empirical Bayes estimation of the multivariate normal covariance matrix'. Ann. Statist., 8, 586-597.

Haff, L.R. (1982). 'Solutions of the Kuler-Lagrange equations for certain multivariate normal estimation problems'. Unpublished manuscript.

James, W. and Stein, C. (1961). 'Estimation with quadratic loss'. Proc. Fourth Berkeley Symp. Math. Statist. Prob., 1, 361-379. University of California Press, Berkeley, CA.

Lin, S.P. and Perlman, M.D. (1985). 'A Monte Carlo comparison of four estimators for a covariance matrix'. In Multivariate Analysis VI (P.R. Krishnaiah, ed.), 411-429. North Holland, Amsterdam.

Muirhead, R.J. (1982). Aspects of Multivariate Statistical Theory. Wiley, New York.

Muirhead, R.J. (1985). 'Estimating a particular function of the multiple correlation coefficient'. J. Amer. Statist. Assoc., 80, 923-925.

Muirhead, R.J. and Leung, P.L. (1985). 'Estimating functions of canonical correlation coefficients'. Linear Algebra and its Applications, 70, 173-183.

Muirhead, R.J. and Leung, P.L. (1986). 'Estimation of parameter matrices and eigenvalues in MANOVA and canonical correlation analysis'. Unpublished manuscript.

Muirhead, R.J. and Verathaworn, T. (1985). 'On estimating the latent roots of $\Sigma_1\Sigma_2^{-1}$'. In Multivariate Analysis VI (P.R. Krishnaiah, ed.), 431-477. North Holland, Amsterdam.

Olkin, I. and Selliah, J.B. (1977). 'Estimating covariances in a multivariate normal distribution'. In Statistical Decision Theory and Related Topics II (S.S, Gupta and D. Moore, eds.), 313-326. Academic Press, New York.

Perlman, M.D. and Rasmussen, V.A. (1975). 'Some remarks on estimating a noncentrality parameter'. Comm. Statist., 4, 455-468.

Stein, C. (1975). 'Estimation of a covariance matrix'. Rietz Lecture, 39th annual meeting IMS. Atlanta, GA.

Stein, C. (1977a). 'Estimation of the parameters of a multivariate normal distribution'. Unpublished notes.

Stein, C. (1977b). 'Lectures on the theory of estimation of many parameters'. (In Russian.) In Studies in the Statistical Theory of Estimation, Part I (I.A. Ibragimov and M.S. Nikulin, eds.), Proceedings of Scientific Seminars of the Steklov Institute, Leningrad Devision, 74, 4-65.

Takemura, A. (1984). 'An orthogonally invariant minimax estimator of the covariance matrix of a multivariate normal population'. Tsukuba J. Math, 8, 367-376.

Verathaworn, T. (1983). Decision Theoretic Approaches to Estimating Covariance Matrices in the Multivariate Normal One-Sample Problems. Ph.D. Dissertation, Department of Statistics, University of Michigan.

B. N. Nagarsenker and P. B. Nagarsenker

ASYMPTOTIC NON-NULL DISTRIBUTIONS OF A STATISTIC FOR TESTING THE EQUALITY OF HERMITIAN COVARIANCE MATRICES IN THE COMPLEX GAUSSIAN CASE

1. INTRODUCTION

Recently there has been a considerable interest in the area of complex multivariate distributions, since these distributions play an important role, particularly in the area of multiple time series. For some discussion about the applications of the complex multivariate distributions in the area of multiple time series, the reader is referred to Goodman and Dubman (1969), Brillinger (1969), Hannan (1970), Wahba (1968, 1971), Priestley, Subha Rao and Tong (1973) and Krishnaiah (1976).

Let $\underset{\sim}{X}_1$: $p x n_1$ and $\underset{\sim}{X}_2$: $p x n_2$, $p \le n_i$ $(i=1,2)$ be independent matrix variates where the columns of $\underset{\sim}{X}_1$ and $\underset{\sim}{X}_2$ are independent and the distributions of $\underset{\sim}{X}_1$ and $\underset{\sim}{X}_2$ are $CN(\underset{\sim}{X}_1;\underset{\sim}{0},\underset{\sim}{\Sigma}_1)$ and $CN(\underset{\sim}{X}_2;\underset{\sim}{0},\underset{\sim}{\Sigma}_2)$ respectively, $CN(\underset{\sim}{X};\underset{\sim}{\mu},\underset{\sim}{\Sigma})$ being the complex multivariate normal distribution defined by

$$CN(\underset{\sim}{X};\underset{\sim}{\mu},\underset{\sim}{\Sigma}) = \pi^{-pn} | \underset{\sim}{\Sigma} |^{-n} \exp[-\mathrm{tr}\,\underset{\sim}{\Sigma}^{-1}(\underset{\sim}{X}-\underset{\sim}{\mu})\overline{(\underset{\sim}{X}-\underset{\sim}{\mu})'}],$$

$\underset{\sim}{X}$ being pxn (see Goodman (1963)). Thus $\underset{\sim}{S}_1 = \underset{\sim}{X}_1\underset{\sim}{X}_1'$ and $\underset{\sim}{S}_2 = \underset{\sim}{X}_2\underset{\sim}{X}_2'$ are independently distributed as $CW(\underset{\sim}{S}_i;p,n_i,\underset{\sim}{\Sigma}_i)$, $i = 1,2$ where $CW(\underset{\sim}{S};p,n,\underset{\sim}{\Sigma})$ is the complex Wishart distribution defined by

$$CN(\underset{\sim}{S};p,n,\underset{\sim}{\Sigma}) = \{\tilde{\Gamma}_p(n)\}^{-1} | \underset{\sim}{\Sigma} |^{-n} | \underset{\sim}{S} |^{n-p} \exp(-\mathrm{tr}\,\underset{\sim}{\Sigma}^{-1}\underset{\sim}{S})$$

where

$$\tilde{\Gamma}_p(n) = \pi^{p(p-1)/2} \prod_{i=1}^{p} \Gamma(n-i+1)$$

A. K. Gupta (ed.), Advances in Multivariate Statistical Analysis, 289–302.

Let $0 < f_1 \leq f_2 \ldots \leq f_p < \infty$ be the characteristic roots of $S_1 S_2^{-1}$ and $0 < \lambda_1 \leq \lambda_2 \ldots \leq \lambda_p < \infty$ be those of $\Sigma_1 \Sigma_2^{-1}$. Then the statistic given by

$$\lambda = \prod_{i=1}^{p} (1-w_i) = |I-W| \tag{1.1}$$

where $W = \text{diag}\,(w_1, w_2, \ldots, w_p)$ and $w_i = f_i/(1+f_i)$,

$(i = 1,2,\ldots,p)$ may be used to test the null hypothesis H_0: $\Sigma_1 = \Sigma_2$ against one sided alternatives K: $\lambda_i \geq 1$ and $\sum_{i=1}^{p} \lambda_i > p$ (see Pillai and Jouris (1971)). We reject H_0 if the observed value of λ is larger than a preassigned constant.

Let $M = I - \Lambda$ where $\Lambda = \text{diag}\,(\lambda_1, \lambda_2, \ldots, \lambda_p)$, $n = n_1 + n_2$ and $n_i/n = t_i$ $(i=1,2)$ where $t_1 + t_2 = 1$. In this paper we obtain the asymptotic expansion of the non-null distribution of $Y = -\sqrt{n} \cdot \log\,(\lambda/t_2^{\,p})$ for the cases (a) when M is fixed and (b) under the sequence of alternatives K_n: $M = P/\sqrt{n}$ where P is a fixed diagonal matrix. Pillai and Nagarsenker (1972) obtained the asymptotic non-null distribution of this statistic under a different sequence of local alternatives for the real multivariate normal distribution using certain identities connnected with Zonal polynomials while in this paper we obtain the asymptotic distributions using a more elegant unified approach by considering the system of partial differential equations obtained by Chikuse (1976) and satisfied by the hypergeometric function ${}_2F_1(a,b;c,A)$ of the Hermitian matrix argument A (see James (1964)).

2. PRELIMINARIES

Lemma 1. The following Kummer transformation formula holds for the hypergeometric function ${}_2F_1(a,b;c;A)$ of the

Hermitian matrix argument $\underset{\sim}{A}$

$${}_2F_1(a,b;c;\underset{\sim}{A}) = |\underset{\sim}{I}-\underset{\sim}{A}|^{-b}\ {}_2F_1(b-a,b;c;-\underset{\sim}{A}(\underset{\sim}{I}-\underset{\sim}{A})^{-1}).$$

(For proof see Sugiama (1972)).

Lemma 2. Let λ be as in (1.1). Then the h-th moment of λ is given by

$$E(\lambda^h) = \left[\frac{\tilde{\Gamma}_p(n)\tilde{\Gamma}_p(n_2+h)}{\tilde{\Gamma}_p(n_2)\tilde{\Gamma}_p(n+h)}\right] \cdot {}_2F_1(h,n_1;n+h;\underset{\sim}{M}) \quad (2.1)$$

where $\underset{\sim}{M} = \underset{\sim}{I} - \underset{\sim}{\Lambda}$.

Proof. From eq. (2.8) of Pillai and Jouris (1971), we have

$$E(\lambda^h) = |\underset{\sim}{\Lambda}|^{-n_1} \frac{\tilde{\Gamma}_p(n)\tilde{\Gamma}_p(n_2+h)}{\tilde{\Gamma}_p(n_2)\tilde{\Gamma}_p(n+h)}\ {}_2F_1(n,n_1;n+h;\underset{\sim}{I}-\underset{\sim}{\Lambda}^{-1}). \quad (2.2)$$

Using lemma 1 in (2.2), we have (2.1).

Lemma 3. The function ${}_2F_1(a,b;c;\underset{\sim}{A})$ is the unique solution of each of the differential equations

$$A_i(1-A_i)\frac{\partial^2 F}{\partial A_i^2} + \left[c-p+1-(a+b-p+2)A_i + \sum_{\substack{j=1\\ j\neq i}}^{p} \frac{A_i(1-A_i)}{A_i-A_j}\right]\frac{\partial F}{\partial A_i} - \sum_{\substack{j=1\\ j\neq i}}^{p} \frac{A_j(1-A_j)}{A_i-A_j}\frac{\partial F}{\partial A_j} = abF$$

$$(i=1,2,\ldots,p) \qquad (2.3)$$

where $A_1, A_2, \ldots, A_p$ are the latent roots of the pxp Hermitian matrix $\underset{\sim}{A}$ and subject to the conditions:

(a) F is a symmetric function of $A_1, A_2, \ldots, A_p$ and

(b) F is analytic about $\underset{\sim}{A} = \underset{\sim}{0}$ and $F(\underset{\sim}{0}) = 1$.

(For proof see Chikuse (1976).)

We also need the asymptotic expansions for large values of n of the hypergeometric functions $F(\underset{\sim}{R}) = {}_2F_1(a,b;c;\underset{\sim}{R}/\sqrt{n})$ and $F_1(\underset{\sim}{R}) = {}_2F_1(a,b;c;\underset{\sim}{R})$ of the Hermitian matrix argument $\underset{\sim}{R}$ where $a = -\sqrt{n}\,T$, $b = nt_1$ and $c = n - \sqrt{n}\,T$. For this let $R_1, R_2, \ldots, R_p$ be the characteristic roots of $\underset{\sim}{R}$. Then by lemma 3, the function F is the unique solution of the following system of partial differential equations:

$$R_i(1-R_i/\sqrt{n})\frac{\partial^2 F}{\partial R_i^2} + \left[c-p+1-(a+b-p+2)\,R_i/\sqrt{n} + \sum_{\substack{j=1\\ j\neq i}}^{p} \frac{R_i(1-R_i/\sqrt{n})}{R_i-R_j}\right]\frac{\partial F}{\partial R_i} - \sum_{\substack{j=1\\ j\neq i}}^{p} \frac{R_j(1-R_j/\sqrt{n})}{R_i-R_j}\frac{\partial F}{\partial R_j} = \frac{ab}{\sqrt{n}}F \qquad (2.4)$$

subject to the conditions (a) and (b) of the lamma 3. Let $H(\underset{\sim}{R}) = \log F(\underset{\sim}{R})$ and so $H(\underset{\sim}{0}) = 0$. Then from (2.4), it follows that H satisfies the following partial differential equation with $a = -\sqrt{n}\,T$, $b = nt_1$ and $c = n - \sqrt{n}\,T$:

$$R_1(1-R_1/\sqrt{n})\left[\frac{\partial^2 H}{\partial R_1^2} + \left(\frac{\partial H}{\partial R_1}\right)^2\right] + \Bigg[n - \sqrt{n}(T+t_1R_1) -$$

$$- (p-1-TR_1) + 1/\sqrt{n} \quad (p-2)R_1 - \sum_{j=2}^{p} \frac{R_1^2}{R_1-R_j} \Bigg\} +$$

$$+ \sum_{j=2}^{p} \frac{R_1}{R_1-R_j} \Bigg] \frac{\partial H}{\partial R_1} - \sum_{j=2}^{p} \frac{R_j}{R_1-R_j} \frac{\partial H}{\partial R_j} +$$

$$+ 1\sqrt{n} \sum_{j=2}^{p} \frac{R_j^2}{R_1-R_j} \frac{\partial H}{\partial R_j} = - nt_1 T \tag{2.5}$$

We now look for a solution of (2.5) of the form

$$H(\underset{\sim}{R}) = \sum_{k=0}^{\infty} \frac{P_k(\underset{\sim}{R})}{n^{k/2}} \tag{2.6}$$

where for all k, $P_k(\underset{\sim}{0}) = 0$ so that $H(\underset{\sim}{0}) = 0$. Substituting (2.6) in (2.5) and equating the coefficients of n on both sides of the resulting equation, we have

$$\frac{\partial P_0}{\partial R_1} = - t_1 T$$

and this implies using the initial conditions that

$$P_0 = P_0(\underset{\sim}{R}) = - t_1 T\sigma_1 \tag{2.7}$$

where $\sigma_i = R_1^i + R_2^i + \dots R_p^i$. Equating the coefficients of $\sqrt{n}$, we have

$$\frac{\partial P_1}{\partial R_1} = (T+t_1R_1) \frac{\partial P_0}{\partial R_1} = - t_1 T(T+t_1R_1)$$

giving

$$P_1 = P_1(R) = - t_1 T(2T\sigma_1+t_1\sigma_2)/2 \tag{2.8}$$

Similarily equating the constant term we have

$$P_2 = P_2(\underset{\sim}{R}) = A_1T + A_2T^2 + A_3T^3 \tag{2.9}$$

where

$$A_1 = -t_1^2(3\sigma_2+2t_1\sigma_3)/6,\ A_2 = t_1\sigma_2(1-2t_1)/2$$

and

$$A_3 = -t_1\sigma_1.$$

Thus we have

Theorem 2.1. An asymptotic solution of the hypergeometric function ${}_2F_1(-\sqrt{n}\ T,\ nt_1;\ n-\sqrt{n}\ T;\ \underset{\sim}{R}/\sqrt{n})$ of the Hermitian matrix argument $\underset{\sim}{R}$ for large values of n is given by

$$\begin{aligned} &{}_2F_1(-\sqrt{n}\ T,\ nt_1;\ n-\sqrt{n}\ T;\ \underset{\sim}{R}/\sqrt{n}) \\ &= e^{-t_1\sigma_1 T}\ [1 + P_1/\sqrt{n} + (P_2+P_1^2/2)/n + \underset{\sim}{O}(n^{-3/2})] \end{aligned} \tag{2.10}$$

where P_1 and P_2 are given by (2.8) and (2.9) respectively.

We shall now obtain the asymptotic expansion of $F_1 = {}_2F_1(-\sqrt{n}\ T,\ nt_1;\ n-\sqrt{n}\ T;\ R)$ for large n. For this let

$$\underset{\sim}{W} = \underset{\sim}{I} - (I-t_1\underset{\sim}{R})^{-1}. \tag{2.11}$$

Then with $a = -\sqrt{n}\ T$, $b = nt_1$ and $c = n-\sqrt{n}\ T$, it follows using lemma 3, that $F_1(\underset{\sim}{W})$ is the unique solution of the following differential equation:

$$\begin{aligned} &w_1E_1(1-w_1)^2\frac{\partial^2F_1}{\partial w_1^{\,2}} + (1-w_1)\Bigg[t_1(1-w_1)(p-1-c) - \\ &- w_1(2E_1+a+b-p+2) + \sum_{j=2}^{p}\frac{w_1(1-w_j)E_1}{w_1-w_j}\Bigg]\frac{\partial F_1}{\partial w_1} - \\ &- (1-w_1)\sum_{j=2}^{p}\frac{w_j(1-w_j)E_j}{w_1-w_j}\frac{\partial F_1}{\partial w_j} = abF_1 \end{aligned} \tag{2.12}$$

where $E_j = -t_1(1-w_j)-w_j$ and $w_1, w_2, \ldots, w_p$ are the latent roots of W. Let

$$F_1(W) = |\underset{\sim}{I}-\underset{\sim}{W}|^a . G(W). \tag{2.13}$$

It can be easily seen from (2.12) that G satisfies the differential equation

$$w_1E_1(1-w_1)^2 \frac{\partial^2 G}{\partial w_1^2} + (1-w_1)[A(w_1)-2(1+a)w_1E_1] \frac{\partial G}{\partial w_1} - $$

$$- \sum_{j=2}^{p} (1-w_j)B(w_j) \frac{\partial G}{\partial w_j}$$

$$= \left[b-(1+a)w_1E_1+A(w_1) - \sum_{j=2}^{p} B(w_j)\right] aG \tag{2.14}$$

where $A(w_1) = \sum_{j=2}^{p} \frac{w_1(1-w_j)E_1}{w_1-w_j} -t_1(1-w_1)(c-p+1) -w_1(a+b-p+2)$,

$B(w_j) = (1-w_1)w_jE_j/(w_1-w_j)$ and subject to the condition that

(a) $G(\underset{\sim}{W})$ is a symmetric function of $w_1, w_2, \ldots, w_p$

and

(b) $G(\underset{\sim}{0}) = 1$.

Finally let $H(\underset{\sim}{W}) = \log G(\underset{\sim}{W})$ and so $H(\underset{\sim}{0}) = 0$. From (2.14), it follows that H satisfies the following partial differential equation:

$$w_1E_1(1-w_1)^2 \left[\frac{\partial^2 H}{\partial w_1^2} + \left(\frac{\partial H}{\partial w_1}\right)^2\right] +$$

$$+ (1-w_1)[A(w_1)-2(1+a)w_1E_1] \frac{\partial H}{\partial w_1} -$$

$$- \sum_{j=2}^{p} (1-w_j)B(w_j) \frac{\partial H}{\partial w_j}$$

$$= a\left[b - (1+a)w_1E_1 + A(w_1) - \sum_{j=2}^{p} B(w_j)\right] \quad (2.15)$$

We now look for a solution of (2.15) of the form

$$H(\underset{\sim}{W}) = \sum_{k=0}^{\infty} \frac{p_k(\underset{\sim}{W})}{n^{k/2}} \quad (2.16)$$

where for all k, $p_k(\underset{\sim}{0}) = 0$ so that $H(\underset{\sim}{0}) = 0$. Using (2.16) in (2.15) and then equating the coefficients of $\sqrt{n}$, n and constant term on both sides of the resulting equation, we have on resubstituting for $a = -\sqrt{n}\,T$, $b = nt_1$ and $c = n - \sqrt{n}\,T$:

$$p_0 = p_0(\underset{\sim}{W}) = T^2(\sigma_1+b_1\sigma_2) \quad (2.17)$$

$$p_1 = p_1(\underset{\sim}{W}) = -T^3G_1 + TG_2 \quad (2.18)$$

and

$$p_2 = p_2(\underset{\sim}{W}) = T^2D_1 + T^4D_2 \quad (2.19)$$

where

$$\sigma_i = tr(\underset{\sim}{W}^i),$$

$$b_1 = t_2/2t_1, G_1 = [6t_1(2t_1-1)\sigma_2-2t_2(1-5t_1)\sigma_3 +$$

$$+ 3t_2^2\sigma_4-6t_1^2\sigma_1]/6t_1^2,\ t_2 = 1-t_1,\ G_2 = -b_1\sigma_1^{\,2},$$

$$D_1 = [t_2^2(4\sigma_1\sigma_3 + \sigma_2^2) - 4\,(1-3t_1)\,t_2\sigma_1\sigma_2-$$

$$- 2t_1(3-4t_1)\sigma_1^{\,2}]/4t_1^2$$

and

$$D_2 = [10t_2^3\sigma_6 - 12t_2^2(1-4t_1)\sigma_5 + 3t_2(31t_1^2 - 17t_1 + 1)\sigma_4 +$$

$$+ 4t_1(23t_1^2 - 21t_1 + 3)\sigma_3 + 6t_1^2(3-8t_1)\sigma_2 + 12t_1^3\sigma_1]/12t_1^3$$

Thus we have

Theorem 2.2. An symptotic sol ion of the hypergeometric function ${}_2F_1(-\sqrt{n}\ T,\ nt_1;\ n-\sqrt{n}\ T;\ \underset{\sim}{R})$ of the Hermitian matrix argument $\underset{\sim}{R}$ for large values of n is given by

$${}_2F_1(-\sqrt{n}\ T,\ nt_1;\ n-\sqrt{n}\ T;\ \underset{\sim}{R})$$

$$= |\underset{\sim}{I}-\underset{\sim}{W}|^{-\sqrt{n}\ T} \exp[p_0 + p_1/n^{1/2} + p_2/n + \underset{\sim}{O}(n^{-3/2})]$$

where $\underset{\sim}{W} = \underset{\sim}{I} - (I - t_1\underset{\sim}{R})^{-1}$ and the coefficients p_0, p_1 and p_2 are given by (2.17), (2.18) and (2.19) respectively.

3. ASYMPOTOTIC DISTRIBUTIONS OF $Y = -\sqrt{n}\ \log(\lambda/t_2^p)$

In this section, we shall obtain the asymptotic expansions of the distribution of Y for the cases (a) and (b) stated in section 1.

Case (a). $\underset{\sim}{M} = \underset{\sim}{I} - \underset{\sim}{\Lambda}$ is a constant matrix.

Let $L_1 = Y - \sqrt{n}\ \log|\underset{\sim}{\Lambda}_1|$ where $\underset{\sim}{\Lambda}_1 = t_1\underset{\sim}{\Lambda} + t_2\underset{\sim}{I}$. Then the characteristic function $\phi(t)$ of L_1, on using lemma 2 is given by

$$\phi(t) = |t_2^{-1}\underset{\sim}{\Lambda}_1|^{-\sqrt{n}\ T} C_1(t)\cdot C_2(t) \qquad (3.1)$$

where

$$T = \sqrt{-1}\ t = it,\quad C_1(t) = \frac{\tilde{\Gamma}_p(n)\ \tilde{\Gamma}_p(nt_2 - \sqrt{n}\ T)}{\tilde{\Gamma}_p(nt_2)\tilde{\Gamma}_p(n-\sqrt{n}\ T)}$$

and

$$c_2(t) = {}_2F_1(-\sqrt{n}\, t,\ nt_1;\ n-\sqrt{n}\, t;\ \underset{\sim}{I}-\underset{\sim}{\Lambda}) \tag{3.2}$$

Now using the well known asymptotic expansion for the logarithm of a gamma function (see Anderson (1984), p. 312), we have

$$C_1(t) = [1 + Q_1/\sqrt{n} + Q_2/n + \underset{\sim}{O}(n^{-3/2})]$$

$$\exp[-pT\sqrt{n}\log t_2 + pa_1T^2/2] \tag{3.3}$$

where $a_r = t_2^{-r}-1$, $Q_1 = pT(3pa_1+a_2T^2)/6$ and $Q_2 = pT^2(3pa_2 + a_3T^2)/12 + Q_1^2/2$

Also using theorem 2.2, we have

$$C_2(t) = |\underset{\sim}{\Lambda}_1|^{\sqrt{n}\,T}[1 + T_1/\sqrt{n} + T_2/n + \underset{\sim}{O}(n^{-3/2})]$$

$$\exp[(\sigma_1+b_1\sigma_2)T^2] \tag{3.4}$$

where $\sigma_1 = \mathrm{tr}(\underset{\sim}{I}-\underset{\sim}{\Lambda}_1^{-1})^i$, $b_1 = 1/2a_1$, $t_1 = p_1$ and $T_2 = p_2 + p_1/2$ as defined in (2.18) and (2.19).

From (3.1), (3.2) and (3.3) we have

$$\phi(t) = e^{(Ta_0)^2/2}[1+\alpha_1/\sqrt{n} + \alpha_2/n + \underset{\sim}{O}(n^{-3/2})] \tag{3.5}$$

where

$$a_0^2 = pa_1 + 2(\sigma_1+b_1\sigma_2),$$

$$\alpha_1 = g_1T^3 + g_2T,\quad g_1 = -G_1 + pa_2/6,\quad g_2 = G_2 + p^2a_1/2,$$

$$\alpha_2 = g_3 T^6 + g_4 T^4 + g_5 T^2,\ g_3 = (p^2 a_2^2 - 12 p a_2 G_1 + 36 G_1^2)/72,$$

$$g_4 = [a_1 a_2 p^3 + p a_3 + 2 p a_2 G_2 - 6 p^2 a_1 G_1 + 12 D_2 - 12 G_1 G_2]/12$$

and

$$g_5 = [a_1^2 p^4 + 2 a_2 p^2 + 4 p^2 a_1 G_2 + 8 D_1 + 4 G_2^2]/8$$

By inverting the characteristic function in (3.5), we have the following theorem:

Theorem 3.1. The non-null distrubtion of L_1 under the assumption the $\underset{\sim}{M}$ is a fixed matrix, has the following asymptotic expansion:

$$P(L_1/a_0 \le x) = \Phi(x) + 1/\sqrt{n}\left[-\frac{g_1}{a_0^3}\Phi^{(3)}(x) - \frac{g_2}{a_0}\Phi^{(1)}(x)\right] + 1/n\left[\frac{g_3}{a_0^6}\Phi^{(6)}(x) + \frac{g_4}{a_0^4}\Phi^{(4)}(x) + \frac{g_5}{a_0^2}\Phi^{(2)}(x)\right] + \underset{\sim}{O}(n^{-3/2})$$

where $\Phi^{(j)}(x)$ is the j-th derivative of the standard normal distribution function $\Phi(x)$.

Case (b). $\underset{\sim}{M} = \underset{\sim}{P}/\sqrt{n}$ where $\underset{\sim}{P}$ is a fixed diagonal matrix as n increases.

Under this sequence of alternatives, the characteristic function $\psi(t)$ of $Y = -\sqrt{n}\log(\lambda/t_2^p)$, using lemma 2, is given by

$$\psi(t) = t_2^{\sqrt{n}\, pT} \cdot C_1(t) \cdot C_2(t) \tag{3.6}$$

where $C_2(t) = {}_2F_1(-\sqrt{n}\,T, nt_1, n-\sqrt{n}\,T, \underset{\sim}{P}/n^{1/2})$, T and $C_1(t)$ are as in (3.1). Taking $\underset{\sim}{R} = \underset{\sim}{P}$ in Theorem 2.1, we have

$$C_2(t) = e^{-t_1\sigma_1 T}\,[1 + P_1/\sqrt{n} + (P_2+P_1^2/2)/n + \\ + \underset{\sim}{O}(n^{-3/2})] \tag{3.7}$$

where $\sigma_i = tr(\underset{\sim}{P}^i)$ and P_1 and P_2 are given in (2.8) and (2.9) respectively. Using (3.3) and (3.7) in (3.6), we have

$$\psi(t) = [1 + \beta_1/\sqrt{n} + \beta_2/n + O(n^{-3/2})] \\ \exp\,[pT^2 a_1/2 - t_1\sigma_1 T]$$

where

$$\beta_1 = \ell_1 T + \ell_2 T^2 + \ell_3 T^3,\ \ell_1 = (p^2 a_1 - t_1^2\sigma_2)/2,$$

$$\ell_2 = -t_1\sigma_1,\ \ell_3 = pa_2/6,$$

$$\beta_2 = h_1 T + h_2 T^2 + h_3 T^3 + h_4 T^4 + h_5 T^5 + h_6 T^6$$

$$h_1 = -t_1^2(3\sigma_2+2t_1\sigma_3)/6,\ h_2 = [p^2(a_1^2 p^2+2a_2) + \\ + t_1\sigma_2(4-2p^2 a_1 t_1 - 8t_1 + t_1^3\sigma_2)]/8,$$

$$h_3 = t_1\sigma_1[-p^2 a_1 - 2 + t_1^2\sigma_2]/2,$$

$$h_4 = [a_1 a_2 p^3 + pa_3 - pa_2 t_1^2\sigma_2 + 6t_1^2\sigma_1^2]/12,$$

$$h_5 = -pa_2 t_1\sigma_1/6,$$

and

$$h_6 = p^2 a_2^2/72.$$

By inverting the characteristic function in (3.8), we have the following theorem:

Theorem 3.2. Under the sequence of local alternatives K_n: $\underset{\sim}{M} = \underset{\sim}{P}/n^{1/2}$, the asyymptotic expansion of the distribution function of the random variable $L_2 = Y + t_1\sigma_1$ is given by

$$P(L_2/b_0 \le x) = \phi(x) + 1/\sqrt{n} \sum_{j=1}^{3} (-1/b_0)^j \ell_j \phi^{(j)}(x) +$$

$$+ 1/n \sum_{j=1}^{6} (1/b_0)^j h_j \phi^{(j)}(x) + \underset{\sim}{O}(n^{-3/2})$$

where $b_0^2 = pa_1/2$ and $\phi(x)$ and $\phi^j(x)$ are as in Theorem 3.1.

Air Force Institute of Technology
Wright-Patterson Air Force Base, Ohio 45433

REFERENCES

Anderson, T. W. (1984). An Introduction to Multivariate Statistical Analysis. New York: John Wiley and Sons.

Brillinger, D. R. (1969). 'The Canonical Analysis of Stationary Time Series,' in Multivariate Analysis - II (P. R. Krishnaiah, Ed.). New York: Academic Press, 331-350.

Chikuse, Y. (1976). 'Partial Differential Equations for Hypergeometric Functions of Complex Matrices and Their Applications.' Ann. Inst. Statist. Math., 28, 187-199.

Goodman, N. R. (1963). 'Statistical Analysis Based on a Certain Multivariate Complex Gaussian Distribution.' Ann. Math. Statist., 34, 152-176.

Goodman, N. R. and Dubman, M.R. (1969). 'Theory of Time-Varying Spectral Analysis and Complex Wishart Matrix

Processes,' in Multivariate Analysis - II (P. R. Krishnaiah, Ed.). New York: Academic Press, 351-366.

Hannan, E. J. (1970). Multiple Time Series, New York: John Wiley and Sons.

James, A. T. (1964). 'Distribution of Matrix Variates and Latent Roots Derived from Normal Samples.' Ann. Math. Statist., 35, 475-501.

Krishnaiah, P. R. (1976). "Some Recent Developments on Complex Multivariate Distributions.' J. Multivariate Analysis, 6, 1-30.

Pillai, K. C. S. and Jouris, G. M. (1971). 'Some Distribution Problems in the Multivariate Complex Gaussian Case.' Ann. Math. Statist., 42, 517-525.

Pillai, K. C. S. and Nagarsenker, B. N. (1972). 'On the Distributions of a Class of Statistics in Multivariate Analysis.' J. Multivariate Analysis, 2, 96-114.

Priestley, M. B., Subha Rao, T. and Tong, H. (1973). 'Identification of the Structure of Multivariate Stochastic Systems' in Multivariate Analysis - III (P. R. Krishnaiah, Ed.). New York: Academic Press.

Sugiama, T. (1972). 'Distributions of the Largest Latent Root of the Multivarate Complex Gaussian Distribution,' Ann. Inst. Statist. Math., 24, 87-94.

Wahba, G. (1968). 'On the Distributions of Some Statistics Useful in the Analysis of Jointly Stationary Time Series.' Ann. Math. Statist., 39, 1849-1862.

Wahba, G. (1971). 'Some Tests of Independence for Stationary Multivariate Time Series.' J. Roy. Statist. Soc. Ser. B., 33, 153-166.

Ingram Olkin and Milton Sobel

A MODEL FOR INTERLABORATORY DIFFERENCES †

ABSTRACT

When a single specimen is divided into samples and sent to different laboratories for analysis, it often happens that a few of the laboratories obtain results that consistently are discrepant from the others. Using an additive model, we propose a new approach based on ordering and selection principles to detect and remove the discrepant laboratories. Parameters are then estimated using the remaining laboratories.

1. INTRODUCTION

When a single specimen is divided into aliquot parts and sent to different laboratories (or given to different technicians) for measurement or analysis, it often happens that a few results are discrepant. Further, the discrepancies may be larger than expected from chance alone. It is generally recognized that certain laboratories (or technicians) tend to overestimate or underestimate their measurements. This situation is exemplified by Figure 1 in which each of eight laboratories made three determinations of the vapor pressure at 35 degrees centigrade of the analyte 2.4–dimitrotoluene.

For such situations a model should include a laboratory error effect and multiple observations. Suppose that n measurements are taken at each of k laboratories. Let x_{ij} denote the i-th measurement from the j-th laboratory $(i = 1, \ldots, n; j = 1, \ldots, k)$. More specifically, assume that

$$x_{ij} = \mu + \theta_j + \varepsilon_{ij}, \tag{1.1}$$

where μ is the true overall mean, θ_j represents the error effect or parameter of the j-th laboratory, and the set of ε_{ij}'s are independently and normally

Key Words: ranking and selection, additive model, subset selection, detection of outliers.

† This work was supported in part by National Science Foundation Grants DMS 84-11411 and MCS 82-02247.

A. K. Gupta (ed.), Advances in Multivariate Statistical Analysis, 303–314.

distributed with a common mean zero and common variance σ^2. Furth suppose that there is sufficient past data to assume that σ^2 is known.

If no assumptions are made about the θ_j, then the true mean μ is no identifiable, and indeed, it may be that all k laboratories underestimate (overestimate) the true value μ. An assumption that $\sum_1^k \theta_j = 0$ is made order to guarantee identifiability. This assumption is equivalent to assumi that the results from the k different laboratories center about the true val μ.

The object is to remove a subset of laboratories that can be regard as discrepant, and use the remaining laboratories to provide an estima (better than the original sample) of the true mean μ. The outliers are th laboratories that correspond to small or large θ-values. We propose a n approach based on ordering and selection principles and methodology designate the discrepant laboratories. These methods are used in Sectic 2 and 3 to set up a probabilistic requirement and to find a least favoral configuration. The rule for selecting discrepant subsets as well as the quired common sample size are provided in Section 2. An improvement the procedure is developed in Section 3.

In a practical application it is often the case that the sample size n each laboratory is determined for reasons of expense, time, equipment, e When the sample size is fixed, the analysis can be inverted to determin P-value which can be regarded as the confidence level associated with rule or as the probability that the outliers are correctly selected.

2. FORMULATION OF THE PROBLEM

The k laboratories measure aliquot parts of a single uniform specimen; he these k samples have a common population mean μ. Although our prim concern lies with the case of an unknown mean μ, comparisons with corresponding problem for known μ may also be of interest in that it provi a check as to whether the proposed rule is efficient in screening out outli

For the model (1.1) we wish to screen out those laboratories with la errors from those with small errors, that is, those for which $|\theta_j| \geq \delta^*$ fr those for which $|\theta_j| < \delta^*$, where $\delta^* > 0$ is specified. The former are m likely to produce sample outliers, whereas the latter will yield a more r able estimate of μ. Our procedure R (defined below) assigns each of th laboratories to one of three disjoint groups defined in the parameter sp by $\theta_j \leq -\delta^*$, $-\delta^* < \theta_j < \delta^*$, and $\theta_j \geq \delta^*$.

In particular, our concern is with the case in which there exist disti indices i and j $(1 \leq i < j \leq k-1)$ such that

$$\theta_{[i+1]} \geq \theta_{[i]} + \delta^*, \quad \theta_{[j+1]} \geq \theta_{[j]} + \delta^*. \qquad (2$$

Such a separation creates at least 3 groups for which the mean in each group is at least δ^* away from the mean in any other group. Of course, there may be more than one such pair (i, j) that creates at least three groups. A separation based on data that is consistent with any such pair will be regarded as a correct assignment of the various populations.

When the separation described by (2.1) holds we want the rule R to have the property that the joint probability that all k assignments are simultaneously correct is at least P^*, where P^* is specified to be close to 1, say 0.90 or 0.95, and considerably larger than $(1/3)^k$.

We say that we have a correct assignment (CA) when all k populations are correctly assigned. This is tantamount to making a correct selection in the usual ranking and selection formulation, since we select those populations in the middle group.

Let $\bar{x}_j$ denote the sample mean from population or laboratory π_j and $\bar{x}$ the overall sample mean. (When μ is known, replace $\bar{x}$ by μ in the procedure below.) The three groups defined by the population parameter above are denoted $\mathcal{G}_1, \mathcal{G}_2, \mathcal{G}_3$.

The assignment procedure R is: For some constant $c > 0$ to be determined,

(i) assign π_j to $\mathcal{G}_1$ if $\bar{x}_j \leq \bar{x} - c$,
(ii) assign π_j to $\mathcal{G}_2$ if $\bar{x} - c < \bar{x}_j < \bar{x} + c$,
(iii) assign π_j to $\mathcal{G}_3$ if $\bar{x} + c \leq \bar{x}_j$.

Those samples from laboratories assigned to group G_2 are then averaged to yield an estimate of μ.

Suppose that the true size of group $\mathcal{G}_i$ is $k_i (i = 1, 2, 3)$, so that

$$k_1 + k_2 + k_3 = k. \tag{2.2}$$

Let $\bar{u}_\alpha, \bar{v}_\beta$ and $\bar{w}_\gamma$ denote the sample means corresponding to those populations from groups $\mathcal{G}_1, \mathcal{G}_2, \mathcal{G}_3$, respectively. Then the overall mean $\bar{x}$ can be written in the form

$$\bar{x} = (\Sigma\bar{u}_\alpha + \Sigma\bar{v}_\beta + \Sigma\bar{w}_\gamma)/k, \tag{2.3}$$

The probability of a correct assignment using procedure R is

$$P\{CA\} = P\{\bar{u}_\alpha \leq \bar{x} - c, \quad \bar{x} - c < \bar{v}_\beta < \bar{x} + c, \quad \bar{x} + c < \bar{w}_\gamma\}, \tag{2.4}$$

where α, β, γ correspond to the groups $\mathcal{G}_1, \mathcal{G}_2, \mathcal{G}_3$, respectively.

To minimize $P\{CA\}$ in the parameter space, set all the population means in $\mathcal{G}_1, \mathcal{G}_2, \mathcal{G}_3$ equal to $\mu - \delta^*$, μ, and $\mu + \delta^*$, respectively. This restriction to three parameter values $\mu - \delta^*$, μ and $\mu + \delta^*$ follows ranking and selection theory methodology (e.g., see Gibbons, Olkin and Sobel, 1977), where $\delta^* > 0$ is specified, and only misclassification of means less than

$\mu - \delta^*$, greater than $\mu + \delta^*$, or equal to μ, are regarded as serious errors. Then

$$\begin{aligned} &E(\bar{u}_\alpha - \bar{x},\ \bar{v}_\beta - \bar{x},\ \bar{w}_\gamma - \bar{x}) \\ &\quad = (-\delta^* + \frac{k_1 - k_3}{k}\delta^*,\ \frac{k_1 - k_3}{k}\delta^*,\ \delta^* + \frac{k_1 - k_3}{k}\delta^*), \end{aligned} \tag{2.5}$$

and

$$\begin{aligned} &\mathrm{Cov}(\bar{u}_\alpha - \bar{x}, \bar{v}_\beta - \bar{x}, \bar{w}_\gamma - \bar{x}) \\ &\quad = \sigma_0^2 \begin{pmatrix} 1 & \rho & \rho \\ & 1 & \rho \\ & & 1 \end{pmatrix}, \end{aligned} \tag{2.6}$$

where

$$\sigma_0^2 = \tfrac{\sigma^2}{n}\tfrac{k-1}{k}, \quad \rho = -\tfrac{1}{k-1}. \tag{2.7}$$

Assume for the moment that k is larger than 5, and preferably larger than 10. In this case, ρ will be close to zero.

Firstly we claim that the worst case in (2.43) occurs when $k_1 = k_3$; we discuss this later at greater length. Secondly, we give an exact expression for $P\{CA\}$ and then expand the result in a Taylor expansion about $\rho = 0$. Because of the weak common correlation ρ, the result for the case of independence is essentially the same as for $\rho = -1/(k-1)$. To establish this we derive and investigate a correction term in the Taylor expansion.

Write

$$\sigma_1 = \sigma\sqrt{\tfrac{k-1}{k}}, \quad a = \tfrac{(k_1 - k_3)\delta^*}{k\sigma_1}, \quad c_1 = \tfrac{\delta^* - c}{\sigma_1}, \quad c_2 = \tfrac{c}{\sigma_1}. \tag{2.8}$$

Under independence, $\rho = 0$, and we obtain from (2.4) and (2.8) that

$$\begin{aligned} P\{CA\} \approx &\Phi^{k_1}((c_1 - a)\sqrt{n})\,[\Phi((c_2 - a)\sqrt{n}) - \\ &- \Phi(-(c_2 + a)\sqrt{n})]^{k_2}\,\bar{\Phi}^{k_3}(-(c_1 + a)\sqrt{n}), \end{aligned} \tag{2.9}$$

where $\Phi(x)$ is the standard normal cumulative distribution function and $\bar{\Phi}(x) = 1 - \Phi(x) = \Phi(-x)$.

That $a = 0$ for the minimizing $P\{CA\}$ can be argued from different points of view. First, recall the assumption (A) $\sum_1^k \theta_j = 0$, and suppose that k is not necessarily an integer. Assumption (A) is consistent with the fact that $k_1 = k_3$ in the least favorable configuration. In fact, if we let the θ_i values in the i-th group be $-\delta^*$, 0, δ^* for $i = 1, 2$, and 3, respectively then it readily follows from assumption (A) that $k_1 = k_3$. Alternatively, to obtain a minimax solution, we can apply the equalizing strategy to the first and third factors in (2.9). This means that we set $k_1 = k_3$, and hence $a = 0$

It then follows that to further minimize (2.9) over pairs (k_1, k_2) with $2k_1 + k_2 = k$, we either have (i) $k_2 = k$ and $k_1 = k_3 = 0$, or (ii) $k_2 = 0$ and $k_1 = k_3 = k/2$. Moreover, to minimize (2.10) for integer values of k, we take $k_1 = k_3 = k/2$ when k is even, and one of k_1, k_3 to be $(k+1)/2$ and the other $(k-1)/2$ when k is odd. The reason for alternatives (i) and (ii) is that although c and c_2 are positive, it is unclear whether c_1 is positive, so that we don't know which of $\Phi(c_1\sqrt{n})$ and $2\Phi(c_2\sqrt{n}) - 1$ is the larger.

The general solution of this problem is obtained as the smallest sample size n such that

$$\min_n\{\Phi(c_1\sqrt{n}),\ 2\Phi(c_2\sqrt{n}) - 1\} \geq (P^*)^{1/k}, \tag{2.10}$$

where $c_1 = (\delta^* - c)/\sigma_1$, $c_2 = c/\sigma_1$.

Closed form expressions can be obtained in certain special cases. In particular, if $c = \delta^*/2$, then $c_1 = c_2 = \delta^*/2\sigma_1$, so that $2\Phi(c_2\sqrt{n}) - 1 < \Phi(c_1\sqrt{n})$. The solution takes the simple form

$$\Phi(c_2\sqrt{n}) = \tfrac{1}{2}\{1 + (P^*)^{1/k}\} \equiv P_1^*, \tag{2.11}$$

or equivalently, the desired common sample size from each of the k laboratories is

$$n = \lceil \tfrac{k-1}{k}\left(\tfrac{2\sigma\Phi^{-1}(P_1^*)}{\delta^*}\right)^2 \rceil, \tag{2.12}$$

where $\lceil m \rceil$ is the smallest integer greater than or equal to m.

Example. Suppose $k = 10$ and we want the probability of a correct assignment to be at least 0.75 when the laboratory effects are at least one standard deviation apart, that is, $\delta^* = \sigma$. For $c = \delta^*/2$ we obtain

$$P_1^* = \tfrac{1}{2}\{1 + (0.75)^{1/10}\} = 0.9858, \quad \Phi^{-1}(P_1^*) = 2.192,$$

$$n = \lceil (\tfrac{9}{10} 4(2.192)^2 \rceil = \lceil 17.3 \rceil = 18,$$

so that 18 observations are required from each of the $k = 10$ laboratories.

Example. Suppose that there are eight laboratories, each providing three measurements:

laboratory	A	B	C	D	E	F	G	H
observations	90	830	85	70	90	110	130	70
	125	930	130	75	70	400	110	80
	85	1300	90	90	100	390	135	70
means	100	1020	102	78	87	300	125	73

Further, suppose that $\sigma = 70$, $\delta^* = 3\sigma = 210$, and $c = \frac{2}{3}\delta^* = 140$. The ru is to assign to group $\mathcal{G}_1$ if

$$\bar{x}_j \leq \bar{x} - c = 235.625 - 140 = 95.6,$$

to assign to group $\mathcal{G}_2$ if

$$95.6 < \bar{x}_j \leq 375.6,$$

and to assign to group $\mathcal{G}_3$ if

$$375.6 < \bar{x}_j.$$

Thus, laboratories D, E, H are assigned to group $\mathcal{G}_1$ laboratories $A,$ (F, G are assigned to group $\mathcal{G}_2$ and laboratory B is assigned to group $\mathcal{G}_3$.

To determine $P\{CA\}$, we first need to compare $\Phi(c_1\sqrt{n})$ with $2\Phi(c_2\sqrt{}$ -1 as in (2.10). From $c_1\sqrt{n} = 1.852$, $c_2\sqrt{n} = 3.703$, so that $\Phi(c_1\sqrt{n})$ 0.968, $2\Phi(c_2\sqrt{n}) - 1 = 0.999$.

Consequently we have the case $k_2 = 0, k_1 = k_3 = 4$, and

$$P\{CA\} \cong \Phi^8(c_1\sqrt{n}) = (0.968)^8 = .77.$$

Thus the probability of a correct assignment is 0.77, which means th when the error parameters θ_j are at least δ^* then with probability 0.77 correctly assign the eight populations to the three groups.

In the next section we obtain an improvement to (2.10) by jointly (termining the value of c and the smallest common sample size n required achieve the P^* condition.

3. A MINIMAX SOLUTION

Let $c = r\delta^*, 0 < r < 1$, and write $\gamma = \delta^*\sqrt{n}/\sigma_1 = \delta^*\sqrt{nk/(k-1)}$, Equating the two expressions in (2.10) and setting the common value eq to $(P^*)^{1/k}$ yields

$$\Phi\left(\frac{(\delta^*-c)\sqrt{n}}{\sigma_1}\right) = 2\Phi\left(\frac{c\sqrt{n}}{\sigma_1}\right) - 1 = (P^*)^{1/k}, \quad (3$$

or equivalently,

$$\Phi(\gamma(1-r)) = 2\Phi(\gamma r) - 1 = (P^*)^{1/k}. \quad (3$$

The two equations in (3.2) are readily solved for γ and r since

$$\Phi^{-1}((P^*)^{1/k}) = \gamma(1-r) \equiv \lambda_1, \quad \Phi^{-1}\left(\frac{(P^*)^{1/k}+1}{2}\right) = \gamma r \equiv \lambda_2.$$

Then

$$\gamma = \lambda_1 + \lambda_2, \quad r = \lambda_2/(\lambda_1 + \lambda_2),$$

$$c = r\delta^*, \; n = \lceil \tfrac{\gamma \sigma_1^2}{\delta^*} \rceil = \lceil (\tfrac{k-1}{k})(\tfrac{\gamma\sigma}{\delta^*})^2 \rceil.$$

Example. Using the data of the example in Section 2,

$$\lambda_1 = 1.905, \; \lambda_2 = 2.192, \; \gamma = 4.097, \; r = 0.535,$$

so that medskip

$$c = 0.535\delta^*, \; n = \lceil \tfrac{9}{10}(4.097)^2 \rceil = \lceil 15.1 \rceil = 16.$$

By jointly determining c and n we obtain $(c, n) = (0.535\delta^*, 16)$ instead of the values $(0.5\delta^*, 18)$. This yields a reduction of 2 observations per population or a total reduction of 20 observations. In other cases the reduction can be larger. For example, if $k = 5, P^* = 0.90$ and $\delta^* = \sigma/2$, the required value of n is 86 per population, whereas for $c = 0.265\sigma$, the value of n is 76, which is a total reduction of 50 observations in this case.

4. A ONE-SIDED ANALOGUE

In the one-sided analogue of our model there are still k laboratories (or populations) but only two groups $\mathcal{G}_1$ and $\mathcal{G}_2$ according to whether the laboratory tends to give correct estimates or tends to overestimate (say) the desired measurement. As before we use $\bar{u}_\alpha$ to denote the sample mean from those populations assigned to group $\mathcal{G}_1$ with true mean μ, and $\bar{v}_\beta$ to denote the sample mean from those populations assigned to group $\mathcal{G}_2$ with true mean $\mu + \delta^*$. The common variance is σ^2 (assumed to be known) for both groups.

Consider the set of random variables

$$\bar{u}_\alpha - \bar{x}, \;\; \bar{v}_\beta - \bar{x}, \;\; \alpha = 1, \ldots, k_1; \; \beta = 1, \ldots, k_2, \tag{4.1}$$

where $\bar{x}$ is the overall sample mean, k_j is the size of group $\mathcal{G}_j, j = 1, 2$, and $k_1 + k_2 = k$. The sampling rule R is to place π_j in the selected (non-discrepant) subset if and only if $\bar{x}_j - \bar{x} < c$, where $\bar{x}_j$ is the sample mean of laboratory j, and $c \geq 0$. The constant c and the common sample size n are both to be determined.

The means, variances and correlations of the random variables (4.1) are

$$E(\bar{u}_\alpha - \bar{x}) = -k_2\delta^*/k, \quad E(\bar{v}_\beta - \bar{x}) = k_1\delta^*/k, \tag{4.2}$$

$$\mathrm{Var}(\bar{u}_\alpha - \bar{x}) = \mathrm{Var}(\bar{v}_\beta - \bar{x}) = \tfrac{\sigma^2}{n}(\tfrac{k-1}{k}) = \tfrac{\sigma_1^2}{n}, \tag{4.3}$$

$$\mathrm{Corr}(u_{\alpha_1}, u_{\alpha_2}) = \mathrm{Corr}(u_\alpha, v_\beta) = \mathrm{Corr}(v_{\beta_1}, v_{\beta_2}) = -\tfrac{1}{k-1}. \tag{4.4}$$

As before the common correlation ρ is weak, and we use independence $(\rho = 0)$ to approximate the desired answer

$$P\{CA|R\} \cong \Phi^{k_1}[(c_2 + a_2)\sqrt{n}]\bar{\Phi}^{k_2}[(c_2 - a_1)\sqrt{n}], \tag{4.5}$$

where $a_i = k_i\delta^*/k\sigma_1 \;\; (i = 1, 2)$, and $c_2 = c/\sigma_1$.

Since the two expectations in (4.2) are functions of k_1, we allow c to depend on k_1 also. In fact, it is highly intuitive that c should be the average of these two expectations, namely,

$$c = \left(\tfrac{k_1-k_2}{2k}\right)\delta^*. \tag{4.6}$$

The right-hand side of (4.5) then reduces to $\Phi^k(\delta^*\sqrt{n}/2\sigma_1)$ and is constant with respect to k_1, so that any value of k_1 will serve to minimize it. However, using (4.6) in the exact expression given below in (A.7), k_1 still appears but only as an exponent, from which it follows that $k_1 = k_2$ yields a minimum.

To see this, rewrite (A.7) with the simplified notation $\tilde{x} = x\sqrt{\rho}/\sqrt{1-\rho}$ and $\tilde{a} = a\sqrt{n}/\sqrt{1-\rho}$, and $k_2 = k - k_1$. Setting the derivative with respect to k_1 of the resulting form of (A.7) equal to zero yields the equation

$$\int_{-\infty}^{\infty} \Phi^{k_1}(\tilde{x}+\tilde{a})\, \bar{\Phi}^{k-k_1}(\tilde{x}-\tilde{a})\, [\log \Phi(\tilde{x}+\tilde{a})]\, \varphi(x)\, dx$$
$$= \int_{-\infty}^{\infty} \Phi^{k_1}(\tilde{x}+\tilde{a})\, \bar{\Phi}^{k-k_1}(\tilde{x}-\tilde{a})\, [\log \bar{\Phi}(\tilde{x}-\tilde{a})]\, \varphi(x)\, dx.$$

The substitution $-\tilde{x}$ for $\tilde{x}$ in the right-hand side yields

$$\int_{-\infty}^{\infty} \bar{\Phi}^{k_1}(\tilde{x}-\tilde{a})\, \Phi^{k-k_1}(\tilde{x}+\tilde{a})\, [\log \Phi(\tilde{x}+\tilde{a})]\, \varphi(x)\, dx,$$

which is equal to the left-hand side when $k_1 = k - k_1$, so that $k_1 = k - k_1$ is a solution. It is then easily seen that this is a minimum and indeed a unique minimum; we omit these details.

Hence $c = 0$ in (4.6) and the approximate solution for n is obtained from

$$\frac{\delta^*\sqrt{n}}{2\sigma_1} = \lambda_1, \tag{4.7}$$

to be

$$n = \left\lceil \left(\tfrac{2\sigma_1\lambda_1}{\delta^*}\right)^2 \right\rceil = \left\lceil 4\left(\tfrac{k-1}{k}\right)\left(\tfrac{\sigma\lambda_1}{\delta^*}\right)^2 \right\rceil, \tag{4.8}$$

where λ_1 is the standard normal quantile associated with $(P^*)^{1/k}$.

Example. If $k = 10, P^* = 0.75$ and $\delta^* = \sigma$ as in the previous example, then $\lambda_1 = 1.905, c = 0$ and

$$n = \lceil 4(\tfrac{9}{10})(1.905)^2 \rceil = \lceil 13.06 \rceil = 14.$$

Thus we require 14 observations from each of the 10 laboratories.

APPENDIX

An exact expression is derived for $P\{CA\}$ as a function of the common known correlation ρ. Unfortunately, the expression is not available as it stands for direct numerical evaluation. For $\rho = 0$ it gives the same answer as that obtained in Section 3. For $\rho \neq 0$ the correction terms indicate the order of magnitude of the correction needed. These correction terms represent the second or third term in a Taylor expansion of $P\{CA\}$ about $\rho = 0$, and are useful for comparing the first term to the value of $P\{CA\}$.

For $k_1 = k_3$ we standardize the variables in (2.4) and (2.5):

$$\begin{aligned}(\bar{u}_\alpha - \bar{x} + \delta^*)\sqrt{n}/\sigma_1 &= Y_\alpha\sqrt{1-\rho} - Y_0\sqrt{\rho},\\ (\bar{v}_\beta - \bar{x})\sqrt{n}/\sigma_1 &= Y_\beta\sqrt{1-\rho} - Y_0\sqrt{\rho},\\ (\bar{w}_\gamma - \bar{x} - \delta^*)\sqrt{n}/\sigma_1 &= Y_\gamma\sqrt{1-\rho} - Y_0\sqrt{\rho},\end{aligned} \tag{A.1}$$

where $Y_0, Y_1, \ldots, Y_k$ are all independently distributed standard normal random variables. It is straightforward to show that the distribution of the $Y's$ is that of the left-hand members in (A.1). The fact that (A.1) contains $\sqrt{\rho}$ may appear to negate the representation in (A.2) for $\rho < 0$. However, this apparent difficulty can be handled as in Steck and Owen (1962) by showing that the imaginary part in (A.2) is zero and that the real part in (A.2) provides the desired answer.

From (A.1), introducing a as in (2.7) and (2.8), and using a standard conditional argument as in Dunnett and Sobel (1954) or in Gibbons, Olkin, Sobel (1977), we obtain for the $P\{CA\}$:

$$\begin{aligned} g(\rho) = \int_{-\infty}^{\infty} &\Phi^{k_1}\Big(\frac{y\sqrt{\rho}+b_1}{\sqrt{1-\rho}}\Big)\Big[\Phi\Big(\frac{y\sqrt{\rho}+b_2}{\sqrt{1-\rho}}\Big) - \\ &- \Phi\Big(\frac{y\sqrt{\rho}-d_2}{\sqrt{1-\rho}}\Big)\Big]^{k_2}\bar{\Phi}^{k_3}\Big(\frac{y\sqrt{\rho}-d_1}{\sqrt{1-\rho}}\Big)\varphi(y)dy,\end{aligned} \tag{A.2}$$

where $\varphi(x)$ is the standard normal density function, and

$$b_1 = c_1\sqrt{n} - a, \quad b_2 = c_2\sqrt{n} - a, \quad d_1 = c_1\sqrt{n} + a, \quad d_2 = c_2\sqrt{n} + a.$$

For ρ close to zero the properties of $g(\rho)$ will not change from those at $\rho = 0$. Hence we claim that a minimum will again be attained by the choice $k_1 = k_3$ (and hence $a = 0$). Further, to obtain a minimax solution we set $(k_1, k_2, k_3) = (0, k, 0)$ or $(k/2, 0, k/2)$. Thus $g(\rho)$ is given by

$$g(\rho) = \min\{g_1(\rho), g_2(\rho)\} \tag{A.3}$$

where

$$g_1(\rho) = \int_{-\infty}^{\infty} \Phi^{k/2}\Big(\frac{y\sqrt{\rho}+c_1\sqrt{n}}{\sqrt{1-\rho}}\Big)\bar{\Phi}^{k/2}\Big(\frac{y\sqrt{\rho}-c_1\sqrt{n}}{\sqrt{1-\rho}}\Big)\varphi(y)dy,$$

$$g_2(\rho) = \int_{-\infty}^{\infty} \Big[\Phi\Big(\frac{y\sqrt{\rho}+c_2\sqrt{n}}{\sqrt{1-\rho}}\Big) - \Phi\Big(\frac{y\sqrt{\rho}-c_2\sqrt{n}}{\sqrt{1-\rho}}\Big)\Big]^k \varphi(y)dy$$

The value of $g(0)$ leads to the solution obtained in Section 3 under th assumption of independence. From (A.3), a rather lengthy exact derivatio of the derivatives $g_1'(0), g_2'(0)$ and $g_2''(0)$ yields

$$\begin{aligned} g_1'(0) &= -\frac{k}{2}\ \varphi^2(c_1\sqrt{n})\ \Phi^{k-2}(c_1\sqrt{n}), \\ g_2'(0) &= 0, \\ g_2''(0) &= -\frac{1}{2}k(k-1)(k-2)c_2\sqrt{n}\ \varphi^3(c_2\sqrt{n})\ \Phi^{k-3}(c_2\sqrt{n}). \end{aligned} \qquad \text{(A.4}$$

Hence the approximation to $P\{CA\}$ with the correction terms based on g_1'(0 and $g_2''(0)$ is

$$P\{CA\} \approx \min\{g_1(0) + \tfrac{1}{k-1}g_1'(0), \quad g_2(0) - \tfrac{1}{(k-1)^2}g_2''(0)\}. \qquad \text{(A.5}$$

Example. Using the same data as before with $k = 10, P^* = 0.75, \delta^* = \sigma$ we insert $c = 0.535\sigma$ and $n = 16$ in the correction terms in (A.5) to obtai 0.001 as the correction for $g_1(\rho)$ and -0.0001 as the correction for $g_2(\rho)$. Thu the error in our earlier solution is negligible for all practical purposes.

For the example of the one-sided case discussed in Section 4, (for whic $c_2 = 0, a_1 = a_2$) the first correction term in (A.6) increases the $P\{CA\}$ valu given by (4.5) and since the value for $n = 14$ obtained in the example is ver close to 0.75 the value for $n = 13$ should be checked. In fact, for $n = 13$ th common value a of a_1 and a_2 is 0.527 and the approximation (A.8) gives

$$\begin{aligned} P\{CA\} &\cong \Phi^{10}(0.527\sqrt{13}) + \frac{5}{9}\varphi^2(0.527\sqrt{13})\Phi^8(0.527\sqrt{13}) \\ &= 0.7493, \end{aligned} \qquad \text{(A.6}$$

which is less than 0.75, though close to it. Consequently, 14 observation are required.

The exact expression that takes into account the common correlatio $\rho = -1/(k-1)$ is given (after setting $c = 0$) by

$$P\{CA|R\} = \int_{-\infty}^{\infty} \Phi^{k_1}\big(\tfrac{x\sqrt{\rho}+a\sqrt{n}}{\sqrt{1-\rho}}\big)\bar{\Phi}^{k_2}\big(\tfrac{x\sqrt{\rho}-a\sqrt{n}}{\sqrt{1-\rho}}\big)\varphi(x)dx, \qquad \text{(A.7}$$

where $a = \delta^*/2\sigma$, and $k_1 = k_2 = k/2$ in the least favorable situation.

Again, it can be shown that for $\rho < 0$ the imaginary part of the righ hand side of (A.7) is zero and that the real part of (A.7) is the desire probability. Since this is exactly the same as $g_1(\rho)$ in (A.3) with c_1 replaced by a, the correction term for $g_1(\rho)$ in (A.5) will also be the same. That is, letting $g(\rho)$ denote the right-hand side of (A.7), the first two terms of a Taylor expansion about $\rho = 0$ yields

$$g(\rho) \approx g(0) + \tfrac{k}{2(k-1)}\varphi^2(a\sqrt{n})\Phi^{k-2}(a\sqrt{n}), \tag{A.8}$$

where we have already set ρ equal to $-1/(k-1)$. This was used in (A.6) where the leading term was calculated to be 0.7474 and the correction term was 0.0019.

Stanford University and University of California, Santa Barbara.

REFERENCES

Dunnett, C. W. and Sobel, M. (1954). 'A bivariate generalization of Student's t distribution, with tables for certain special cases'. *Biometrika*, **41**, 153–169.

Gibbons, J. D., Olkin, I. and Sobel, M. (1977). *Selecting and Ordering Populations: A New Statistical Methodology.* New York: Wiley.

Steck, G. P. and Owen, D. B. (1962). 'A note on the equicorrelated multivariate normal distribution'. *Biometrika* **49**, 269–271.

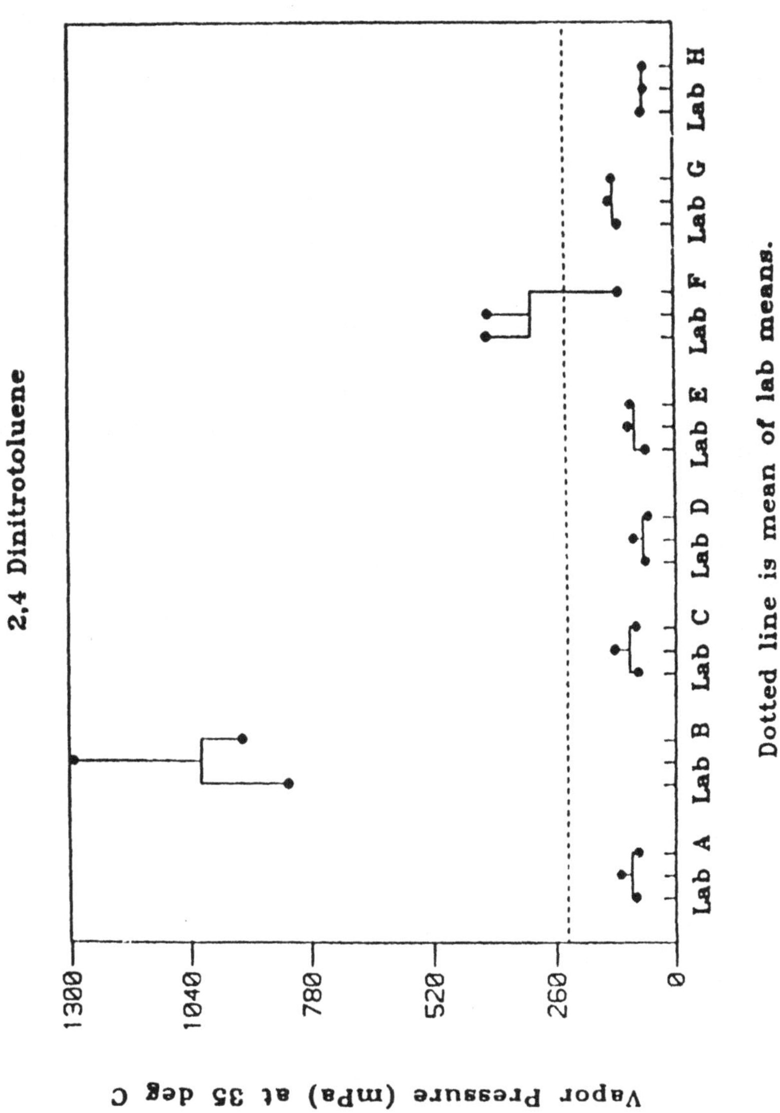
2,4 Dinitrotoluene
Vapor Pressure (mPa) at 35 deg C
1300
1040
780
520
260
0
Lab A
Lab B
Lab C
Lab D
Lab E
Lab F
Lab G
Lab H
Dotted line is mean of lab means.

FIGURE 1

Andrew L. Rukhin

BAYES ESTIMATORS IN LOGNORMAL REGRESSION MODEL

1. INTRODUCTION

In this paper we consider a lognormal regression model which is often used to describe a production process in economics (see Dhrymes (1962), Goldberger (1968), Zellner (1971)). Namely, we assume that

$$z_i = \log y_i = \theta_1 v_{i1} + \dots + \theta_k v_{ik} + \varepsilon_i, \quad i = 1,\dots,n \qquad (1.1)$$

where v's are given values of indpendent (explanatory) variables, θ's are unknown regression coefficients, and ε's are independent normal errors each with zero mean and unknown variance σ^2. The observations y's in (1.1) typically represent economic output at n succesive time moments.

Under this model the minimum variance linear unbiased estimate of θ is the least squares regression estimator $\tilde{\theta} = (V^TV)^{-1}V^TZ$. We assume that the design matrix V has full rank k. It is well-known that $\tilde{\theta}$ has multivariate normal distribution with mean θ and covariance matrix $\sigma^2\Sigma$ where $\Sigma = (V^TV)^{-1}$. Also the scaled residual sum of squares

$$(Z - V\tilde{\theta})^T(Z - V\tilde{\theta})/\sigma^2 = S^2/\sigma^2$$

has χ^2 distribution with $n - k$ degrees of freedom.

In the problem of predicting the next value of y at, say, $v = d$ or when evaluating the expected value of y, it is of interest to estimate

$$E\exp\{<\tilde{\theta},d>\} = E\exp\{\sum_{i=1}^{k}\tilde{\theta}_i d_i\}$$

$$= \exp\{<\theta,d> + b\sigma^2\}$$

A. K. Gupta (ed.), Advances in Multivariate Statistical Analysis, 315–325.

for some vector d, rather than $E\langle\tilde{\theta},d\rangle = \langle\theta,d\rangle$. Thus we study here the statistical estimation problem of the parametric function $g(\theta,\sigma) = E\exp\langle\tilde{\theta},d\rangle$ or $\log g(\theta,\sigma) = \langle\theta,d\rangle + b\sigma^2$ where d is a given vector and b is a given real constant. It is assumed that available data consists of a normal vector X with mean θ and covariance matrix $\sigma^2\Sigma$ where Σ is fixed and σ^2 is unknown, and an independent statistic S^2 such that S^2/σ^2 has χ^2-distribution with $m-1$ degrees of freedom.

Traditional estimators δ have the form

$$\delta(X,S) = \langle X,d\rangle + cS^2 \tag{1.2}$$

for some c. For instance the best unbiased estimator δ_U, is

$$\delta_U(X,S) = \langle X,d\rangle + bS^2/(m-1)$$

and the maximum likelihood estimator $\hat{\delta}$, is

$$\hat{\delta}(X,S) = \langle X,d\rangle + bS^2/m \ .$$

The loss function considered in this paper has the form

$$L(\theta,\sigma;\delta) = [\langle\theta,d\rangle + b\sigma^2 - \delta]^2/\sigma^4 \ . \tag{1.3}$$

This choice is motivated by the fact that under more traditional quadratic loss, $(g - e^{\delta})^2$, all estimators above have infinite risk for large values σ^2. Under (1.3) the risk of any estimator is finite,

$$EL(\theta,\sigma;\delta) = E\langle X-\theta,d\rangle^2/\sigma^4 + E(cS^2 - b\sigma^2)^2/\sigma^4$$

$$= \langle\Sigma d,d\rangle/\sigma^2 + [c^2(m^2-1) - 2cb(m-1) + b^2].$$

Therefore there exists the optimal choice of the constant c

$$c = c_0 = b/(m+1).$$

The corresponding estimator

$$\delta_0(X,S) = \langle X,d\rangle + bS^2/(m+1) \tag{1.4}$$

improves upon δ_U and $\hat{\delta}$.

Notice that if $d = 0$ our problem reduces to the estimation of unknown variance σ^2. In this case δ_0 is the best equivariant estimator of σ^2 and under (1.3) it has a constant risk. (This fact explains the presence of the rescaling factor σ^{-4} in (1.3)).

In this paper we shall investigate the behavior of δ_0. For this purpose we can and will assume that $\sum = I$. Indeed it suffices to replace X by $\sum^{-1/2}X$ with corresponding change of d. Also by performing a suitable orthogonal transformation one can always assume that d is proportional to the first basis vector, so that $\log g = a\theta_1 + b\sigma^2$. Here a and b are given constants and θ_1 is the first coordinate of θ. The estimator (1.4) takes the form

$$\delta_0(X,S) = aX_1 + bS^2/(m+1) . \tag{1.5}$$

In Section 2 we prove the admissibility of δ_0 in the class of scale equivariant estimators if $a \neq 0$. This should be contrasted with the inadmissibility of δ_0 in this class when $a = 0$ (see Stein (1964), Brown (1968), Brewster and Zidek (1974), Strawderman (1974)). In Section 3 we determine the form of generalized Bayes estimators and their relationship to Bayes solutions in the case $a = 0$. This relationship leads to a natural extension of the Brewster-Zidek admissible and minimax estimator of normal variance to general situation $(a \neq 0)$.

2. SCALE EQUIVARIANT ESTIMATORS

In this section we consider scale equivariant estimators of the form

$$\delta(X,S) = aU(1 - h(U/S)) + bS^2(1 - f(U/S))/(m+1). \tag{2.1}$$

Here $U = X_1$ and h and f are measurable functions of $z = U/S$ such that $f(-z) = f(z)$.

Theorem 1. Estimator (1.5) is admissible in the class of all procedures (2.1) for $a \neq 0$.

Proof. Assume that for some h and f

$$\begin{aligned} R(\theta_1,\sigma;\delta) &= E[\delta(X,S) - a\theta_1 - b\sigma^2]^2\sigma^{-4} \\ &\leq a^2E(U - \theta_1)^2\sigma^{-4} + b^2E(S^2/(m+1) - \sigma^2)^2\sigma^{-4} \\ &= a^2\sigma^{-2} + 2b^2/(m+1) \end{aligned} \tag{2.2}$$

with strict inequality for some θ_1 and σ.

One has with $\xi = \theta_1/\sigma$

$$\begin{aligned} R(\xi,\sigma;\delta) &= a^2E[U(1-h) - \xi]^2\sigma^{-2} \\ &+ 2abE[U(1-h) - \xi][S^2(1-f)/(m+1)-1]\sigma^{-1} \\ &+ b^2E[S^2(1-f)/(m+1) - 1]^2. \end{aligned}$$

Here the expected values are calculated for $\sigma = 1$.

Inequality (2.2) implies that

$$E[U(1-h) - \xi]^2 \leq E[U - \xi]^2$$

and because of the admissibility of U as the estimator of ξ (cf. Brown (1973)) one concludes that $h_1 = 0$.

It follows that for all ξ

$$\begin{aligned} &abE[U - \xi][S^2(1-f)/(m+1) - 1] \\ &\qquad = -abE[U - \xi]S^2f/(m+1) \leq 0 \,. \end{aligned} \tag{2.3}$$

Because of our assumption the last expected value in (2.3) is an odd function of ξ. Thus (2.3) can hold only if $f = 0$ which proves Theorem 1.

Remark. Inequality (2.2) implies that the estimator $S^2(1-f)/(m+1)$ improves upon the best equivariant estimator $S^2/(m+1)$. If f is any function for which this is the case then

$$\tilde{f}(z) = [f(-z) + f(z)]/2$$

also possesses this property. Thus the symmetry condition on f is not really restrictive.

The admissibility of δ_0 can be also proven for classes of estimators considered by Olkin and Selliah (1977).

For $a = 0$ one can find functions f such that estimator (2.1) is better than δ_0. Some of these estimators are generalized Bayes with respect to prior density of the form $\sigma^{-\alpha}\lambda(\theta_1/\sigma)$. These prior densities in the case $a \neq 0$ lead exactly to the class of estimators (2.1). In the next section we study the general form of Bayes procedures.

3. BAYES ESTIMATORS

Let $\lambda(\theta,\sigma)$ be a density of a generalized prior distribution with respect to invariant "uniform" measure $d\theta d\sigma/\sigma$. The Bayes estimator for loss (1.3) has the form

$$\delta_B(X,S) = \iint [a\theta_1 + b\sigma^2]\exp\{-[\|X - \theta\|^2 + S^2]/2\sigma^2\} \lambda\sigma^{-m-k-4}d\theta d\sigma \Big/ \iint \exp\{-[\|X-\theta\|^2 + S^2]/2\sigma^2\} \lambda\sigma^{-m-k-4}d\theta d\sigma \; .$$

Assume that the following integrations by parts are legitimate:

$$\int_{-\infty}^{\infty} (\theta_1 - X_1)\lambda\exp\{-(\theta_1 - X_1)^2/2\sigma^2\}d\theta_1 = \sigma^2\int_{-\infty}^{\infty} \lambda_1\exp\{-(\theta_1 - X_1)^2/2\sigma^2\}d\theta_1, \tag{3.1}$$

$$\int_{-\infty}^{\infty} (\theta_i - X_i)^2\lambda\exp\{-(\theta_i - X_i)^2/2\sigma^2\}d\theta_i = \sigma^2\int_{-\infty}^{\infty} [\lambda + \sigma^2\lambda_{ii}]\exp\{-(\theta_i - X_i)^2/2\sigma^2\}d\theta_i, \tag{3.2}$$

$i = 1,\dots,k$ and

$$[\|X-\theta\|^2+S^2]\int_0^\infty \lambda\sigma^{-m-5}\exp\{-[\|X-\theta\|^2+S^2]/2\sigma^2\}d\sigma$$

$$=\int_0^\infty [(m+2)\lambda-\sigma\lambda_\sigma]\sigma^{-m-3}\exp\{-[\|X-\theta\|^2+S^2]2\sigma^2\}d\sigma. \qquad (3.3)$$

Here $\lambda_1 = \partial\lambda/\partial\theta_1$, $\lambda_\sigma = \partial\lambda/\partial\sigma$ and $\lambda_{ii} = \partial^2\lambda/\partial\theta_i^2$. Combining these formulae we obtain the following representation of the Bayes estimator

$$\delta_B(X,S) - aX_1 - bS^2/(m+1)$$

$$= b(m+1)^{-1}\iint[\mathcal{D}\lambda]\sigma^{-m-k}\exp\{-[\|X-\theta\|^2+S^2]/2\sigma^2\}d\theta d\sigma$$

$$/\iint\lambda\sigma^{-m-k-4}\exp\{-[\|X-\theta\|^2+S^2]/2\sigma^2\}d\theta d\sigma, \qquad (3.4)$$

where for $b \neq 0$

$$\mathcal{D}\lambda = \Delta\lambda + \lambda_\sigma/\sigma + a(m+1)\lambda_1/(b\sigma^2),$$

and

$$\Delta\lambda = \sum_i \lambda_{ii} \; .$$

It follows that δ_0 is a generalized Bayes estimator against "non-informative" density $\lambda \equiv 1$. Also it is evident that $\delta_B = \delta_0$ if and only if the density λ is a solution of the following parabolic differential equation

$$\mathcal{D}\lambda = 0 \; . \qquad (3.5)$$

Notice that if $a = 0$, (3.5) takes a particularly simple form

$$L\lambda = \Delta\lambda + \lambda_\sigma/\sigma = 0 \; . \qquad (3.6)$$

Equation (3.6) is closely related to the adjoint heat equation (cf. Widder (1975)). Its typical solutions are of the form $\sigma^{-k}\exp\{\|\xi\|^2/2\sigma^2\}$ or convolutions of these kernels with functions representing initial values ($\sigma = 0$). Therefore these solutions look even less like probability densities than $\lambda \equiv 1$, and they do not admit a good approximation by proper densities. This condition is exactly responsible for the inadmissibility of the corresponding generalized Bayes estimator (in this case $S^2/(m+1)$ as an estimator of σ^2). In fact $S^2/(m+1)$ is known to be inadmissible and smooth improvements over it are known. In particular the generalized Bayes estimator $\tilde{\delta}$ with respect to the prior density

$$\tilde{\lambda}(\theta,\sigma) = \int_0^{\infty}\exp\{-t\|\theta\|^2/2\sigma^2\}t^{k/2-1}(1+t)^{-1}dt/\sigma^k \quad (3.7)$$

is uniformly better than $S^2/(m+1)$ and possesses other optimality properties (cf. Brewster and Zidek (1974)).

An easy calculation shows that $L\tilde{\lambda}$ as a function of θ is proportional to the Dirac delta function so that $\tilde{\lambda}$ coincides with a particular value of Green's function for equation (3.6). Thus in the case $a = 0$, $b = 1$

$$\iint[\mathcal{D}\tilde{\lambda}]\sigma^{-m-k}\exp\{-[\|X-\theta\|^2+S^2]/2\sigma^2\}d\theta d\sigma$$

$$= -2(2\pi)^{k/2}\int\sigma^{-m-k-2}\exp\{-[\|X\|^2+S^2]/2\sigma^2\}d\sigma$$

$$= -2(2\pi)^{k/2}d_{m+k}[\|X\|^2+S^2]^{-(m+k+1)/2},$$

where $$d_n = \int_0^{\infty}t^n\exp\{-t^2/2\}dt = 2^{(n-1)/2}\Gamma((n+1)/2).$$

Similarly

$$\iint\tilde{\lambda}\sigma^{-m-k-4}\exp\{-[\|X-\theta\|^2+S^2]/2\sigma^2\}d\theta d\sigma$$

$$= (2\pi)^{k/2}\iint\sigma^{-m-k-4}\exp\{-[t\|X\|^2(1+t)^{-1}+S^2]/2\sigma^2\}$$

$$\times\, t^{k/2-1}(1+t)^{-k-1}dtd\sigma$$

$$= 2(2\pi)^{k/2}d_{m+k+2}\int_0^1 u^{k-1}[S^2+u^2\|X\|^2]^{-(m+k+3)/2}du$$

$$= (2\pi)^{k/2}d_{m+k+2}S^{-m-k-3}[y^{-1}-1]^{k/2}\int_0^y t^{k/2-1}(1-t)^{(m+1)/2}dt$$

with $y = \|X\|^2/[\|X\|^2 + S^2]$.

Therefore

$$\tilde{\delta}(X,S) = S^2[1 - 2y^{k/2}(1-y)^{(m+k+1)/2}(m+k+1)^{-1} \Big/ \int_0^y t^{k/2-1}(1-t)^{(m+1)/2}dt]/(m+1). \quad (3.8)$$

Thus $\tilde{\delta}$ is a simple function of an incomplete beta function and can be calculated from the existing tables (Pearson (1968)) for any dimension k. It can also be always expressed in terms of elementary functions.

Notice that

$$\tilde{\delta}(0,S) = S^2/(m+k+1)$$

so that the maximal amount of shrinkage provided by $\tilde{\delta}$ is $k/(m+k+1)$ (which tends to one as k increases).

Now we turn to the case $a \neq 0$. Let

$$\lambda(\theta,\sigma) = \rho(\theta_1 - a(m+1)\log\sigma/b, \theta_2, \ldots, \theta_k, \sigma) . \quad (3.9)$$

Then

$$\mathcal{D}\lambda = \Delta\rho + \rho_\sigma/\sigma = L\rho . \quad (3.10)$$

Because of the discussion above, the equation $L\rho = 0$ does not have solutions ρ which are approximable by proper prior densities. In particular δ_0 must be inadmissible. In the case $k = 1$ this has been proven in Rukhin (1986) and in the case $k \geq 2$ the inadmissibility of δ_0 can be proved by considering alternative estimators of the form

$$\delta = aX_1 + bS^2(1 - f(X_2/S,\dots,X_k/S))/(m+1).$$

We summarize these results.

Theorem 2. If for a prior density λ integrations by parts in (3.1) - (3.3) are legitimate then the Bayes estimator δ_B has form (3.4). In particular $\delta_B = \delta_0$ if and only if λ satisfies (3.5). Estimator δ_0 is inadmissible. If $a = 0$ a better (Brewster-Zidek) estimator which is generalized Bayes against density (3.7), has form (3.8).

Because of (3.10) it is interesting to determine the Bayes estimator for λ defined by (3.9) with $\rho = \tilde{\lambda}$, when $a \neq 0$. One has with $c = a(m+1)/b$

$$\iint [\mathcal{D}\lambda]\sigma^{-m-k}\exp\{-[\|X - \theta\|^2+S^2]/2\sigma^2\}d\theta d\sigma$$

$$= 2(2\pi)^{k/2}\int \sigma^{-m-k-2}\exp\{-[(X_1-c\log\sigma)^2 + \sum_2^k X_j^2+S^2]/2\sigma^2\}d\sigma$$

and

$$\iint \lambda\sigma^{-m-k-4}\exp\{-[\|X-\theta\|^2+S^2]/2\sigma^2\}d\theta d\sigma$$

$$= (2\pi)^{k/2}\int_0^\infty\int_0^1 u^{k-1}\exp\{-[u^2((X_1-c\log\sigma)^2+\sum_2^k X_j^2)+S^2]/2\sigma^2\}\sigma^{-m-k-4}du d\sigma .$$

Thus the corresponding estimator δ has a considerably more complicated structure than the Brewster-Zidek estimator $\tilde{\delta}$. There remains an open problem if δ improves on δ_0. For $k = 1$ the numerical results support the conjecture that the estimator

$$\delta(X,S) = aX + bS^2/(m+1)$$

$$-2b(m+1)^{-1}\int_0^\infty t^{m+k}\exp\{-t^2[(X + c\log t)^2+S^2]/2\}dt$$

$$/\int_0^\infty\int_0^1 t^{m+k+2}\exp\{-t^2[u^2(X + c\log t)^2+S^2]/2\}dudt,$$

which can be easily evaluated by Gaussian quadratures, indeed is better than δ_0.

Andrew L. Rukhin
Department of Mathematics and Statistics
University of Massachusetts
Amherst, MA 01003

References

Brewster, J. F., and Zidek, J. V. (1974). Improving on equivariant estimators. Ann. Statist. 2, 21-38.

Brown, L. D. (1968). Inadmissibility of the usual estimators of scale parameters in problems with unknown location and scale parameters. Ann. Math. Statist. 39,29-48.

Brown, L. D. (1973). Estimation with incompletely specified loss functions, J. Amer. Statist. Assoc., 70, 417-427.

Dhrymes, P. J. (1962). On devising unbiased estimators for the parameters of the Cobb-Douglas production function. Econometrica, 30, 297-304.

Goldberger, A. S. (1963). The interpretation and estimation of Cobb-Douglas functions. Econometrica 36, 464-472.

Olkin, I., and Selliah (1977). Estimating covariances in a multivariate normal distribution, in Statistical Decision Theory Related Topics, S. Gupta and D. Moore, eds., Academic Press, N.Y.

Pearson, K. (1968). Tables of the Incomplete Beta-Function, University Press, Cambridge.

Rukhin, A. (1986). Estimating a linear function of the normal mean and variance. Sankhya. Ser A, 48.

Stein, C. (1964). Inadmissibility of the usual estimator for the variance of a normal distribution with unknown mean. Ann. Inst. Statist. Math. 16, 155-160.

Stein, C. (1965). Approximation of improper measures by prior probability measures. Bernoulli, Bayes, Laplace Anniversary Volume, 217-241, Springer-Verlag, N.Y.

Strawderman, W. E. (1974). Minimax estimation of powers of the variance of a normal population under squared error loss. Ann. Statist. 2, 190-198.

Widder, D. V. (1975). The Heat Equation, Academic Press, N.Y.

Zellner, A. (1971). Bayesian and non-Bayesian analysis of the lognormal distribution and lognormal regression. J. Amer. Statist. Assoc. 66, 327-330.

Minoru Siotani

MULTIVARIATE BEHRENS-FISHER PROBLEM BY HETEROSCEDASTIC METHOD

SUMMARY

The Behrens-Fisher problem is considered by using the so called heteroscedastic method. This leads to a test procedure for testing the hypothesis of equality of two normal mean vectors with a given power for a given value of the distance between mean vectors, when population covariance matrices are different. Example is discussed by using Fisher's iris data.

1. INTRODUCTION

Let $\{\underset{\sim}{x}_r^{(i)}\}$, $r = 1,\ldots,N_i$, $i = 1,2$ be random samples from $N_p(\underset{\sim}{\mu}_i,\Sigma_i)$, $\Sigma_i > 0$, where Σ_1 and Σ_2 are unequal and unknown. Let $\bar{\underset{\sim}{x}}_i$ and $S_i \equiv V_i/n_i$ be the sample mean vector and unbiased covariance matrix with n_i degrees of freedom (d.f.) in the i-th sample, respectively and $n_i = N_i-1$, N_i being the sample size (i = 1,2). We wish here to test the hypothesis H: $\underset{\sim}{\mu}_1 = \underset{\sim}{\mu}_2$ against K: $\underset{\sim}{\mu}_1 \neq \underset{\sim}{\mu}_2$.

This multivariate version of the Behrens-Fisher problem has been discussed by several authors; for example, Bennett (1951), James (1954), Yao (1965), Eaton (1969), Ito (1969), and Scheffé (1970). In this paper, we discuss the problem based on the heteroscedastic method, which was proposed by Dudewicz and Bishop (1979) and basically depends on the two-stage sampling scheme due to Stein (1945) and Chatterjee (1959). We obtain the test procedure for testing H with a significance level α and with a given value of power function for a preassigned value of $\delta^2 = (\underset{\sim}{\mu}_1-\underset{\sim}{\mu}_2)'(\underset{\sim}{\mu}_1-\underset{\sim}{\mu}_2)$. It needs to show how we control the sample sizes N_i, $i = 1,2$, so as to satisfy the requirement for the test. Note that this is impossible under a single-stage sampling scheme.

A numerical example is given to explain the test proce-

A. K. Gupta (ed.), Advances in Multivariate Statistical Analysis, 327–340.

dure using the iris data discussed by Fisher (1936), which wa used in the construction of a set of simultaneous confidence intervals with a given length by Hyakutake and Siotani (1986

2. CONSTRUCTION OF THE TEST STATISTICS

We first take random sample of a given size N_o; $\underset{\sim}{x}_1^{(i)}, \ldots, \underset{\sim}{x}_{N_o}^{(i)}$ (i = 1,2) from each population and compute $\bar{\underset{\sim}{x}}_i$ and S_i or V_i, where

$$\bar{\underset{\sim}{x}}_i = \frac{1}{N_o}\sum_{r=1}^{N_o}\underset{\sim}{x}_r^{(i)}, \quad V_i = \sum_{r=1}^{N_o}(\underset{\sim}{x}_r^{(i)}-\bar{\underset{\sim}{x}}_i)(\underset{\sim}{x}_r^{(i)}-\bar{\underset{\sim}{x}}_i)', \qquad S_i = V_i/\nu, \tag{1}$$

where $\nu = N_o-1 \geq p$. Then N_i (i = 1,2) is, according to Stei (1945) and Chatterjee (1959), defined as

$$N_i = \max\{N_o+p^2, [c\cdot tr(TS_i)]+1\}, \ (i=1,2) \tag{2}$$

where c is a given positive constant, T:p×p is a given positive definite (p.d.) symmetric matrix [$T = I_p$, if means of components of $\underset{\sim}{x}$ are equally important.] and symbol [a] stand for the greatest integer not greater than real number a. Now we take N_i-N_o additional observations

$$\underset{\sim}{x}_{N_o+1}^{(i)}, \ldots, \underset{\sim}{x}_{N_i}^{(i)}$$

from each population $N_p(\underset{\sim}{\mu}_i, \Sigma_i)$ (i = 1,2). Let

$$X^{(i)}:p\times N_i = [\underset{\sim}{x}_1^{(i)},\ldots,\underset{\sim}{x}_{N_o}^{(i)},\underset{\sim}{x}_{N_o+1}^{(i)},\ldots,\underset{\sim}{x}_{N_i}^{(i)}] \tag{3}$$

and we form p matrices

$$A_k^{(i)}:p\times N_i = [\underset{\sim}{a}_{k1}^{(i)},\ldots,\underset{\sim}{a}_{k,N_o}^{(i)},\underset{\sim}{a}_{k,N_o+1}^{(i)},\ldots,\underset{\sim}{a}_{k,N_i}^{(i)}],$$

k = 1,...,p; i = 1,2, satisfying the following three conditions:

(1) $\underset{\sim}{a}_{k1}^{(i)} = \cdots = \underset{\sim}{a}_{k,N_o}^{(i)}$,

(2) $A_k^{(i)} \underset{\sim}{j}_{N_i} = \underset{\sim}{e}_k$, where $j_{N_i}:N_i\times 1 = (1,1,\ldots,1)'$ and $\underset{\sim}{e}_k:p\times 1 = (0,\ldots,0, \underset{(k)}{1}, 0,\ldots,0)'$,

(3) $A^{(i)}A^{(i)'} = \frac{1}{c}T^{-1} \otimes S_i^{-1}$, where $\otimes$ indicates the direct product of two matrices and $A^{(i)}:p^2\times N_i = [A_1^{(i)'},\ldots, A_p^{(i)'}]'$.

Based on $A_k^{(i)}$ and $X^{(i)}$, new random vectors $\underset{\sim}{z}_i$, i = 1,2, are defined by

$$\underset{\sim}{z}_i:p\times 1 = [\mathrm{tr}(A_1^{(i)}X^{(i)'}),\ldots,\mathrm{tr}(A_p^{(i)}X^{(i)'})]'. \tag{4}$$

The test procedure for testing H: $\underset{\sim}{\mu}_1 = \underset{\sim}{\mu}_2$ by the heteroscedastic method is formed in the following way [see Dudewicz and Bishop (1979) or Siotani et al. (1985)]: Suppose we have two normal populations $N_p(\underset{\sim}{\mu}_1,\Sigma)$ and $N_p(\underset{\sim}{\mu}_2,\Sigma)$, Σ known and sampling is made in single stage with the same size N from each population. Let $\bar{\underset{\sim}{y}}_i$, i = 1,2, be sample mean vectors in this case. Then the test procedure for the same testing problem as the original can be easily obtained as the one based on

$$\frac{1}{2}(\bar{\underset{\sim}{y}}_1-\bar{\underset{\sim}{y}}_2)'(\Sigma/N)^{-1}(\bar{\underset{\sim}{y}}_1-\bar{\underset{\sim}{y}}_2)$$

which is a χ^2-variate with p d.f. After this, we replace $\bar{\underset{\sim}{y}}_1$, $\bar{\underset{\sim}{y}}_2$, and Σ/N with $\underset{\sim}{z}_1$, $\underset{\sim}{z}_2$, and $(cT)^{-1}$, respectively, to have the new statistic

$$w = \frac{c}{2p}(\underset{\sim}{z}_1-\underset{\sim}{z}_2)'T(\underset{\sim}{z}_1-\underset{\sim}{z}_2). \tag{5}$$

The critical region is $[w > w(\alpha)]$ for a given significance level α, $(1 > \alpha > 0)$, where $w(\alpha)$ is the upper 100α % point of w. Thus the test procedure derived by the heteroscedastic method is based on the statistic w defined by (5).

The distribution of w has been studied by Hyakutake and Siotani (1986) and Hyakutake et al. (1986) in both null and nonnull cases. It is completely free from Σ_1 and Σ_2, and the nonnull distribution function of w depends on $\underset{\sim}{\mu}_1$ and $\underset{\sim}{\mu}_2$ only through the noncentrality parameter $\theta = (c/2p)\delta^2$, where $\delta^2 = (\underset{\sim}{\mu}_1-\underset{\sim}{\mu}_2)'T(\underset{\sim}{\mu}_1-\underset{\sim}{\mu}_2)$. Tables of $w(\alpha)$ were provided by Hyakutake et al. (1986).

To design the procedure concretely, we wish to determine the constant c so that the power of test has a preassigned value P_o for a given value δ_o^2 of δ^2; that is,

$$\Pr\{w > w(\alpha) \mid \neq H,\ \delta_o^2\} = P_o. \tag{6}$$

Then for this choice of c, we compute N_i (i = 1,2) from the expression (2) and take additional observations from each population. For convenience, we reproduce the distribution function of w in the next section.

3. NONNULL DISTRIBUTION FUNCTION w

The nonnull distribution function of w is reproduced from Hyakutake et al. (1986) as follows:

When p = 1,

$$F_1(w;\Delta^2) = \int_0^w f_1(z;\Delta^2)\,dz \tag{7}$$

where $\Delta^2 = c(\mu_1-\mu_2)^2$, (T = 1), and

$$\begin{aligned} f_1(z;\Delta^2) = {} & [\sqrt{\nu}2^{\nu-\frac{1}{2}}\{\Gamma(\tfrac{\nu}{2})\}^2]^{-1}\cdot \\ & \cdot\sum_{j=0}^{\infty}\sum_{\ell=0}^{\infty}\frac{(\frac{1}{2}\Delta^2)^j}{j!\ell!}\,\frac{1}{(2\nu)^{2j}}\cdot \\ & \cdot\frac{\Gamma(\frac{1}{2}+\ell)\Gamma(\frac{\nu+1}{2}+2j)\Gamma(\nu+\frac{1}{2}+2j+\ell)}{\Gamma(\frac{1}{2}+j)\Gamma(\frac{\nu}{2}+1+2j+\ell)}\cdot \\ & \cdot w^{j-\frac{1}{2}}\left(\frac{w^*}{2\nu}\right)^{\ell}\left(1+\frac{w^*}{2\nu}\right)^{-(\nu+\frac{1}{2}+2j+\ell)}, \end{aligned} \tag{8}$$

where $w^* = w + \frac{1}{2}\Delta^2$.

When $p \geq 2$, the exact formula is not available and an asymptotic expansion formula is given as follows:

$$
\begin{aligned}
F_p(w;\theta) = {} & G_p^* + \frac{1}{8\nu}[k_2 G_{p+4}^* + 2a_1 G_{p+2}^* - a_o G_p^*] + \\
& + \frac{1}{384p\nu^2}[3k_2k_4k_6 G_{p+8}^* + 4k_2k_4b_3 G_{p+6}^* + \\
& \qquad + 6k_2b_2 G_{p+4}^* - 12b_1 G_{p+2}^* + \\
& \qquad + b_o G_p^*] + \\
& + \frac{\theta}{8p\nu}[\theta G_{p+8}^* - 2(2\theta - k_2) G_{p+6}^* + \\
& \qquad +2(3\theta + a_1 - 2k_2) G_{p+4}^* - 2(2\theta + 2a_1 - k_2) G_{p+2}^* + \\
& \qquad + (\theta + 2a_1) G_p^*] + \\
& + \frac{\theta}{384p^2\nu^2}[3\theta^3 G_{p+16}^* - 12\theta^2(2\theta - k_6) G_{p+14}^* + \\
& \qquad +2\theta\{42\theta^2 + 2\theta(b_3 - 18k_6) + 9k_4k_6\} G_{p+12}^* - \\
& \qquad -12\{14\theta^3 - \theta^2(15k_6 - 2b_3) - \theta k_4(b_3 - 6k_6) - \\
& \qquad\quad -k_2k_4k_6\} G_{p+10}^* + \\
& \qquad +6\{35\theta^3 + 10\theta^2(b_3 - 4k_6) + \theta(b_2 - 8k_4b_3 + \\
& \qquad\qquad +18k_4k_6) + \\
& \qquad\quad +2k_2k_4(b_3 - 2k_6)\} G_{p+8}^* - \\
& \qquad -4\{42\theta^3 + \theta^2(20b_3 - 45k_6) + 6\theta(c_1 - k_2b_3 - \\
& \qquad\qquad -2k_6b_3 + 3k_4k_6 - 8) - \\
& \qquad\quad -3k_2(b_2 - 2k_4b_3 + k_4k_6\} G_{p+6}^* + \\
& \qquad +6\{14\theta^3 + 2\theta^2(5b_3 - 6k_6) +
\end{aligned}
$$

$$+\theta(2b_2+4d_o-8k_6b_3+3k_4k_6-88)-$$

$$-2(b_1+2k_2c_1-k_2k_8b_3-16k_2)\}\cdot$$

$$\cdot G^*_{p+4} -$$

$$-4\{6\theta^3+3\theta^2(2b_3-k_6)+3\theta(2c_1-k_8b_3$$

$$-16)-$$

$$-(3pb_2+2c_o+12d_o-32d_1)\}G^*_{p+2}+$$

$$+(3\theta^3+4\theta^2b_3+6\theta b_2-12b_1)G^*_p] +$$

$$+ O(\nu^{-3}), \qquad (9$$

where $\theta = \Delta^2/2p$, $\Delta^2 = c\cdot(\underset{\sim}{\mu}_1-\underset{\sim}{\mu}_2)'T(\underset{\sim}{\mu}_1-\underset{\sim}{\mu}_2) \equiv c\cdot\delta^2$ $(\theta = \frac{c}{2p}\delta^2)$, $G^*_m \equiv G^*_m(w;\theta)$ is the distribution function of noncentral χ^2-distribution with m d.f. and noncentrality parameter θ, and

$$k_i = p+i,\ i = 2,4,6,8,$$

$$a_1 = 2p^2+p-2, \qquad a_o = 4p^2+3p-2,$$

$$b_3 = 6p^2+3p-10, \qquad b_2 = 8p^4+4p^3-17p^2-2p+8,$$

$$b_1 = 8p^5+26p^4-49p^3+48p^2+12p-16,$$

$$b_o = 48p^5+168p^4-693p^3+532p^2+132p-112,$$

$$c_1 = 8p^4+4p^3-5p^2+4p-4,$$

$$c_o = 24p^5+54p^4-159p^3+147p^2+18p-28,$$

$$d_1 = 6p^2+3p-4, \qquad d_o = 8p^4+4p^3+7p^2+10p-10.$$

It is noted here that, when $w = \infty$, i.e., when all $G^*_m = 1$, th coefficients of $1/\nu$ and $1/\nu^2$ become zero, which gives us assurance for our computation.

The null distribution function is obtained just by putting $\theta = 0$, i.e.,

$$F_p(w;0) = G_p + \frac{1}{8\nu}[(p+2)G_{p+4}+2(2p^2+p-2)G_{p+2}- -(4p^2+3p-2)G_p] + + \frac{1}{384p\nu^2}[3(p+2)(p+4)(p+6)G_{p+8}+4(p+2)(p+4)\cdot \cdot(6p^2+3p-10)G_{p+6}+ +6(p+2)(8p^4+4p^3-17p^2-2p+8)G_{p+4}- -12(8p^5+26p^4-49p^3+48p^2+12p- -16)G_{p+2}+ +(48p^5+168p^4-693p^3+532p^2+ +132p-112)G_p] + O(\nu^{-3}), \tag{10}$$

where $G_m \equiv G_m(w)$ is the distribution function of central χ^2-distribution with m d.f.

Let χ_α^2 be the upper 100α % point of the χ^2-distribution with p d.f. Then we obtain an asymptotic expansion formula for $w(\alpha)$ as

$$w(\alpha) = \chi_\alpha^2 + \frac{1}{4p\nu}\chi_\sigma^2(\chi_\alpha^2+4p^2+3p-2) + + \frac{1}{96p^2\nu^2}\chi_\alpha^2[4\chi_\alpha^4+(72p^2+61p-50)\chi_\alpha^2+ +(408p^3-305p^2-96p+68)] + O(\nu^{-3}). \tag{11}$$

For the numerical values of $G_m^*(w;\theta)$, we can make use in part of Biometrika Tables edited by Pearson and Hartley (1972, Table 25), and tables computed by Haynam et al. (1973) together with some interporations and by drawing graphs.

4. A NUMERICAL EXAMPLE

To illustrate the test procedure by heteroscedastic method, we give a numerical example based on the iris data used by Fisher (1936) for the discrimination problem. Suppose we have two iris populations; versicolor and virginica and measurements are made on sepal length (x_1), sepal width (x_2), petal length (x_3), and petal width (x_4). Suspecting the equality of population covariance matrices, we use the heteroscedastic method to test the equality of population mean vectors. Suppose we have the following data (Table 1) for the first-stage samples of common size $N_o = \nu+1 = 21$.

Table 1

First-stage samples drawn from two populations of Iris species

r	Iris Versicolor				Iris Virginica			
	x_1	x_2	x_3	x_4	x_1	x_2	x_3	x_4
1	7.0	3.2	4.7	1.4	6.3	3.3	6.0	2.5
2	6.4	3.2	4.5	1.5	5.8	2.7	5.1	1.9
3	6.9	3.1	4.9	1.5	7.1	3.0	5.9	2.1
4	5.5	2.3	4.0	1.3	6.3	2.9	5.6	1.8
5	6.5	2.8	4.6	1.5	6.5	3.0	5.8	2.2
6	5.7	2.8	4,5	1.3	7.6	3.0	6.6	2.1
7	6.3	3.3	4.7	1.6	4.9	2.5	4.5	1.7
8	4.0	2.4	3.3	1.0	7.3	2.9	6.3	1.8
9	6.6	2.9	4.6	1.3	6.7	2.5	5.8	1.8
10	5.2	2.7	3.9	1.4	7.2	3.6	6.1	2.5
11	5.0	2.0	3.5	1.0	6.5	3.2	5.1	2.0
12	5.9	3.0	4.2	1.5	6.4	2.7	5.3	1.9
13	6.0	2.2	4.0	1.0	6.8	3.0	5.5	2.1
14	6.1	2.9	4.7	1.4	5.7	2.5	5.0	2.0
15	5.6	2.9	3.6	1.3	5.8	2.8	5.1	2.4
16	6.7	3.1	4.4	1.4	6.4	3.2	5.3	2.3
17	5.6	3.0	4.5	1.5	6.5	3.0	5.5	1.8

18	5.8	2.7	4.1	1.0	7.7	3.8	6.7	2.2
19	6.2	2.2	4.5	1.5	7.7	2.6	6.9	2.3
20	5.6	2.5	3.9	1.1	6.0	2.2	5.0	1.5
21	5.9	3.2	4.8	1.8	6.9	3.2	5.7	2.3

Then sample covariance matrices are computed as

$$S_1 = \begin{bmatrix} 0.3461 & 0.1329 & 0.2139 & 0.0619 \\ 0.1329 & 0.1446 & 0.1146 & 0.0585 \\ 0.2139 & 0.1146 & 0.2056 & 0.0745 \\ 0.0619 & 0.0585 & 0.0745 & 0.0486 \end{bmatrix},$$

$$S_2 = \begin{bmatrix} 0.5139 & 0.1413 & 0.4123 & 0.0709 \\ 0.1413 & 0.1453 & 0.1110 & 0.0680 \\ 0.4123 & 0.1110 & 0.3896 & 0.0721 \\ 0.0709 & 0.0680 & 0.0721 & 0.0746 \end{bmatrix}.$$

Note. It has shown in Siotani et al. (1985, pp.350-355) that, under the assumption of normality, the hypothesis of equality of population covariance matrices is rejected with 1% significance level for both irises.

We assume that the population of Iris Versicolor is $N_4(\underset{\sim}{\mu}_1, \Sigma_1)$ and that of Iris Virginica is $N_4(\underset{\sim}{\mu}_2, \Sigma_2)$, where $\Sigma_1 \neq \Sigma_2$. We wish to test the hypothesis $H: \underset{\sim}{\mu}_1 = \underset{\sim}{\mu}_2$ against $K: \underset{\sim}{\mu}_1 \neq \underset{\sim}{\mu}_2$ with significance level $\alpha = 0.05$ and with the power $P_o = 0.90$ for $\delta_o^2 = 6.5$. We now have to determine the value of c so as to satisfy this requirement. Since the critical value w(0.05) is found from Hyakutake et al. (1986) to be w(0.05) = 12.59 (for $\alpha = 0.05$, $p = 4$, $\nu = 20$), we compute θ satisfying

$$0.10 = \Pr\{w \leq 12.59 | \theta\} \qquad \theta = \frac{c}{2p}\delta_o^2 = \frac{c}{8}(6.5)$$

based on the formula (9), which is found to be $\theta \doteq 20.46$. Hence

$$c = (8)(20.46)/6.5 = 25.18.$$

With this choice of c and $T = I_4$, total sample sizes N_1 and N_2 are obtained, by (2), as $N_1 = N_2 = 37$, so that $N_i - N_o = 16$

(for both i = 1,2) additional observations are taken from respective populations, which are given in Table 2 below:

Table 2

Additional observations in the second-stage

r	Iris Versicolor x_1	x_2	x_3	x_4	Iris Virginica x_1	x_2	x_3	x_4
22	6.1	2.8	4.0	1.3	5.6	2.8	4.9	2.0
23	6.3	2.5	4.9	1.5	7.7	2.8	6.7	2.0
24	6.1	2.8	4.7	1.2	6.3	2.7	4.9	1.8
25	6.4	2.9	4.3	1.3	6.7	3.3	5.7	2.1
26	6.6	3.0	4.4	1.4	7.2	3.2	6.0	1.8
27	6.8	2.8	4.8	1.4	6.2	2.8	4.8	1.8
28	6.7	3.0	5.0	1.7	6.1	3.0	4.9	1.8
29	6.0	2.9	4.5	1.5	6.4	2.8	5.6	2.1
30	5.7	2.6	3.5	1.0	7.2	3.0	5.8	1.6
31	5.5	2.4	3.8	1.1	7.4	2.8	6.1	1.9
32	5.5	2.4	3.7	1.0	7.9	3.8	6.4	2.0
33	5.8	2.7	3.9	1.2	6.4	2.8	5.6	2.2
34	6.0	2.7	5.1	1.6	6.3	2.8	5.1	1.5
35	5.4	3.0	4.5	1.5	6.1	2.6	5.6	1.4
36	6.0	3.4	4.5	1.6	7.7	3.0	6.1	2.3
37	6.7	3.1	4.7	1.5	6.3	3.4	5.6	2.4

By applying Hyakutake's method [Hyakutake (1985)], we compute $A^{(i)}: 4^2 \times 37$, i = 1,2, which are shown in Table 3. Then basic random vector variates $\underset{\sim}{z}_1$ and $\underset{\sim}{z}_2$ are given by

$$\underset{\sim}{z}_1' = (6.197,\ 3.140,\ 4.318,\ 0.991),$$

$$\underset{\sim}{z}_2' = (6.264,\ 3.027,\ 6.234,\ 2.483),$$

and
$$\underset{\sim}{z}_1' - \underset{\sim}{z}_2' = (-0.067,\ 0.113,\ -1.916,\ -1.492).$$

Now observed value of the statistic w (with $T = I_4$) is found

Table 3

Matrices $A^{(i)}$, i=1,2 for p=4 and c=25.18.

$A^{(1)} =$

-.0009	...	-.0009	.6134	.0270	.0270	.0270	.0270	.0270	.0270	.0270	.0270	.0270	.0270	.0270	.0270	.0270	.0270	.0270
-.0225	...	-.0225	-.2753	.7476	0	0	0	0	0	0	0	0	0	0	0	0	0	0
.0197	...	.0197	-.7695	-.2498	.6066	0	0	0	0	0	0	0	0	0	0	0	0	0
.0037	...	.0037	.6748	-.5667	-1.041	.8559	0	0	0	0	0	0	0	0	0	0	0	0
-.0244	...	-.0244	-.0244	-.0244	-.0244	-.0244	.6102	0	0	0	0	0	0	0	0	0	0	0
.0156	...	.0156	-.0304	-.0014	-.0498	-.0389	-.2346	.7569	.0270	.0270	.0270	.0270	.0270	.0270	.0270	.0270	.0270	.0270
.0135	...	.0135	.0135	.0135	.0135	.0135	-.7449	-.2368	.6444	0	0	0	0	0	0	0	0	0
.0052	...	.0052	.0052	.0052	.0052	.0052	.6501	-.6010	-1.036	.8578	0	0	0	0	0	0	0	0
-.0211	...	-.0211	-.0211	-.0211	-.0211	-.0211	-.0211	-.0211	-.0211	-.0211	.6118	0	0	0	0	0	0	0
-.0171	...	-.0171	-.0171	-.0171	-.0171	-.0171	-.0171	-.0171	-.0171	-.0171	-.2594	.7547	0	0	0	0	0	0
.0508	...	.0508	.0047	.0338	-.0147	-.0038	.0434	-.0050	.0256	-.0120	-.7171	-.1998	.6440	.0270	.0270	.0270	.0270	.0270
.0059	...	.0059	.0059	.0059	.0059	.0059	.0059	.0059	.0059	.0059	.6492	-.5828	-1.068	.8308	0	0	0	0
-.0186	...	-.0186	-.0186	-.0186	-.0186	-.0186	-.0186	-.0186	-.0186	-.0186	-.0186	-.0186	-.0186	-.0186	.6131	0	0	0
-.0151	...	-.0151	-.0151	-.0151	-.0151	-.0151	-.0151	-.0151	-.0151	-.0151	-.0151	-.0151	-.0151	-.0151	-.2569	.7562	0	0
.0097	...	.0097	.0097	.0097	.0097	.0097	.0097	.0097	.0097	.0097	.0097	.0097	.0097	.0097	-.7451	-.2259	.6492	0
.0461	...	.0461	0	.0291	-.0194	-.0085	.0387	-.0097	.0209	-.0167	.0345	.0345	-.0313	-.0501	.6719	-.5602	-.9909	.8900

(first 21 columns bracketed: 21)

$A^{(2)} =$

-.0081	...	-.0081	.7640	.0270	.0270	.0270	.0270	.0270	.0270	.0270	.0270	.0270	.0270	.0270	.0270	.0270	.0270	.0270
-.0185	...	-.0185	-.3126	.7002	0	0	0	0	0	0	0	0	0	0	0	0	0	0
.0300	...	.0300	-.7977	-.1271	.2955	0	0	0	0	0	0	0	0	0	0	0	0	0
-.0012	...	-.0012	.2977	-.5538	-.4091	.6910	0	0	0	0	0	0	0	0	0	0	0	0
-.0303	...	-.0303	-.0303	-.0303	-.0303	-.0303	.7570	0	0	0	0	0	0	0	0	0	0	0
.0208	...	.0208	-.0159	.0044	-.0852	-.0393	-.2744	.7032	.0270	.0270	.0270	.0270	.0270	.0270	.0270	.0270	.0270	.0270
.0222	...	.0222	.0222	.0222	.0222	.0222	-.7835	-.1169	.3447	0	0	0	0	0	0	0	0	0
-.0010	...	-.0010	-.0010	-.0010	-.0010	-.0010	.2900	-.5817	-.3742	.6910	0	0	0	0	0	0	0	0
-.0262	...	-.0262	-.0262	-.0262	-.0262	-.0262	-.0262	-.0262	-.0262	-.0262	.7590	0	0	0	0	0	0	0
-.0140	...	-.0140	-.0140	-.0140	-.0140	-.0140	-.0140	-.0140	-.0140	-.0140	-.2997	.7069	0	0	0	0	0	0
.0622	...	.0622	.0256	.0459	-.0438	.0021	.0535	-.0080	.0326	-.0096	-.7551	-.0817	.2967	.0270	.0270	.0270	.0270	.0270
.0017	...	.0017	.0017	.0017	.0017	.0017	.0017	.0017	.0017	.0017	.2919	-.5548	-.4518	.6655	0	0	0	0
-.0231	...	-.0231	-.0231	-.0231	-.0231	-.0231	-.0231	-.0231	-.0231	-.0231	-.0231	-.0231	-.0231	-.0231	.7606	0	0	0
-.0125	...	-.0125	-.0125	-.0125	-.0125	-.0125	-.0125	-.0125	-.0125	-.0125	-.0125	-.0125	-.0125	-.0125	-.2975	.7082	0	0
.0166	...	.0166	.0166	.0166	.0166	.0166	.0166	.0166	.0166	.0166	.0166	.0166	.0166	.0166	-.7854	-.1102	.3473	0
.0492	...	.0492	.0125	.0328	-.0569	-.0110	.0404	-.0211	.0195	-.0227	.0354	.0354	-.1465	-.0881	.3158	-.5305	-.3269	.6798

(first 21 columns bracketed: 21)

$$w = \frac{c}{2p}(\underset{\sim}{z}_1-\underset{\sim}{z}_2)'(\underset{\sim}{z}_1-\underset{\sim}{z}_2)$$

$$= \frac{25.18}{8}[(-0.067)^2+(0.113)^2+(-1.916)^2+(-1.492)^2]$$

$$= 18.616,$$

which is greater than $w(0.05) = 12.59$; hence the hypothesis H: $\underset{\sim}{\mu}_1 = \underset{\sim}{\mu}_2$ is rejected at 5% significance level and the power of this test is greater than or equal to $P_0 = 0.90$ for $\delta \geq \delta_0 = \sqrt{6.5} = 2.55$.

5. COMMENTS

In this paper we have treated the case of two samples. However, the heteroscedastic method can be applied to the case of k samples; see for example, Dudewicz and Taneja (1981). This is essentially based on the fact that the distribution of $\underset{\sim}{z}_i$ defined by (4) is completely independent of unknown population covariance matrix Σ_i. Also asymptotic expansions for distributions of test statistics obtained for the case of k samples can be derived by applying Lemma 2 and its Corollary in Hyakutake and Siotani (1986), although we may need a large N_0, a common size of first-stage samples from each of k populations.

The method described in this paper or the multivariate heteroscedastic method may be not appropriate to use in practice for large p, dimensionality; for if $p = 10$, one has to take at least $p^2 = 100$ additional observations because of (2) which seems to be too big for experimenters.

There is an important problem left to be investigated in future, that is, a comparison of the test procedure in this paper with other methods mentioned in Section 1 with respect topower of test or sample sizes attaining a specified value of power. However, since the power functions of James' and Yao's tests are quite complicated, we are forced to employ the Monte Carlo experiment.

Department of Applied Mathematics
Science University of Tokyo
Kagurazaka, Shinjuku-ku, Japan

REFERENCES

Bennett, B.M. (1951). 'Note on a solution of the generalized Behrens-Fisher problem.' Ann. Inst. Statist. Math., 2, 87-90.

Chatterjee, S.K. (1959). 'On an extension of Stein's two-sample procedure to multi-normal problem.' Culcutta Statist. Assoc., Bull., 8, 121-148.

Dudewicz, E.J., and Bishop, T.A. (1979). 'The heteroscedastic method.' Optimizing Methods in Statistics (J.S. Rustagi, ed.), Academic Press, New York, 183-203.

Dudewicz, E.J., and Taneja, V.S. (1981). 'A multivariate solution of the multivariate ranking and selection problem.' Commun. Statist.-Theor. Meth., A10(18), 1849-1868.

Eaton, M.L. (1969). 'Some remarks on Scheffé's solution to the Behrens-Fisher problem.' J. Amer. Statist. Assoc., 64, 1318-1322.

Fisher, R.A. (1936). 'The use of multiple measurements in taxonomic problem.' Ann. Eugen., 7, 179-188.

Haynam, G.E., Govindarajulu, Z., and Leone, F.C. (1970). 'Tables of the cumulative non-central chi-square distribution.' Selected Tables in Mathematical Statistics (edited by the Inst. Math. Statist.), Vol. 1, Amer. Math. Soc., Providence, Rhode Island.

Hyakutake, H. (1985). 'A construction method of certain matrices required in the multivariate heteroscedastic method.' Tech. Repo., 149, Statistical Research Group, Hiroshima Univ., Hiroshima, Japan.

Hyakutake, H., and Siotani, M. (1986). 'The multivariate heteroscedastic method: distributions of statistics and an application.' Amer. J. Math. Manage. Sci., 7 (in print)

Hyakutake, H., Siotani, M., Li, Chu-yu, and Mustafid (1986). 'Distributions of some statistics in heteroscedastic inference method —power functions and percentage points—.' J. Japan Statist. Soci., 16, 7-20.

Ito, K. (1969). 'On the effect of heteroscedasticity and nonnormality upon some multivariate test procedures.' Multivariate Analysis II (P.R. Krishnaiah ed.), Academic Press, New York, 87-120.

James, G.S. (1954). 'Tests of linear hypotheses in univariate and multivariate analysis when the ratios of the population variances are unknown.' Biometrika, 41, 19-43.

Pearson, E.S., and Hartley, H.O. (eds.) (1972). Biometrika Tables for Statisticians, Volume 2, Cambridge University Press.

Scheffé, H. (1970). 'Practical solutions of the Behrens-Fisher problem.' J. Amer. Statist. Assoc., 65, 1501-1508.

Siotani, M., Hayakawa, T., and Fujikoshi, Y. (1985). Modern Multivariate Statistical Analysis: A Graduate Course and Handbook. American Sciences Press, Columbus, Ohio.

Stein, C. (1945). 'A two-sample test for a linear hypothesis whose power is independent of the variance.' Ann. Math. Statist., 16, 243-258.

Yao, Y. (1965). 'An approximate degrees of freedom solution to the multivariate Behrens-Fisher problem.' Biometrika, 52, 139-147.

This research was partially supported by Grant-in-Aid for Scientific Research of the Ministry of Education under Contract Number 321-6009-61530017, and also in part by Research Grant of Science University of Tokyo under Contract Number 86-1001.

M.S. Srivastava

TESTS FOR COVARIANCE STRUCTURE IN FAMILIAL DATA AND PRINCIPAL COMPONENT

1. INTRODUCTION

Let $\underline{x}_1,\ldots,\underline{x}_n$ be independently and identically distributed as $N_p(\underline{\mu},\Sigma)$, where $N_p(\underline{\mu},\Sigma)$ denotes the distribution of a p-dimensional normal random vector with mean vector $\underline{\mu}$ and covariance matrix $\Sigma = (\sigma_{ij})$. In the familial data analysis, certain structure on the covariance matrix is assumed. It would thus be desirable to test whether this model is correct. Similarly, in principal component analysis, it is sometimes desirable to test if some given set of orthogonal vectors are eigenvectors of the unknown covariance matrix. In this note, we consider these two problems and give likelihood ratio tests along with their asymptotic distributions; the second problem was considered by Mallows (1961). Bootstrap methods are given when normality is suspected.

2. TESTING FOR COVARIANCE STRUCTURE IN FAMILIAL DATA

Suppose the random vector $\underline{x} = (x_1,\ldots,x_p)'$ records parallel observations on a mother and her p-1 children, x_1 being the observation on the mother, $\{x_i, 2 \le i \le p\}$ being the observation on the siblings. A simple model for such familial data postulates that the covariance matrix of x has the following structure:

A. K. Gupta (ed.), Advances in Multivariate Statistical Analysis, 341–352.

$$\text{var}(x_1) = \sigma_m^2, \quad \text{var}(x_i) = \sigma_c^2, \quad 2 \le i \le p ,$$

$$\text{cov}(x_1, x_i) = \sigma_{mc} = \rho_{mc}\, \sigma_m\, \sigma_c, \quad 2 \le i \le p ,$$

$$\text{cov}(x_i, x_j) = \sigma_{cc} = \rho_{cc}\, \sigma_c^2, \quad 2 \le i \ne j \le p .$$

The values of σ_m^2, σ_{mc}, σ_c^2, and σ_{cc} are unknown but $\underline{x}$ is assumed to be normally distributed as $N_p(\underline{\mu}, \Sigma)$. We wish to test the fit of this model to a sample of N familial data vectors, $\underline{x}_1, \ldots, \underline{x}_N$, all of the same dimension p. Thus, we wish to test the hypothesis that

$$H\colon \Sigma = \Sigma_o \quad \text{vs} \quad A\colon \Sigma \ne \Sigma_o ,$$

where

$$\Sigma_o = \begin{bmatrix} \sigma_m^2 & \sigma_{mc} & \sigma_{mc} & \cdots & \sigma_{mc} \\ \sigma_{mc} & \sigma_c^2 & \sigma_{cc} & \cdots & \sigma_{cc} \\ \sigma_{mc} & \sigma_{cc} & \sigma_c^2 & & \sigma_{cc} \\ \vdots & \vdots & \vdots & \ddots & \\ \sigma_{mc} & \sigma_{cc} & \sigma_{cc} & & \sigma_c^2 \end{bmatrix}$$

Consider, as in Srivastava (1984), a known $(p-1)\times(p-1)$ nonsingular matrix Γ given by $\Gamma' = [(p-1)^{-1}\, \underline{e}_{p-1}, C']$, where $\underline{e}_r = (1, \ldots, 1)'$, is an r-vector of ones and C is of the order $(p-2)\times(p-1)$ and is such that $C\underline{e} = \underline{0}$, $CC' = I_{p-2}$. Let

$$\underline{y}_i = \begin{bmatrix} 1 & \underline{0}' \\ \underline{0} & \Gamma \end{bmatrix} \underline{x}_i = \tilde{\Gamma}\, \underline{x}_i, \quad i = 1, \ldots, N .$$

Then, with

$$\underline{\nu} = \tilde{\Gamma}\,\underline{\mu},\ \Omega = \tilde{\Gamma}\,\Sigma\,\tilde{\Gamma}' ,$$

$$\Omega_{11} = \begin{bmatrix} \sigma_m^2 & \sigma_{mc} \\ \sigma_{mc} & \eta^2 \end{bmatrix},\quad \gamma_c^2 = \sigma_c^2(1-\rho_{cc}) , \tag{2.1}$$

$$\eta^2 = (p-1)^{-1}\,\{1+(p-2)\rho_{cc}\}\sigma_c^2 ,$$

$$\Omega_o = \mathrm{diag}(\Omega_{11},\ \gamma_c^2\, I_{p-2}) , \tag{2.2}$$

we find that $\underline{y}_i$'s are iid $N_p(\underline{\nu},\Omega_o)$ under the hypothesis H. Since, the matrix $\tilde{\Gamma}$ is known and the transformation from $\underline{x}_i$ to $\underline{y}_i$ is one to one, the hypothesis H is equivalent to testing the hypothesis

$$H:\ \Omega = \Omega_o \quad \text{vs} \quad A:\ \Omega \neq \Omega_o . \tag{2.3}$$

Let

$$S = \sum_{i=1}^{N} (\underline{x}_i - \bar{\underline{x}})(\underline{x}_i - \bar{\underline{x}})',\ \bar{\underline{x}} = N^{-1} \sum_{i=1}^{N} \underline{x}_i .$$

$$V = \tilde{\Gamma}\, S\, \tilde{\Gamma}' = \begin{bmatrix} V_{11} & V_{12} \\ V_{12}' & V_{22} \end{bmatrix} .$$

Then, the likelihood ratio test rejects the hypothesis H for large values of

$$\lambda = [(p-2)^{-1}\,\mathrm{tr}\, V_{22}]^{(p-2)}/|V_{2.1}|, \tag{2.4}$$

where

$$V_{2.1} = V_{22} - V_{12}' V_{11}^{-1} V_{12} .$$

Letting

$$P = V_{22}^{-\frac{1}{2}} V_{12}' V_{11}^{-1} V_{12} V_{22}^{-\frac{1}{2}} ,$$

we find that

$$\lambda = [\{(p-2)^{-1} \text{ tr } V_{22}\}^{p-2}/|V_{22}|] \; [|I-P|^{-1}] ,$$

$$= \lambda_1 \lambda_2 , \quad \text{say} ,$$

where P and V_{22} are independently distributed under the hypothesis H, see Theorem 3.6.9 of Srivastava and Khatri (1979). Hence, since $V \sim W_p(\Omega_o, n)$, $n = N-1$, under H, we get

$$E(\lambda_1^h) = s^{-sh} \frac{\Gamma_s(\frac{n}{2} - h)\Gamma(\frac{ns}{2})}{\Gamma_s(\frac{n}{2})\Gamma(\frac{ns}{2} - sh)} , \quad s = p-2 ,$$

and

$$E(\lambda_2^h) = E\,[|I - P|^{-h}]$$

$$= \frac{\Gamma_s(\frac{n}{2}) \; \Gamma_s(\frac{n-2}{2} - h)}{\Gamma_s(\frac{n}{2} - h) \; \Gamma_s(\frac{n-2}{2})}$$

Using multiplicative formula

$$\Gamma(s\,\alpha) = (2\pi)^{-\frac{s-1}{2}} \Gamma_s(s\alpha - \tfrac{1}{2}) \prod_{j=0}^{s-1} \Gamma(\alpha + s^{-1} j) ,$$

we find that

$$E(\lambda^h) = \prod_{j=0}^{s-1} \frac{\Gamma(\frac{n}{2} + s^{-1}j)\Gamma(\frac{n-r-j}{2} - h)}{\Gamma(\frac{n}{2} - h + s^{-1}j)\Gamma(\frac{n-r-j}{2})} \tag{2.5}$$

Hence

$$P\{r \log \lambda \leq z\} = P\{\chi_f^2 \leq z\} + \frac{a}{r^2} [P(\chi_{f+4}^2 \leq z) - P(\chi_f^2 \leq z)] + 0(r^{-3}) , \tag{2.6}$$

where

$$f = \frac{(p-2)(p-1)}{2} + 2(p-2) - 1 ,$$

$$r = n - 2\alpha,$$

$$\alpha = \frac{1}{4f}\left[\frac{(p-4)(p-3)(2p-5)p}{6(p-2)} + (p-3)p+2(p-2)(p+1)\right],$$

$$a = \frac{p(p-3)}{48(p-2)} [(p-3)\{(p-2)^2 - 2p+8\}-2(2\alpha-1)(p-4)(2p-5)$$
$$+ 4(p-2)(6\alpha^2-6\alpha+1)]$$
$$+ \frac{p-2}{12} [8+6(p-3)+(p-3)(2p-5)-6(2\alpha-1)(p-1)$$
$$+ 4(6\alpha^2-6\alpha+1)] .$$

3. TESTING FOR SPECIFIED EIGENVECTORS

Let $\underline{x} \sim N(\underline{\mu}, \Sigma)$. Let B be a $p \times k$ matrix whose columns are orthonormal. Suppose the null hypothesis asserts that the columns of B are eigenvectors of the unknown covariance matrix Σ. Let C be any $p\times(p-k)$ matrix such that $\Gamma = (B \vdots C)$ is orthogonal. Then the null hypothesis asserts that the variates forming the elements of $B'\underline{x}$ are uncorrelated with each other and with the elements of $C'\underline{x}$; i.e., that they are the k principal components of the basic distribution. Since $B'B = I_k$, $C'C = I_{p-k}$, $C'B = 0$, we get under the hypothesis

$$\Sigma B = B\Lambda_1, \quad B'\Sigma B = \Lambda_1, \quad C'\Sigma B = 0 , \tag{3.1}$$

where Λ_1 is a $k \times k$ diagonal matrix whose diagonal elements are unknown. Thus, if $\Omega = \Gamma'\Sigma\Gamma$, we wish to test the hypothesis

$$H: \Omega = \Gamma'\Sigma\Gamma = \begin{pmatrix} \Lambda_1 & 0 \\ 0 & C'\Sigma C \end{pmatrix} = \begin{pmatrix} \Lambda_1 & 0 \\ 0 & \Omega_2 \end{pmatrix} \tag{3.2}$$

against the alternative $A: \Omega > 0$.

Let $\underline{x}_1, \ldots, \underline{x}_N$ be iid $N_p(\underline{\mu}, \Sigma)$. Let

$$S = \sum_{i=1}^{N} (\underline{x}_i - \bar{\underline{x}})(\underline{x} - \bar{\underline{x}})', \quad \bar{\underline{x}} = N^{-1} \sum_{i=1}^{N} \underline{x}_i , \tag{3.3}$$

$$V = (v_{ij}) = \Gamma'S\Gamma = \begin{pmatrix} B'SB & B'SC \\ C'SB & C'SC \end{pmatrix} \equiv \begin{pmatrix} V_{11} & V_{12} \\ V_{12}' & V_{22} \end{pmatrix} , \tag{3.4}$$

where V_{11}: $k \times k$, V_{12}: $k \times (p-k)$, V_{22}: $(p-k) \times (p-k)$. Then $V \sim W_p(\Omega, n)$, where $n = N - 1$. The likelihood ratio test is based on the statistic

$$\lambda = \left[\frac{\left(\prod_{i=1}^{k} v_{ii}\right) |V_{22}|}{|V|} \right] = \frac{\left(\prod_{i=1}^{k} v_{ii}\right)}{|V_{11} - V_{12} V_{22}^{-1} V_{12}'|} \tag{3.5}$$

see Mallows (1961); the hypothesis is rejected for large

values of λ. We shall now show that the test statistic λ does not depend on C. We have

$$V_{11} - V_{12} V_{22}^{-1} V_{22}'$$

$$= B'SB - (B'SC)(C'SC)^{-1}(C'SB)$$

$$= B'[S - SC(C'SC)^{-1}C'S]B$$

$$= B'[B(B'S^{-1}B)^{-1}B']B$$

$$= (B'S^{-1} B)^{-1} ,$$

from Corollary 1.9.2, page 19 of Srivastava and Khatri (1979). Hence

$$\lambda = [\prod_{i=1}^{k} (B'SB)_{ii}] \, |(B'S^{-1}B)| ,$$

where $(B'SB)_{ii}$ denotes the ith diagonal element of the matrix $(B'SB)$.

Next, we find the hth moment of λ under the hypothesis H. Let

$$D_{v^{1/2}} = \text{diag}\,(v_{11}^{1/2},\ldots,v_{kk}^{1/2}),$$

$$L = D_{v^{-1/2}} V_{11} D_{v^{-1/2}} ,$$

$$R = V_{11}^{-1/2} V_{12} V_{22}^{-1} V_{12}' V_{11}^{-1/2} .$$

Then

$$\lambda = \{|L| \; |I - R|\}^{-1} .$$

From a slight modification of Problem 3.1, Srivastava

and Khatri (1979, p. 97), L and R are independently distributed under the hypothesis H. Assuming, without loss of generality, that $k \leq m$, the distribution of R is given by

$$f(R) = \frac{\Gamma_m(\tfrac{1}{2}n)\, \Gamma_k(\tfrac{1}{2}n)}{\Gamma_p(\tfrac{1}{2}n)} \frac{\pi^{\frac{1}{2}km}}{\Gamma_k(\tfrac{1}{2}m)} |R|^{\frac{1}{2}(m-k-1)} |I-R|^{\frac{1}{2}(n-p-1)}, \tag{3.7}$$

where $m = p - k$, and $\Gamma_r(\tfrac{1}{2}n) = \pi^{\frac{1}{4}r(n-1)} \prod_{i=1}^{r} \Gamma(\frac{n-i+1}{2})$. And, from Theorem 3.5.1, Srivastava and Khatri (1979), p. 87, the pdf of L under H is given by

$$f(L) = \frac{[\Gamma(\tfrac{1}{2}n)]^k}{\Gamma_k(\tfrac{1}{2}n)} |L|^{\frac{1}{2}(n-k-1)} . \tag{3.8}$$

Hence, after some simplification,

$$E(\lambda^h) = \left[\frac{\Gamma(\tfrac{1}{2}n)}{\Gamma(\tfrac{1}{2}n-h)}\right]^k \frac{\Gamma_k(\frac{n-m}{2} - h)}{\Gamma_k(\frac{n-m}{2})} . \tag{3.9}$$

Thus, with

$$r = n - 2\alpha ,$$

where

$$2\alpha = \frac{6m(p+1) + (2k+5)}{6(p+m-1)} ,$$

we get

$$P\{r \ln \lambda \leq z\} = P\{\chi_f^2 \leq z\} - \frac{2\gamma_2}{3r^2} P[(\chi_{f+4}^2 \leq z) - P(\chi_f^2 \leq z)] + 0(r^{-3}), \tag{3.10}$$

where

$$f = \frac{k(p+m-1)}{2} ,$$

$$\gamma_2 = \frac{k^2}{128f} [6k^4 + 2(10m+7)k^3 + (24m^2 + 32m - 1)k^2$$
$$+ 4(2m^3 + 3m^2 + 5m - 2)k$$
$$+ (4m^4 - 8m^3 - 4m^2 + 24m - 11)]$$

4. BOOTSTRAPPING

When normality is suspected, the above distributional results become invalid. Even the asymptotic distribution theory for these tests becomes complex. As shown in Beran and Srivastava (1985), the nonparametric bootstrap methods offer an attractive alternative approach. The only assumption required in carrying out the bootstrap method is the existence of fourth moments for the distribution F from which n independent samples $\underline{y}_1, \ldots, \underline{y}_n$ have been drawn. Let Ω_F denote the covariance of $\underline{y}_i$'s.

For any $p \times p$ symmetric matrix $A = (a_{ij})$, let $u\,\text{vec}\,(A)$ denote the $p(p+1)/2$ dimensional column vector $\{a_{ij};\ 1 \le i \le j \le p\}$ formed from the elements in the upper half of A, including the diagonal elements. Suppose $Z_F = \{\underline{Z}_{F,ij}\}$ is a symmetric $p \times p$ random matrix such that the distribution of $u\,\text{vec}\,(Z_F)$ is normal with mean vector zero and covariance matrix Λ_F. Let

$$W_n = (n-1)^{-1} V, \text{ where } V = \Sigma_{i=1}^{n} (\underline{y}_i - \bar{\underline{y}})(\underline{y}_i - \bar{\underline{y}})' . \qquad (4.1)$$

That is W_n is the sample covariance matrix of the $\underline{y}_i$'s. Suppose $T_n = nh(W_n)$ is a test statistic for the null hypothesis

H: the $\{\underset{\sim}{y}_i, i \geq 1\}$ are independently identically distributed $p \times 1$ random vectors with unknown cdf F_o, which has finite fourth moments; F_o is such that $\Omega_{F_o} = \pi(\Omega_{F_o})$, (4.2)

where π is a linear projection, not the identity map, which takes any $p \times p$ nonsingular matrix into a $p \times p$ covariance matrix. The function h defining the test statistic T_n is twice continuously differentiable at u vec (Ω_{F_o}), with h and the first derivative of h vanishing there for every possible choice of Ω_{F_o} satisfying the hypothesis H. The test rejects H for large values of T_n. Let $\ddot{h}$ denote the second derivative of h, and let $\underline{z}_{F_o}$ = u vec (Z_{F_o}). Then, denoting by '$\mathcal{L}$' the distribution 'of',

$$\mathcal{L}(T_n | F_o) \Longrightarrow \mathcal{L}(\underline{z}'_{F_o}\, \ddot{h}\, (\Omega_{F_o})\, \underline{z}_{F_o}) , \tag{4.3}$$

as n tends to infinity. Approximate critical values for the test can be obtained by computing the upper quantiles of the limiting distribution on the right side of (4.3), after first estimating the unknown cdf F_o by the empirical cdf.

Alternatively, we can construct a bootstrap estimate for the distribution of $\mathcal{L}(T_n | F_o)$, as follows. Let

$$\underline{u} = [\pi(w_n)]^{\frac{1}{2}}\, w_n^{-\frac{1}{2}}\, \underline{y}_i, \quad 1 \leq i \leq n .$$

Let $\{u_i^*; 1 \leq i \leq n\}$ be iid random vectors whose cdf is the realized empirical cdf of the $\{\underline{u}_i; 1 \leq i \leq n\}$. Let

$W^*_{n,\underline{u}}$ be the sample covariance matrix of the $\underline{u}^*_i$. Then with probability one $\mathcal{L}[nh(W^*_{n,\underline{u}})] \Longrightarrow \mathcal{L}(\underline{z}'_{F_o} \ddot{h}(\Sigma_{F_o})\underline{z}_{F_o})$ from Corollary 3 of Beran and Srivastava (1985). The distribution of $nh(W^*_{n,\underline{u}})$ can be approximated by Monte Carlo methods.

The null hypothesis (2.3) for the familial data problem is of the form (4.2) where π is a linear projection defined by

$$\pi(\Omega_o) = \begin{pmatrix} \Omega & 0 \\ 0' & [(p-2)^{-1}\mathrm{tr}(\Omega_{22})I_{p-2}] \end{pmatrix},$$

for any $p \times p$ covariance matrix Ω_o partitioned as in (2.2). The function h defined by (see (2.4)

$$h(W_n) = (p-2)\log[(p-2)^{-1}\mathrm{tr}\ w_{n,22}] - \log\det(w_{n,22} - w'_{n,12}\ w^{-1}_{n,11}\ w_{n,12}),$$

where

$$W_n = \begin{pmatrix} w_{n,11} & w_{n,12} \\ w'_{n,12} & w_{n,22} \end{pmatrix},$$

is twice continuously differnetiable at $\pi(\Omega_o)$ with both h and the first derivative of h vanishing there. Thus, the bootstrap test for the familial data model based on the statistic (4.5) has asymptotic size α by (4.3) if the hypothesis H is rejected when $nh(w_n) \geq d^*_{n,\alpha}$, where $d^*_{n,\alpha}$ is the upper $100 \times \alpha$ % point of the bootstrap distribution obtained by Monte Carlo methods.

The bootstrapping of the second problem has been discussed in Beran and Srivastava (1985).

5. ACKNOWLEDGEMENT

The research was supported by Natural Sciences and Engineering Council of Canada.

Department of Statistics
University of Toronto
Toronto, Ontario
M5S 1A1

REFERENCES

Beran, R. and Srivastava, M.S. (1985). Bootstrap tests and confidence for functions of a covariance matrix. *Ann. Statist.* 13, 95-115.

Mallows, C.L. (1961). Latent vectors of random symmetric matrices. *Biom trika*, 48, 133-149.

Srivastava, M.S. (1984). Estimation of interclass correlations in familial data. *Biometrika*, 71, 177-185.

Srivastava, M.S. and Khatri, C.G. (1979). An introduction to multivariate statistics. Elsevier North Holland, New York.

Nariaki Sugiura and Yoshihiko Konno

RISK OF IMPROVED ESTIMATORS FOR GENERALIZED VARIANCE AND PRECISION

Summary

Let the distributions of observed random matrices $X(p\times r)$ and $S(p\times p)$ be $N(\xi,\Sigma\otimes I_r)$ and $W_p(n,\Sigma)$ respectively. Assume that they are independent. The risk of improved estimators for $|\Sigma|$ or $|\Sigma^{-1}|$ based on X and S under squared loss is evaluated in terms of incomplete beta function of matrix argument. Numerical comparison for the reduction of risk over the best equivariant estimators is given.

1. INTRODUCTION

Suppose that observed random matrix $X(p\times r)$ has normal distribution $N(\xi,\Sigma\otimes I_r)$ and that $S(p\times p)$ has Wishart distribution $W_p(n,\Sigma)$, where $n\geqq p$ and the matrix ξ of mean vectors and the covariance matrix Σ are unknown. Shorrock and Zidek (1976) have shown that the best affine equivariant estimator (=best scalar multiple of $|S|$) for generalized variance $|\Sigma|$ under the squared loss

$$L_1(d;|\Sigma|) = \left(\frac{d}{|\Sigma|}-1\right)^2 \tag{1.1}$$

is given by

$$d_1(S) = \frac{(n-p+2)!}{(n+2)!}|S| \tag{1.2}$$

with constant risk $R(d_1)=p(2n-p+3)/\{(n+1)(n+2)\}$. Then using

This research was supported by University of Tsukuba Project Research 1986.

A. K. Gupta (ed.), Advances in Multivariate Statistical Analysis, 353–371.

zonal polynomials of matrix argument developed by James (1964) and Constantine (1963), they proved that $d_1(S)$ is dominated by the improved estimator

$$d_1^*(X,S) = \mathrm{Min}\{d_1(S), \frac{(n+r-p+2)!}{(n+r+2)!}|S+XX^t|\}. \qquad (1.3)$$

The result is an extension of Stein (1964) who proved the case of p=1. Sinha (1976) also gave another proof without using zonal polynomials. If the loss $L_2(d;|\Sigma|)=d/|\Sigma|-\log(d/|\Sigma|)-1$ is used instead of (1.1), the best affine equivariant estimator is given by $d_2(S)=\{(n-p)!/n!\}|S|$, which is dominated by

$$d_2^*(X,S) = \mathrm{Min}\{d_2(S), \frac{(n+r-p)!}{(n+r)!}|S+XX^t|\}.$$

This fact was pointed out by Sinha and Ghosh (1985).

To see the reduction of risk of $d_1^*(X,S)$ over $d_1(S)$, we first tried simulation study. However we found that it is not sufficiently precise to detect such a delicate difference as this. Simulation sometimes gave us decreased values where it should be increased. In Section 2 of this paper we shall give an expression of risk of $d_1^*(X,S)$ useful for numerical computation, though it contains incomplete beta functions of matrix argument.

To estimate generalized precision $|\Sigma^{-1}|$ under the squared loss

$$L_1^*(e;|\Sigma^{-1}|) = (\frac{e}{|\Sigma^{-1}|}-1)^2, \qquad (1.4)$$

affine equivariant estimator e(X,S) is defined by

$$e(AXH+B,ASA^t) = |A|^{-2}e(X,S)$$

for any pxp nonsingular matrix A, any rxr orthogonal matrix H, and any pxr matrix B. The best affine equivariant estimator is given by

$$e_1(S) = \frac{(n-4)!}{(n-p-4)!}|S^{-1}| \tag{1.5}$$

when $n \geq p+4$ and has constant risk $R(e_1)=p(2n-p-5)/\{(n-2)(n-3)\}$. In Section 3, we show that $e_1(S)$ is dominated by

$$e_1^*(X,S) = \mathrm{Max}\{e_1(S), \frac{(n+r-4)!}{(n+r-p-4)!}|S+XX^t|^{-1}\} \tag{1.6}$$

when $n \geq p+4$, following the same argument as in Shorrock and Zidek (1976) and give an expression of risk of $e_1^*(X,S)$ suitable for numerical computations. We note that if the loss $L_2^*(e;|\Sigma^{-1}|)=e|\Sigma|-\log(e|\Sigma|)-1$ is used instead of (1.4), the best affine equivariant estimator is given by $e_2(S)=\{(n-2)!/(n-p-2)!\}\ |S^{-1}|$ in case of $n \geq p+2$, which is dominated by

$$e_2^*(X,S) = \mathrm{Max}\{e_2(S), \frac{(n+r-2)!}{(n+r-p-2)!}|S+XX^t|^{-1}\}. \tag{1.7}$$

In Section 4 numerical comparison of reduction of risk of $d_1^*(X,S)$ or $e_1^*(X,S)$ is given. It is shown that the reduction is more remarkable for estimating generalized precision than for estimating generalized variance.

2. ESTIMATION OF GENERALIZED VARIANCE

In order to derive a useful representation of the risk of the improved estimator $d_1^*(X,S)$ under L_1 loss defined by (1.3) and (1.1), we shall define a multivariate incomplete beta function, which is also used in Section 3.

Definition 2.1. Let k be a nonnegative integer and let $\kappa=\{k_1,\ldots,k_r\}$ be a partition of k into not more than r parts where $k=k_1+\ldots+k_r$ and $k_1 \geq \ldots \geq k_r \geq 0$. Let $C_\kappa(Z)$ be the zonal polynomial of positive definite matrix Z of order r corresponding to κ. Define an r-variate incomplete beta function with parameters α, β corresponding to partition κ by

$$I_a^{(r)}(\alpha,\beta;\kappa) = \frac{(\alpha+\beta)_\kappa}{(\beta)_\kappa C_\kappa(I)} \int_{\substack{|Z|<a \\ 0<Z<I}} \frac{|Z|^{\alpha-\frac{r+1}{2}} |I-Z|^{\beta-\frac{r+1}{2}}}{B_r(\alpha,\beta)} C_\kappa(I-Z)dZ \tag{2.1}$$

where $0<Z<I$ in the domain of integration stands for Z and $I-Z$ being positive definite and $(\alpha)_\kappa = \Pi_{i=1}^r (\alpha-(i-1)/2)_{k_i}$ with $(b)_0 = 1$ and $(b)_m = b(b+1)\ldots(b+m-1)$.

The constant factor in (2.1) is so chosen that for $a=1$, $I_a^{(r)}(\alpha,\beta;\kappa)=1$ by the formula (22) in Constantine (1963). For the case of $r=1$

$$I_a^{(1)}(\alpha,\beta;\kappa) = \int_0^a \frac{z^{\alpha-1}(1-z)^{\beta+k-1}}{B(\alpha,\beta+k)} dz$$

which is the usual incomplete beta function denoted by $I_a(\alpha,\beta+k)$. We now prove the following theorem.

<u>Theorem 2.1.</u> Assume that $n\geq p$. Put $\Lambda = \xi^t\Sigma^{-1}\xi$ and $a=\{(n+2)!(n+r-p+2)!\}/\{(n+2-p)!(n+r+2)!\}$. Then under the L_1 loss, the risk of the improved estimator $d_1^*(X,S)$ is given by

(i) when $r\geq p$,

$$\Sigma_{k=0}^\infty \Sigma_{(\kappa)} \text{etr}(-\frac{\Lambda}{2}) \frac{C_\kappa(\frac{\Lambda}{2})}{k!} [\frac{(n-p+1)(n-p+2)}{(n+1)(n+2)} \{I_a^{(p)}(\frac{n+4}{2},\frac{r}{2};\kappa)$$

$$- 2I_a^{(p)}(\frac{n+2}{2},\frac{r}{2};\kappa)\} + \Pi_{i=1}^p (\frac{n+r+2k_i-i+1}{n+r-i+3})$$

$$\times\{\Pi_{i=1}^p (\frac{n+r+2k_i-i+3}{n+r-i+3}) - 2\}\{1 - I_a^{(p)}(\frac{n}{2},\frac{r}{2};\kappa)\} + 1]$$

(ii) when $r<p$,

$$\Sigma_{k=0}^{\infty}\Sigma_{(\kappa)} etr(-\frac{\Lambda}{2})\frac{C_{\kappa}(\frac{\Lambda}{2})}{k!}[\frac{(n-p+1)(n-p+2)}{(n+1)(n+2)}$$

$$\{I_a^{(r)}(\frac{n+r-p+4}{2},\frac{p}{2};\kappa) - 2I_a^{(r)}(\frac{n+r-p+2}{2},\frac{p}{2};\kappa)\}$$

$$+ \frac{(n+r-p+1)(n+r-p+2)}{(n+r+1)(n+r+2)}\Pi_{i=1}^{r}(\frac{n+r+2k_i-i+1}{n+r-i+1})$$

$$x\{\Pi_{i=1}^{r}(\frac{n+r+2k_i-i+3}{n+r-i+3}) - 2\}\{1 - I_a^{(r)}(\frac{n+r-p}{2},\frac{p}{2};\kappa)\} + 1]$$

where $\Sigma_{(\kappa)}$ denotes sum of all possible partition κ of k.

<u>Corollary 2.1.</u> When $p=1$, $a=(n+2)/(n+r+2)$ and the risk is given by

$$\Sigma_{k=0}^{\infty} etr(-\frac{\Lambda}{2})\frac{(tr\Lambda/2)^k}{k!}[\frac{n}{n+2}\{I_a(\frac{n}{2}+2,\frac{r}{2}+k)$$

$$- 2I_a(\frac{n}{2}+1,\frac{r}{2}+k)\} - \frac{(n+r+2k)(n+r+2-2k)}{(n+r+2)^2}\{1 - I_a(\frac{n}{2},\frac{r}{2}+k)\} + 1].$$

<u>Corollary 2.2.</u> When $r=1$, $a=(n-p+3)/(n+3)$ and the risk is given by

$$\Sigma_{k=0}^{\infty} \exp(-\frac{\Lambda}{2})\frac{(\Lambda/2)^k}{k!}[\frac{(n-p+1)(n-p+2)}{(n+1)(n+2)}\{I_a(\frac{n-p+5}{2},\frac{p}{2}+k)$$

$$- 2I_a(\frac{n-p+3}{2},\frac{p}{2}+k)\} - \frac{(n-p+2)(n-p+3)(n+2k+1)(n-2k+3)}{(n+1)(n+2)(n+3)^2}$$

$$x\{1 - I_a(\frac{n-p+1}{2},\frac{p}{2}+k)\} + 1].$$

The risk of Stein's improved estimator (Stein (1964)) is given by Corollary 2.1 and the risk of the improved estimator based on p-variate one-sample problem of size n+1 is given by Corollary 2.2. They are the easiest cases to be computed, since only the usual incomplete beta functions are involved. To prove Theorem 2.1, we first note that the joint density function of $V=\chi_m^2+\chi_n^2(\lambda)$ and $T=\chi_m^2/(\chi_m^2+\chi_n^2(\lambda))$ where χ_m^2 and $\chi_n^2(\lambda)$ stand for chi-square variate with m degrees of freedom and noncentral chi-square variate with n degrees of freedom and noncentral parameter λ respectively and they are assumed to be independent, is given by

$$\Sigma_{k=0}^{\infty}\exp(-\frac{\lambda}{2})\frac{(\lambda/2)^k}{k!}\frac{v^{\frac{m+n}{2}+k-1}\,e^{-\frac{v}{2}}\,t^{\frac{m}{2}-1}\,(1-t)^{\frac{n}{2}+k-1}}{2^{(m+n)/2+k}\,\Gamma(\frac{m+n}{2}+k)\,B(\frac{m}{2},\frac{n}{2}+k)}. \quad (2.2)$$

Define an auxiliary Poisson random variable with parameter $\lambda/2$ by K. Then an equivalent expression to (2.2) is that for any function g(v) and h(t), assuming the expectation in LHS is finite,

$$E_\lambda[g(V)h(T)]$$
$$= E_\lambda^K[\frac{E_{\lambda=0}[g(V)V^K|K]}{E_{\lambda=0}[V^K|K]}\,\frac{E_{\lambda=0}[h(T)(1-T)^K|K]}{E_{\lambda=0}[(1-T)^K|K]}], \quad (2.3)$$

where conditional distribution of V for given K is the chi-square distribution obtained from $\chi_m^2+\chi_n^2(0)$ and the conditional distribution of T for given K is the beta distribution with parameters (m/2,n/2) obtained from $\chi_m^2/(\chi_m^2+\chi_n^2(0))$. V and T are conditionally independent for given K. Note that this gives a mixture representation of noncentral chi-square distribution due to Robbins and Pitman (1949) by putting h(t)=1 and that of noncentral beta distribution by putting g(v)=1. Multivariate extension of (2.3) was obtained by Shorrock and Zidek (1976), though they considered $\chi_n^2(\lambda)/(\chi_m^2+\chi_n^2(\lambda))$ instead of $\chi_m^2/(\chi_m^2+\chi_n^2(\lambda))$. The result is summarized in the following lemma which is essential for the proof of Theorem 2.1. Zidek (1978) may also be noted in this direction.

Lemma 2.1 (Shorrock and Zidek). Let S have $W_p(n,I)$ and let $X(p\times r)$ have $N(\xi, I_p \otimes I_r)$. Assume that S and X are independent. Put $V=S+XX^t$ and $T=(S+XX^t)^{-1/2}S(S+XX^t)^{-1/2}$ when $r \geqq p$ and $T=I-X^t(S+XX^t)^{-1}X$ when $r<p$. Further put $\Lambda=\xi\xi^t$ and let an auxiliary random variable K be a random partition of nonnegative integers into $s=\mathrm{Min}\{p,r\}$ parts with probability $P(K=\kappa)=\mathrm{etr}(-\frac{\Lambda}{2})C_\kappa(\frac{\Lambda}{2})\ /k!$ for $\kappa=\{k_1,\dots,k_s\}$ $k_1 \geqq \dots \geqq k_s \geqq 0$ and $k_1+\dots+k_s=k$. Then for any functions $g(V)$ and $h(T)$ invariant under the transformations $V \to H_1VH_1^t$ and $T \to H_2TH_2^t$ for any orthogonal matrices H_1 and H_2,

$$E_\Lambda[g(V)h(T)] \tag{2.4}$$
$$= E_\Lambda^K\left[\frac{E_{\Lambda=0}[g(V)C_K(V)\,|\,K]}{E_{\Lambda=0}[C_K(V)\,|\,K]}\ \frac{E_{\Lambda=0}[h(T)C_K(I-T)\,|\,K]}{E_{\Lambda=0}[C_K(I-T)\,|\,K]}\right]$$

where the expectation in LHS is assumed to be finite and the conditional distribution of V for given K is $W_p(n+r,I)$ obtained from $S+XX^t$ under $\Lambda=0$ and the conditional distribution of T for given K is a p-variate beta distribution with parameter $(n/2,r/2)$ obtained from $(S+XX^t)^{-1/2}S(S+XX^t)^{-1/2}$ under $\Lambda=0$ when $r \geqq p$ and an r-variate beta distribution with parameter $((n+r-p)/2,p/2)$ obtained from $I-X^t(S+XX^t)^{-1}X$ under $\Lambda=0$ when $r<p$. The V and T are conditionally independent for given K.

Proof of Theorem 2.1. Since the loss function (1.1) is invariant under the transformation $(X,S) \to (AX,ASA^t)$ and $(\xi,\Sigma) \to (A\xi,A\Sigma A^t)$ for any nonsingular matrix A, we may assume that $\Sigma=I$. Using Lemma 2.1, we can write $d_1^*(X,S)=|V|\,h(T)$ where $h(T)=\{(n-p+2)!/(n+2)!\}\mathrm{Min}\{|T|,a\}$ and get the risk of d_1^* as

$$R(d_1^*,\Lambda) = E_\Lambda[\{|V|h(T) - 1\}^2]$$

$$= E_\Lambda[|V|^2 h(T)^2 - 2|V|h(T) + 1] \qquad (2.5)$$

$$= E_\Lambda^K[\frac{E_0[|V|^2 C_K(V)|K]}{E_0[C_K(V)|K]} \frac{E_0[h(T)^2 C_K(I-T)|K]}{E_0[C_K(I-T)|K]}]$$

$$- 2E_\Lambda^K[\frac{E_0[|V|C_K(V)|K]}{E_0[C_K(V)|K]} \frac{E_0[h(T)C_K(I-T)|K]}{E_0[C_K(I-T)|K]}] + 1$$

where E_0 stands for $E_{\Lambda=0}$. Reproductive property of zonal polynomials $E_0[C_\kappa(V)]=2^k((n+r)/2)_\kappa C_\kappa(I_p)$ yields

$$\frac{E_0[|V|^2 C_\kappa(V)|K=\kappa]}{E_0[C_\kappa(V)|K=\kappa]} = 2^{2p} \frac{\Gamma_p((n+r+4)/2)}{\Gamma_p((n+r)/2)} \frac{((n+r+4)/2)_\kappa}{((n+r)/2)_\kappa}. \qquad (2.6)$$

By Lemma 2.1 we get, when $r \geqq p$,

$$E_0[h(T)^2 C_\kappa(I-T)|K=\kappa]$$

$$= \{\frac{(n+2-p)!}{(n+2)!}\}^2 \{\int_{|T|<a} |T|^{\frac{n+3-p}{2}} |I-T|^{\frac{r-p-1}{2}} C_\kappa(I-T)dT$$

$$+ a^2 \int_{|T|>a} |T|^{\frac{n-p-1}{2}} |I-T|^{\frac{r-p-1}{2}} C_\kappa(I-T)dT\} \qquad (2.7)$$

which can be expressed by incomplete beta function defined by (2.1). After simplification we get

$$\frac{E_0[h(T)^2 C_\kappa(I-T)|K=\kappa]}{E_0[C_\kappa(I-T)|K=\kappa]}$$

$$= \{\frac{(n+2-p)!}{(n+2)!}\}^2 \frac{(\frac{n+r}{2})_\kappa \ \Gamma_p(\frac{n+4}{2}) \ \Gamma_p(\frac{n+r}{2})}{(\frac{n+r+4}{2})_\kappa \Gamma_p(\frac{n+r+4}{2}) \ \Gamma_p(\frac{n}{2})} I_a^{(p)}(\frac{n+4}{2},\frac{r}{2};\kappa)$$

$$+ \{\frac{(n+r-p+2)!}{(n+r+2)!}\}^2 \{1 - I_a^{(p)}(\frac{n}{2},\frac{r}{2};\kappa)\} \qquad (2.8)$$

where we used $E_0[C_\kappa(I-T)|K=\kappa]=\{(\frac{r}{2})_\kappa/(\frac{n+r}{2})_\kappa\}C_\kappa(I)$ by Constantine (1963). Multiplication of (2.6) and (2.8) with the help of formulas $\Gamma_p(\frac{n+2}{2})/\Gamma_p(\frac{n}{2})=2^{-p}n!/(n-p)!$ and $(\frac{n+2}{2}-\frac{i-1}{2})_{k_i}/(\frac{n}{2}-\frac{i-1}{2})_{k_i}=\frac{n+1-i+2k_i}{n+1-i}$ yields the first term in RHS of (2.5) as

$$\frac{(n-p+1)(n-p+2)}{(n+1)(n+2)} I_a^{(p)}(\frac{n+4}{2},\frac{r}{2};\kappa) \qquad (2.9)$$

$$+ \{\Pi_{i=1}^{p} \frac{n+r+2k_i-i+3}{n+r+3-i} \ \frac{n+r+2k_i-i+1}{n+r+3-i}\}\{1 - I_a^{(p)}(\frac{n}{2},\frac{r}{2};\kappa)\}.$$

Similarly the second term in RHS of (2.5) is given by

$$\frac{(n-p+1)(n-p+2)}{(n+1)(n+2)} I_a^{(p)}(\frac{n+2}{2},\frac{r}{2};\kappa)$$

$$+ \{\Pi_{i=1}^{p} \frac{n+r+2k_i-i+1}{n+r+3-i}\}\{1 - I_a^{(p)}(\frac{n}{2},\frac{r}{2};\kappa)\}. \qquad (2.10)$$

Combined with (2.9) and (2.10), we get the formula (i).

When p>r, RHS of (2.8) is changed to

$$\{\frac{(n-p+2)!}{(n+2)!}\}^2 \frac{(\frac{n+r}{2})_\kappa}{(\frac{n+r+4}{2})_\kappa} \frac{\Gamma_r(\frac{n+r-p+4}{2})\Gamma_r(\frac{n+r}{2})}{\Gamma_r(\frac{n+r-p}{2})\Gamma_r(\frac{n+r+4}{2})}$$

$$\times I_a^{(r)}(\frac{n+r-p+4}{2},\frac{p}{2};\kappa) + \{\frac{(n+r-p+2)!}{(n+r+2)!}\}^2\{1 -$$

$$- I_a^{(r)}\left(\frac{n+r-p}{2},\frac{p}{2};\kappa\right)\}. \tag{2.11}$$

Combined with (2.6) yields the first term in RHS of (2.5) as

$$\frac{(n-p+1)(n-p+2)}{(n+1)(n+2)} I_a^{(r)}\left(\frac{n+r-p+4}{2},\frac{p}{2};\kappa\right)$$

$$+ \frac{(n+r-p+1)(n+r-p+2)}{(n+r+1)(n+r+2)}\{\Pi_{i=1}^{r} \frac{n+r+2k_i-i+3}{n+r+3-i}\,\frac{n+r+2k_i-i+1}{n+r+1-i}\}\{1$$

$$- I_a^{(r)}\left(\frac{n+r-p}{2},\frac{p}{2};\kappa\right)\}. \tag{2.12}$$

Similarly the second term in RHS of (2.5)is computed which gives the formula (ii).

Our numerical study shows that the minimum value of $R(d_1^*,\Lambda)$ is obtained at $\Lambda=0$ given by

$$R(d_1^*,0)$$

$$= \frac{(n-p+1)(n-p+2)}{(n+1)(n+2)}\{I_a^{(p)}\left(\frac{n+4}{2},\frac{r}{2}\right) -2I_a^{(p)}\left(\frac{n+2}{2},\frac{r}{2}\right)\}$$

$$- \frac{(n+r-p+1)(n+r-p+2)}{(n+r+1)(n+r+2)}\{1 - I_a^{(p)}\left(\frac{n}{2},\frac{r}{2}\right)\} + 1 \tag{2.13}$$

for $r \geqq p$ and

$$R(d_1^*,0)$$

$$= \frac{(n-p+1)(n-p+2)}{(n+1)(n+2)}\{I_a^{(r)}\left(\frac{n+r-p+4}{2},\frac{p}{2}\right)$$

$$- 2I_a^{(r)}\left(\frac{n+r-p+2}{2},\frac{p}{2}\right)\} - \frac{(n+r-p+1)(n+r-p+2)}{(n+r+1)(n+r+2)}\{1$$

$$- I_a^{(r)}\left(\frac{n+r-p}{2},\frac{p}{2}\right)\} + 1 \tag{2.14}$$

for $r<p$, where

$$l_a^{(p)}(\alpha,\beta) = \frac{1}{B_p(\alpha,\beta)}\int_{|T|<a,0<T<I} |T|^{\alpha-(p+1)/2}|I - T|^{\beta-(p+1)/2}dT. \quad (2.15)$$

The maximum rate of reduction of risk $\{R(d_1)-R(d_1^*,\Lambda)\}/R(d_1)$ is obtained when $\Lambda=0$.

3. ESTIMATION OF GENERALIZED PRECISION

To estimate generalized precision $|\Sigma^{-1}|$ based on X and S under squared loss given in (1.4), we shall consider, following Stein (1964) and Shorrock and Zidek (1976), equivariant estimator $\phi(X,S)$ under the subgroup of affine transformations, defined by $\phi(AXH,ASA^t)=|A|^{-2}\phi(X,S)$ for any nonsingular matrix of order p and any orthogonal matrix H of order r. Then ϕ must be of the form $\phi(X,S)=|V|^{-1}\psi(T)$ where $V=S+XX^t$ and $T=(S+XX^t)^{-1/2}S(S+XX^t)^{-1/2}$ when $r\geqq p$ and $T=I-X^t(S+XX^t)^{-1}X$ when $r<p$. We now prove the following theorem.

Theorem 3.1. Let $n\geqq p+4$. Then the estimator $e_1^*(X,S)$ defined by (1.6) dominates the best affine equivariant estimator $e_1(S)$ defined by (1.5).

Proof. By Lemma 2.1 the risk of equivariant estimator is expressed by

$$E\,[\{|V|^{-1}\psi(T) - 1\}^2]$$

$$= E_\Lambda^K[E_0\{\frac{E_0[|V|^{-2}C_K(V)|K]}{E_0[C_K(V)|K]}\psi(T)^2 - \tag{3.1}$$

$$- 2\frac{E_0[|V|^{-1}C_K(V)|K]}{E_0[C_K(V)|K]}\psi(T) + 1\}\frac{C_K(I-T)}{E_0[C_K(I-T)|K]}|K].$$

Hence the minimum of $R(\phi,\Lambda)$ with respect to ψ for given K is obtained by

$$\psi_\kappa(T) = E_0[|V|^{-1}C_\kappa(V)|K=\kappa]/E_0[|V|^{-2}C_\kappa(V)|K=\kappa]$$

$$= 2^p \frac{\Gamma_p(\frac{n+r-2}{2})(\frac{n+r-2}{2})_\kappa}{\Gamma_p(\frac{n+r-4}{2})(\frac{n+r-4}{2})_\kappa}$$

$$= \frac{(n+r-4)!}{(n+r-p-4)!} \Pi_{i=1}^p \frac{n+r-3-i+2k_i}{n+r-3-i}$$

$$\geqq \frac{(n+r-4)!}{(n+r-p-4)!}.$$

If we put $\psi_0(T)=\mathrm{Max}\{\psi(T),(n+r-4)!/(n+r-p-4)!\}$, then the estimator $\phi(X,S)=|V|^{-1}\psi(T)$ is dominated by $\phi_0(X,S)=|V|^{-1}\psi_0(T)$. Taking $\psi(T)=|T|^{-1}(n-4)!/(n-4-p)!$ gives the conclusion.

By the same argument it is shown that under the loss $e|\Sigma|-\log(e|\Sigma|)-1$, the best affine equivariant estimator $\{(n-2)!/(n-p-2)!\}|S^{-1}|$ is dominated by $e_2^*(X,S)$ given by (1.7). The similar argument as in the proof of Theorem 2.1 gives the following mixture representation of the risk of $e_1^*(X,S)$.

<u>Theorem 3.2.</u> Assume that $n\geqq p+4$. Put $\Lambda=\xi^t\Sigma^{-1}\xi$ and

$a=\{(n-4)!(n+r-p-4)!\}/\{(n-p-4)!(n+r-4)!\}$. Under the L_1^* loss given in (1.4), the risk of the improved estimator $e_1^*(X,S)$ is given by

(i) when $r \geqq p$,

$$\Sigma_{k=0}^{\infty}\Sigma_{(\kappa)}\,\mathrm{etr}(-\frac{\Lambda}{2})\frac{C_\kappa(\frac{\Lambda}{2})}{k!}[\frac{(n-p-2)(n-p-3)}{(n-2)(n-3)}$$

$$\times\{I_a^{(p)}(\frac{n-4}{2},\frac{r}{2};\kappa) - 2I_a^{(p)}(\frac{n-2}{2},\frac{r}{2};\kappa)\} + \Pi_{i=1}^{p}(\frac{n+r-i-3}{n+r+2k_i-i-1})$$

$$\times\{\Pi_{i=1}^{p}(\frac{n+r-i-3}{n+r+2k_i-i-3}) - 2\}\{1 - I_a^{(p)}(\frac{n}{2},\frac{r}{2};\kappa)\} + 1]$$

(ii) when $r<p$,

$$\Sigma_{k=0}^{\infty}\Sigma_{(\kappa)}\,\mathrm{etr}(-\frac{\Lambda}{2})\frac{C_\kappa(\frac{\Lambda}{2})}{k!}[\frac{(n-p-2)(n-p-3)}{(n-2)(n-3)}$$

$$\times\{I_a^{(r)}(\frac{n+r-p-4}{2},\frac{p}{2};\kappa) - 2I_a^{(r)}(\frac{n+r-p-2}{2},\frac{p}{2};\kappa)\}$$

$$+\frac{(n+r-p-2)(n+r-p-3)}{(n+r-2)(n+r-3)}\Pi_{i=1}^{r}(\frac{n+r-i-1}{n+r+2k_i-i-1})$$

$$\times\{\Pi_{i=1}^{r}(\frac{n+r-i-3}{n+r+2k_i-i-3}) - 2\}\{1 - I_a^{(r)}(\frac{n+r-p}{2},\frac{p}{2};\kappa)\} + 1].$$

<u>Corollary 3.1.</u> When $p=1$, $a=(n-4)/(n+r-4)$ and the risk of $e_1^*(X,S)$ is given by

$$\Sigma_{k=0}^{\infty}\mathrm{etr}(-\frac{\Lambda}{2})\frac{(\mathrm{tr}\,\Lambda/2)^k}{k!}[\frac{n-4}{n-2}\{I_a(\frac{n}{2}-2,\frac{r}{2}+k) - 2I_a(\frac{n}{2}-1,\frac{r}{2}+k)\}$$

$$-\frac{(n+r-4)(n+r-4+4k)}{(n+r-2+2k)(n+r-4+2k)}\{1- I_a(\frac{n}{2},\frac{r}{2}+k)\} + 1].$$

Corollary 3.2. When $r=1$, $a=(n-p-3)/(n-3)$ and the risk of $e_1^*(X,S)$ is given by

$$\sum_{k=0}^{\infty}\exp(-\frac{\Lambda}{2})\frac{(\Lambda/2)^k}{k!}[\frac{(n-p-2)(n-p-3)}{(n-2)(n-3)}\{I_a(\frac{n-p-3}{2},\frac{p}{2}+k)$$

$$-2I_a(\frac{n-p-1}{2},\frac{p}{2}+k)\} - \frac{(n-p-1)(n-p-2)(n+4k-3)}{(n-2)(n+2k-1)(n+2k-3)}$$

$$\times\{1 - I_a(\frac{n-p+1}{2},\frac{p}{2}+k)\} + 1] .$$

Corollaries 3.1 and 3.2 are useful for the numerical computation in Section 4, which shows that the minimum risk of $R(e_1^*,\Lambda)$ with respect to Λ is obtained when $\Lambda=0$. It is given by

$$R(e_1^*,0)$$

$$= \frac{(n-p-2)(n-p-3)}{(n-2)(n-3)}\{I_a^{(p)}(\frac{n}{2}-2,\frac{r}{2}) - 2I_a^{(p)}(\frac{n}{2}-1,\frac{r}{2})\}$$

$$- \frac{(n+r-p-2)(n+r-p-3)}{(n+r-2)(n+r-3)}\{1 - I_a^{(p)}(\frac{n}{2},\frac{r}{2})\} + 1$$

for $r\geqq p$ and

$$R(e_1^*,0)$$

$$= \frac{(n-p-2)(n-p-3)}{(n-2)(n-3)}\{I_a^{(r)}(\frac{n+r-p-4}{2},\frac{p}{2})$$

$$- 2I_a^{(r)}(\frac{n+r-p-2}{2},\frac{p}{2})\} - \frac{(n+r-p-2)(n+r-p-3)}{(n+r-2)(n+r-3)}$$

$$\cdot\{1 - I_a^{(r)}(\frac{n+r-p}{2},\frac{p}{2})\} + 1$$

for $r<p$, where r-variate incomplete beta function $I_a^{(r)}(\alpha,\beta)$

corresponding to the partition $\kappa=(0)$ is given in (2.15). The maximum rate of reduction of risk $\{R(e_1)-R(e_1^*,\Lambda)\}/R(e_1)$ is obtained at $\Lambda=0$.

4. NUMERICAL RESULTS

Tables 1 and 2 give the values of risk of improved estimator d_1^* defined by (1.3) when p=1, based on Corollary 2.1 and when r=1, based on Corollary 2.2. The maximum rate of reduction of risk $100\times\{R(d_1)-R(d_1^*,0)\}/R(d_1)$ is shown in the parentheses. We can see from Table 1 that the rate of reduction increases as r increases and it decreases as $\lambda=\mathrm{tr}\Lambda$ increases. The maximum rate of reduction is only 1.7% when n=5 and p=r=1. From Table 2 we can see that the rate of reduction increases slightly and then decreases as p increases.

To obtain the risk of $d_1^*(X,S)$ when p=2 or r=2, based on Theorem 2.1, we need zonal polynomials. We made use of Sugiyama (1979) to get the coefficients of zonal polynomials. The results are shown in Tables 3 and 4. We can see from Table 3 that the risk is monotonically increasing with respect to each latent root of Λ.

Table 1

Risk of $d_1^*(X,S)$ when n=5 (upper) and n=10 (lower) for p=1.

r	$\lambda=0$		$\lambda=1$	$\lambda=2$	$\lambda=5$
	.2809	(1.7%)	.282	.283	.285
1	.1646	(1.2%)	.165	.166	.167
	.2749	(3.8%)	.277	.279	.284
2	.1617	(3.0%)	.163	.164	.166
	.2697	(5.6%)	.272	.275	.283
3	.1590	(4.6%)	.160	.162	.166
	.2616	(8.4%)	.264	.268	.279
5	.1543	(7.4%)	.156	.158	.164
	.2498	(12.6%)	.251	.255	.268
10	.1461	(12.3%)	.147	.150	.158

Table 2

Risk of $d_1^*(X,S)$ when n=10 and r=1.

p	λ=0	λ=1	λ=2	λ=5
2	.3135 (1.5%)	.314	.316	.318
3	.4477 (1.5%)	.449	.450	.454
4	.5675 (1.4%)	.569	.570	.574
5	.6730 (1.3%)	.674	.695	.680
6	.7641 (1.1%)	.765	.766	.770
10	.9829 (.2%)	.983	.983	.984

Table 3

Risk of $d_1^*(X,S)$ when p=r=2, n=10 for Λ=diag(λ_1,λ_2).

λ_2	λ_1=0	λ_1=1	λ_1=2
0	.3079 (3.2%)	.309	.311
1		.311	.313
2			.315

Table 4 shows the increase of the maximum rate of reduction for increasing r but with respect to p, the rate rises shortly and then falls down as in the case of p=1 or r=1 in Tables 1 and 2.

Table 4

Risk of $d_1^*(X,S)$ at Λ=0 when n=10 and the maximum rate of reduction of risk $100\times\{R(d_1)-R(d_1^*,0)\}/R(d_1)$.

p	r=2	r	p=2
1	.1617 (3.0%)	3	.3031 (4.7%)
2	.3079 (3.2%)	5	.2952 (7.2%)
3	.4405 (3.1%)	10	.2822 (11.3%)
4	.5595 (2.8%)	20	.2696 (15.3%)
6	.7567 (2.1%)		
10	.9816 (.3%)		

Corresponding to Tables 1 and 2, the risk of the

improved estimator $e_1^*(X,S)$ defined by (1.6) is shown in Tables 5 and 6. They are computed by Corollaries 3.1 and 3.2. We can see that the risk of $e_1^*(X,S)$ has the same tendency as that of $d_1^*(X,S)$. However the maximum rates of reduction of risk shown in the parentheses are considerably higher than those of $d_1^*(X,S)$. The maximum rate of reduction for $e_1^*(X,S)$ is 12.5% when n=5 and p=r=1, whereas the corresponding maximum rate for $d_1^*(X,S)$ is 1.7%.

Table 5

Risk of $e_1^*(X,S)$ when n=5 (upper) n=10 (lower) for p=1.

r	$\lambda=0$	$\lambda=1$	$\lambda=2$	$\lambda=5$
	.5833 (12.5%)	.596	.617	.654
1	.2424 (3.0%)	.244	.246	.249
	.5026 (24.6%)	.521	.552	.624
2	.2316 (7.4%)	.235	.239	.248
	.4444 (33.3%)	.463	.498	.588
3	.2213 (11.5%)	.225	.232	.245
	.3689 (44.7%)	.385	.418	.519
5	.2038 (18.5%)	.208	.216	.237
	.2792 (58.1%)	.289	.311	.397
10	.1742 (30.3%)	.178	.186	.215

Table 6

Risk of $e_1^*(X,S)$ when n=10 and r=1.

p	$\lambda=0$	$\lambda=1$	$\lambda=2$	$\lambda=5$
2	.4460 (3.9%)	.449	.454	.462
3	.6153 (4.3%)	.619	.625	.637
4	.7517 (4.3%)	.755	.761	.776
5	.8562 (4.1%)	.859	.864	.878
6	.9296 (3.6%)	.931	.934	.945

Corresponding to Tables 3 and 4, the risk of $e_1^*(X,S)$

for generalized precision is computed by Theorem 3.2 and is shown in Tables 7 and 8. The rates of reduction are almost three times more than those of $d_1^*(X,S)$.

Table 7

Risk of $e_1^*(X,S)$ when r=p=2 and n=10 for $\Lambda=\mathrm{diag}(\lambda_1,\lambda_2)$.

λ_2	$\lambda_1=0$	$\lambda_1=1$	$\lambda_1=2$
0	.4223 (9.1%)	.427	.435
1		.436	.444
2			.451

Table 8

Risk of $e_1^*(X,S)$ at $\Lambda=0$ when n=10 and the maximum rate of reduction of risk $100\mathrm{x}\{R(e_1)-R(e_1^*,0)\}/R(e_1)$.

p	r=2	r	p=2
1	.2316 (7.4%)	3	.4002 (13.8%)
2	.4223 (9.0%)	5	.3629 (21.8%)
3	.5805 (9.7%)	10	.2999 (35.4%)
4	.7088 (9.8%)	20	.2366 (49.0%)
5	.8089 (9.4%)		
6	.8828 (8.4%)		

Nariaki Sugiura and Yoshihiko Konno
Department of Mathematics
University of Tsukuba
Sakura mura, Ibaraki 305
Japan

References

Constantine, A. G. (1963). 'Some non-central distribution problems in multivariate analysis.' Ann. Math. Statist. 34, 1270-1285.

James, A. T. (1964) 'Distributions of matrix variates and latent roots derived from normal samples.' Ann. Math. Statist. 35, 475-501.

Robbins, H. and Pitman, E. J. G. (1949). 'Application of the method of mixtures to quadratic forms in normal variates.' Ann. Math. Statist. 20, 552-560.

Shorrock, R. W. and Zidek, J. V. (1976). 'An improved estimator of the generalized variance.' Ann. Statist. 4, 629-638.

Sinha, B. K. (1976). 'On improved estimators of the generalized variance.' J. Multivariate Anal. 6, 617-625.

Sinha, B. K. and Ghosh, M. (1985). 'Inadmissibility of the best equivariant estimators of the variance-covariance matrix and the generalized variance under entropy loss.' Univ. of Pittsburgh Technical Report No. 85-27.

Stein, C. (1964) 'Inadmissibility of the usual estimator for the variance of a normal distribution with unknown mean.' Ann. Inst. Statist. Math. 16, 155-160.

Sugiyama, T. (1979). 'Coeffecient of zonal polynomials of order two.' Computer Science Monographs No. 12, Publication of Inst. Statist. Math.

Zidek, J. V. (1978). 'Deriving unbiased risk estimators of multinormal mean and regression coefficient estimators using zonal polynomials.' Ann. Statist. 6, 769-782.

W.Y. Tan

SAMPLING DISTRIBUTIONS OF DEPENDENT QUADRATIC FORMS FROM NORMAL AND NONNORMAL UNIVERSES

Abstract. This paper derives the sampling distribution of dependent quadratic forms from normal and nonnormal universes. It is shown that when all the population cumulants are finite, under some conditions, the joint distribution of dependent quadratic forms can be expressed as infinite series involving Laquerre polynomials.

1. Introduction

For testing several dependent linear hypotheses or for performing simultaneously several dependent goodness of fit test, it is usually desirable to find the joint distribution of several dependent quadratic forms (for illustration, see Jensen, 1970a,b). Under normality assumption for the random variables, this gives rise to the so-called multivariate gamma or multivariate chi-square distributions. Now, in many practical situations, the populations are in general not normal (see Geary 1947, for example). For evaluating the robustness of the testing procedures based on normality assumption, the purpose of this paper is to derive the sampling distributions of dependent quadratic forms from nonnormal universes.

2. Sampling Distribution Of Dependent Quadratic Forms From Normal Universes

Consider k quadratic forms in the random variables $\underset{\sim}{\epsilon}' = (\epsilon_1, \epsilon_2, \dots \epsilon_n)$:

$$Q_i = \underset{\sim}{\epsilon}' A_i \underset{\sim}{\epsilon}, \qquad i = 1, \dots, k,$$

where the A_i's are nxn symmetric matrices of real numbers.

Assume that $\underset{\sim}{\epsilon}$ is normal with expectation $E(\underset{\sim}{\epsilon}) = \underset{\sim}{0}$ and positive definite variance- covariance matrix $V_{\underset{\sim}{\epsilon}} = \Sigma$; assume further that A_i is positive semi-definite with rank $A_i = r_i, r = \sum_{i=1}^{k} r_i \leq n$.

A. K. Gupta (ed.), Advances in Multivariate Statistical Analysis, 373–378.

THEOREM (2.1). *If $A_j\Sigma A_j = A_j$ for $j = 1,\ldots,k$, then the joint characteristic function (c.f.) $\phi(t_1,\ldots,t_k)$ of $Q_j, j = 1,\ldots,k$ is given by:*

$$\phi(t_1,\ldots,t_k) = \{\Pi_{j=1}^k(1-2it_j)^{-\frac{r_j}{2}}\}\exp\{g(w_1,\ldots,w_k)\} \tag{2.1}$$

where $i = \sqrt{-1}$, $w_j = 2it_j/(1-2it_j)$ and $g(w_1,\ldots,w_k)$ is an analytic function of $(w_1,\ldots,w_k)$.

PROOF: Let $\underset{\sim}{e} = \underset{\sim}{\epsilon} - E(\underset{\sim}{\epsilon}) = \underset{\sim}{\epsilon} - \underset{\sim}{\theta}$. Then, $\underset{\sim}{e}$ is normal with $E(\underset{\sim}{e}) = \underset{\sim}{0}$ and $V_{\underset{\sim}{e}} = \Sigma$ and Q_j can be written as

$$Q_j = \underset{\sim}{e}'A_j\underset{\sim}{e} + 2\underset{\sim}{\theta}'A_j\underset{\sim}{e} + \underset{\sim}{\theta}'A_j\underset{\sim}{\theta}.$$

The joint c.f. $Q(t_1,\ldots,t_k)$ is then given by:

$$Q(t_1,\ldots,t_k) = \exp[i\sum_{j=1}^k t_j\underset{\sim}{\theta}'A_j\underset{\sim}{\theta}]E\exp[i\sum_{j=1}^k t_j(\underset{\sim}{e}'A_j\underset{\sim}{e}) + 2i\underset{\sim}{\theta}'(\sum_{j=1}^k t_jA_j)\underset{\sim}{e}], i = \sqrt{-1}$$

Let $H = \Sigma^{-1} - 2i\sum_{i=1}^k t_jA_j$, and let $|A|$ denote the determinant of A. Then, it can easily be shown that

$$Q(t_1,\ldots,t_k) = |I_n - 2i(\sum_{j=1}^k t_jA_j)\Sigma|^{-\frac{1}{2}} \times \exp\{-\frac{1}{2}\underset{\sim}{\theta}'\Sigma^{-1}\underset{\sim}{\theta} + \frac{1}{2}\underset{\sim}{\theta}'\Sigma^{-1}\left[(I_n - 2i\Sigma(\sum_{j=1}^k t_jA_j)\right]^{-1}\underset{\sim}{\theta}\}. \tag{2.2}$$

Write A_j as $A_j = A_{j1}A'_{j1}$, where A_{j1} is $n \times r_j$ with rank $A_{j1} = r_j$, and put: $P = (A_{11}, A_{21},\ldots,A_{k1})$ and $T = \sum_{j=1}^{k\oplus} t_jI_{rj}$, the direct sum of matrices t_jI_{rj}, (see Searle 1971, p231). Then, $\sum_{j=1}^k t_jA_j = PTP'$; further $A_j\Sigma A_j = A_j$ implies $A_{j1}\Sigma A'_{j1} = I_{rj}$ for all $j = 1,\ldots,k$. Let $B_{uv} = A_{u1}\Sigma A'_{v1}, u, v = 1,\ldots,k$, and $W = \left[(\frac{\delta uv - 2it_u}{1-2it_u})B_{uv}\right]$, where δ_{uv} is the Kronecker's δ, and note the matrix results: $|I_n + BA| = |I_p + AB|$ and $(I_n + BA)^{-1} = I_n - B(I_p + AB)^{-1}A$ for any $n\times p$ matrix B and $p\times n$ matrix A. Then one has:

$$\begin{aligned}|I_n - 2i(\sum_{j=1}^k t_jA_j)\Sigma|^{-\frac{1}{2}} &= |I_n - 2iPTP'\Sigma|^{-\frac{1}{2}}\\ &= |I_r - 2iTP'\Sigma P|^{-\frac{1}{2}}\\ &= \{\Pi_{j=1}^k(1-2it_j)^{-\frac{r_j}{2}}\}|W|^{-\frac{1}{2}};\end{aligned} \tag{2.3}$$

$$\Sigma^{-1}\left[I_n - 2i\Sigma(\sum_{j=1}^{k} t_j A_j)\right]^{-1} = \Sigma^{-1}\left[I_n - 2i\Sigma PTP'\right]^{-1}$$
$$= \Sigma^{-1}\{I_n + 2i\Sigma P(I_r - 2iTP'\Sigma P)^{-1}TP'\}$$
$$= \Sigma^{-1} + 2iP(I_r - 2iTP'\Sigma P)^{-1}TP'$$
$$= \Sigma^{-1} + PW^{-1}\left[\sum_{j=1}^{k\oplus} w_j I_{r_j}\right]P', \tag{2.4}$$

where $w_j = 2it_j/(1-2it_j)$,
On substituting (2.3) and (2.4) into (2.2) one has:

$$Q(t_1,\ldots,t_k) = \Pi_{j=1}^{k}(1-2it_j)^{-\frac{r_j}{2}} \times \{\exp\{-\frac{1}{2}\log|W| + \frac{1}{2}\underset{\sim}{\theta}'PW^{-1}(\sum_{j=1}^{k\oplus} w_j I_{r_j})P'\underset{\sim}{\theta}\}$$

Since the elements of W depend on $(t_1,\ldots,t_k)$ only through $w_j = 2it_j/(1-2it_j)$, hence, $-\frac{1}{2}\log|W| + \frac{1}{2}\underset{\sim}{\theta}'PW^{-1}(\sum_{j=1}^{k\oplus} w_j I_{r_j})P'\underset{\sim}{\theta} = g(w_1,\ldots,w_k)$. Obviously, $g(w_1,\ldots,w_k)$ is an analytic function of $(w_1,\ldots,w_k)$. ∎

Note that if we fix j, and put $t_u = 0$ for all $u \neq j$, then $|W| = 1$ and $PW^{-1}(\sum_{j=1}^{k\oplus} w_j I_{r_j})P' = w_j A_{j1}A'_{j1}$. It follows that the c.f. of Q_j is

$$Q_j(t_j) = (1-2it_j)^{-\frac{r_j}{2}}\exp\{\frac{1}{2}\Delta_j^2(\frac{2it_j}{1-2it_j})\},$$
$$\text{where } \Delta_j^2 = \underset{\sim}{\theta}'A_{j1}A'_{j1}\underset{\sim}{\theta} = \underset{\sim}{\theta}'A_j\underset{\sim}{\theta}.$$

That is, Q_j is distributed as a noncental chi-square distribution with degrees of freedom r_j and noncentrality Δ_j^2.

Let $z_j = it_j/(1-it_j), j = 1,\ldots,k$. From Theorem (2.1), the joint c.f. of $(Q_j/2, j = 1,\ldots,k)$ is :

$$\psi(t_1,\ldots,t_k) = \{\Pi_{j=1}^{k}(1-it_j)^{-\alpha_j}\}\exp\{g(z_1,\ldots,z_k)\},$$

where $\alpha_j = r_j/2$

Expanding $\exp\{g(z_1,\ldots,z_k)\}$ in Taylor series around $(0,\ldots,0)$ with respect to $(z_1,\ldots,z_k)$ and using the Lemma given below, one obtains the joint distribution of $(X_j = Q_j/2, \quad j = 1,\ldots,k)$ as:

$$f(x_1,\ldots,x_k) = \sum_{i_1=0}^{\infty}\cdots\sum_{i_k=0}^{\infty} C_{i_1}\cdots_{i_k}\Pi_{j=1}^{k}\{L_{i_j}^{(\alpha_j)}(x_j)P_{\alpha_j}(x_j)\}, \tag{2.5}$$

where $L_j^{\alpha}(x)$ is the jth degree Laguerre polynomial with parameter α, and

$$C_{i_1} \cdots {}_{i_k} = E\{\Pi_{j=1}^k L_{i_j}^{(\alpha_j)}(X_j)\}\Pi_{j=1}^k \binom{\alpha_j + i_j - 1}{i_j},$$

and where

$$P_\alpha(x) = \frac{1}{\Gamma(\alpha)} x^{\alpha-1} e^{-x}, \quad \alpha > 0.$$

LEMMA. *Let $\alpha > 0$ and $i = \sqrt{-1}$. Then,*

$$\frac{1}{2\pi} \int_{-\infty}^{\infty} (\frac{-it}{1-it})^n (1-it)^{-\alpha} \exp(-itx) dt$$
$$= \frac{(n!)\Gamma(\alpha)}{\Gamma(\alpha+n)} L_n^{\alpha}(x) P_\alpha(x), \tag{2.6}$$

PROOF. *By the inversion formula, we have:*

$$P_{a+j}(x) = \frac{1}{2\pi} \int_{-\infty}^{\infty} (1-it)^{-(\alpha+j)} \exp(-itx) dt.$$

Hence

$$\begin{aligned}
&\frac{1}{2\pi} \int_{-\infty}^{\infty} (\frac{-it}{1-it})^n (1-it)^{-\alpha} \exp(-itx) dt \\
&= \frac{1}{2\pi} \int_{-\infty}^{\infty} (-it)^n (1-it)^{-(\alpha+n)} \exp(-itx) dt \\
&= (\frac{d}{dx})^n \frac{1}{2\pi} \int_{-\infty}^{\infty} (1-it)^{-(\alpha+n)} \exp(-itx) dt \\
&= (\frac{d}{dx})^n P_{\alpha+n}(x) = \frac{(n!)}{\Gamma(\alpha+n)} x^{\alpha-1} e^{-x} L_n^{(\alpha)}(x) \\
&= \frac{(n!)\Gamma(\alpha)}{\Gamma(\alpha+n)} L_n^{(\alpha)}(x) P_\alpha(x),
\end{aligned}$$

∎

3. SAMPLING DISTRIBUTION OF DEPENDENT QUADRATIC FORMS FROM NONNORMAL UNIVERSES

To find the joint sampling distribution of $(X_j = Q_j/2, \quad j = 1, \ldots, k)$ from nonnormal universes, we assume that $(\epsilon_1, \ldots, \epsilon_n)$ have finite higher cumulants $\kappa(r_1,\ldots,r_n)$ for all $r_1+\cdots+r_n \geq 3$. Then, following Davis (1976), the joint distribution of $(X_j, j=1,2,\ldots,n)$ is obtained by the following two procedures:

(a) Assume that $\underset{\sim}{\epsilon} \sim N(\underset{\sim}{\theta}+\underset{\sim}{Z},\Sigma)$ and obtain the conditional distribution of $(X_j, j=1,\ldots,k)$ given $\underset{\sim}{Z}' = (Z_1,\ldots,Z_n)$.

(b) Assume that the Z_i's are pseudo random variables having the same finite higher cumulants as those of $\underset{\sim}{\epsilon}$ but $E\underset{\sim}{Z} = \underset{\sim}{0}$ and $E(Z_iZ_j)=0$ for all i and j. The sampling distribution of $(X_j, j=1,\ldots,k)$ is obtained from the conditional distribution of $(X_j, j=1,\ldots,k)$ by taking expectation over the Z_i's.

Put $\underset{\sim}{\phi} = \underset{\sim}{\theta} + \underset{\sim}{Z}$. From the previous section, the conditional distribution of $(X_j, j=1,\ldots,k)$ given $\underset{\sim}{Z}$ is then given by:

$$f(x_1,\ldots,x_n|\underset{\sim}{Z}) = \sum_{i_1=0}^{\infty}\cdots\sum_{i_k=0}^{\infty} d_{i_1\ldots i_k}(\underset{\sim}{Z})\Pi_{j=1}^{k}\{L_{i_j}^{(\alpha_j)}(x_j)P_{\alpha_j}(x_j)\},$$

where only the coefficients $d_{i_1\ldots i_k}(\underset{\sim}{Z})$ are functions of $\underset{\sim}{Z}$.

Taking expectation over the Z_j's, the distribution of $(X_j, j=1,\ldots,k)$ is then given by:

$$f(x_1,\ldots,x_n) = \sum_{i_1=0}^{\infty}\cdots\sum_{i_k=0}^{\infty} C_{i_1}\cdots_{i_k}\Pi_{j=1}^{k}\{L_{i_j}^{(\alpha_j)}(x_j)P_{\alpha_j}(x_j)\}, \tag{3.1}$$

where

$$C_{i_1}\cdots_{i_k} = E\{\Pi_{j=1}^{k}L_{i_j}^{(\alpha_j)}(X_j)\}\Pi_{j=1}^{k}\binom{\alpha_j+i_j-1}{i_j}.$$

4. Laguerre Polynomial Approximation To The Distribution Of Dependent Quadratic Forms From Normal And Nonnormal Distribution

From (3.1) it is apparent that one may use finite Laguerre polynomial series to approximate the distribution of $(X_j, j=1,\ldots,k)$ from normal and nonnormal universes. For the normal universes, this approximation has already been derived by Tan and Wong (1978). The approximation results for nonnormal universes and its application to robustness studies will be derived in another paper; thus we are not going any further here.

Department of Mathematical Sciences
Memphis State University
Memphis, TN 38152

REFERENCES

Davis, A. W., 'Statistical distributions in univariate and mulivariate Edgeworth populations.,' *Biometrika* **63** (1976), 661-670.

Jensen, D.R., 'The joint distribution of quadratic forms and related distributions,' *Aust. Jour. Statist.* **12** (1970a), 13-22.

Jensen, D.R., 'The joint distribution of traces of Wishart matrices and their applications,' *Ann. Math. Statist.* **41** (1970b), 133-145.

Geary, R.C., 'Testing for normality,' *Biometrika* **34** (1947), 209-242.

Searle, S.R., *Linear Models*, John Wiley and Sons, New York, 1971.

Tan W.Y. and Wong, S.P., 'On approximating the central and noncentral multivariate gamma distributions,' *Comm. in Statist* **B7** (1978), 227-243.

BIBLIOGRAPHY OF WORKS BY K. C. S. PILLAI

1. Trend analyzer. *Proc. Indian Academy of Science*, 1943.
2. Confidence interval for the correlation coefficient. *Sankhya* (1946), 415-422.
3. A note on ordered samples. *Sankhya 8*(1948), 375-380.
4. On the distribution of midrange and semirange in samples form a normal population. *Ann. Math. Statist.* *21*(1950), 100-105.
5. Some notes on ordered samples from a normal population. *Sankhya 11*(1951), 23-28.
6. On the distribution of an anlogue of Student's t. *Ann. Math. Statist. 22*(1951), 467-472.
7. On the distribution of Studentized range. *Biometrika 39*(1952), 194-195.
8. On the distribution of the ratio of the ith observation in an ordered sample from a normal population to an independent estimate of the standard deviation. *Ann. Math. Statist. 25*(1954), 569-572 (with K. V. Ramachandran).
9. Some new test criteria in multivariate analysis. *Ann. Math. Statist. 26*(1955), 117-121.
10. On the distribution of the largest or smallest root of a matrix in multivariate analysis. *Biometrika 43*(1956), 122-127.
11. Some results useful in multivariate analysis. *Ann. Math. Statist. 27*(1956), 1106-1114.
12. Sample surveys in the Philippines. *Statistical Reporter*, Manila, *2*(1958), No. 2, 160-168.
13. On Hotelling's generalization of T^2. *Biometrika 46* (1959), 160-168 (with Pablo Samson, Jr.).
14. On the distribution of the largest of six roots of a matrix in multivariate analysis. *Biometrika 46*(1959), 237-240 (with Celia G. Bantegui).
15. On the distribution of the extreme Studentized deviate from the sample mean. *Biometrika 46*(1959), 467-472 (with Benjamin P. Tienzo).
16. On the moments of the trace of a matrix and approximations to its distribution. *Ann. Math. Statist. 30*(1959), 1135-1140 (with Tito A. Mijares).
17. Upper percentage points of the extreme Studentized deviate from the sample mean. *Biometrika 46*(1959), 195-196.

18. Upper percentage points of a substitute F-ratio using ranges. *Biometrika 48*(1961), 195-196 (with Angeles R. Buenaventura).
19. On the distribution of linear functions and ratios of linear functions of ordered correlated normal random variables with emphasis on range. *Biometrika 51*(1964), 143-151 (with Shanti S. Gupta and G. P. Steck).
20. On the distribution of the largest of seven roots of a matrix in multivariate analysis. *Biometrika 51*(1964), 270-275.
21. On the moments of elementary symmetric functions of the roots of two matrices. *Ann. Math. Statist. 35* (1964), 1704-1712.
22. On linear functions of ordered correlated normal random variables. *Biometrika 52*(1965), 365-379 (with Shanti S. Gupta).
23. On the distribution of the largest characteristic root of a matrix in multivariabe analysis. *Biometrika 52* (1965), 405-411.
24. On elementary symmetric functions of the roots of two matrices in multivariate analysis. *Biometrika 52* (1965), 499-506.
25. Some results on the non-central multivariate beta distribution and moments of traces of two matrices. *Ann. Math. Statist. 36*(1965), 1511-1520 (with C. G. Khatri).
26. On the moments of the trace of a matrix and approximations to its non-central distribution. *Ann. Math. Statist. 37*(1966), 1311-1318 (with C. G. Khatri).
27. On the non-central multivariate beta distribution and the moments of traces of some matrices. *Proc. International Symposium on Multivariate Analysis*, Dayton, Ohio (1966).
28. On the distribution of the second elementary symmetric function of the roots of a matrix. *Ann. Inst. Statist. Math. 19*(1967), 167-179 (with A. K. Gupta).
29. On the moments of traces of two matrices in multivariate analysis. *Ann. Inst. Statist. Math. 19*(1967), 143-156 (with C. G. Khatri).
30. On the distribution of the largest root of a matrix in multivariate analysis. *Ann. Math. Statist. 38*(1967), 616-617.
31. Upper percentage points of the largest root of a matrix in multivariate analysis. *Biometrika 54*(1967), 189-194.

32. Power comparisons of tests of two multivariate hypotheses based on four criteria. *Biometrika 54*(1967), 195-210 (with Kanta Jayachandran).
33. On the non-central distributions of two test criteria in multivariate analysis of variance. *Ann. Math. Statist. 39*(1968), 215-226 (with C. G. Khatri).
34. On the non-central distribution of the second elementary symmetric function of the roots of a matrix. *Ann. Math. Statist. 39*(1968), 833-839 (with A. K. Gupta).
35. On the moment generating function of Pillai's $V^{(s)}$ criterion. *Ann. Math. Statist. 39*(1968), 877-880.
36. Power comparisons of tests of equality of two covariance matrices based on four criteria. *Biometrika 55*(1968), 315-322 (with K. Jayachandran).
37. On the moments of elementary symmetric functions of the roots of two matrices and approximations to a distribution. *Ann. Math. Statist. 39*(1968), 1274-1281 (with C. G. Khatri).
38. Power comparisons of tests of two multivariate hypotheses based on individual characteristic roots. *Ann. Inst. Statist. Math. 21*(1969), 109-118 (with C. O. Dobson).
39. On the exact distribution of Wilks' criterion. *Biometrika 56*(1969), 109-118 (with A. K. Gupta).
40. On the moments of elementary symmetric functions of the roots of two matrices. *Ann. Inst. Statist. Math. 21* (1969), 309-320 (with G. M. Jouris).
41. Noncentral distributions of the smallest and second smallest roots of matrices in multivariate analysis. *Proc. Sixth Arab Science Congress* (1969) (with S. Al-Ani).
42. Non-central distributions of the largest latent roots of three matrices in multivariate analysis. *Ann. Inst. Statist. Math. 21*(1969), 321-327 (with T. Sugiyama).
43. Distributions of vectors corresponding to the largest roots of three matrices. *Multivariate Analysis II* (1969), Academic Press, New York, Ed. P. R. Krishnaiah (with C. G. Khatri).
44. On the distributions of the ratios of the roots of a covariance matrix and Wilks' criterion for tests of three hypotheses. *Ann. Math. Statist. 40*(1969), 2033-2040 (with S. Al-Ani and G. M. Jouris).

45. Power comparisons of tests of equality of two covariance matrices based on individual characteristic roots. *J. Amer. Statist. Asso.* *65*(1970), 438-446 (with S. Al-Ani).
46. On the exact distribution of Pillai's $V^{(s)}$ criterion. *J. Amer. Statist. Asso.* *65*(1970), 447-454 (with K. Jayachandran).
47. An approximation to the CDF of the largest root of a covariance matrix. *Ann. Inst. Statist. Math., Supplement* *6*(1970), 115-124 (with T. C. Chang).
48. Monotonicity of power functions of some tests of hypotheses concerning multivariate complex normal distributions. *Ann. Inst. Statist. Math.* *22*(1970), 307-318 (with H. C. Li).
49. Asymptotic expansions for distributions of the roots of two matrices from classical and complex Gaussian populations. *Ann. Math. Statist.* *41*(1970), 1541-1556 (with H. C. Li and T. C. Chang).
50. On the non-central distributions of the largest roots of two matrices in multivariate analysis. *Probability and Statistics* (1970), The University of North Carolina Press, Chapel Hill, Chapter 29.
51. Some distribution problems in the multivariate complex Gaussian case. *Ann. Math. Statist.* *42*(1971), 517-525 (with G. M. Jouris).
52. On the distribution of the sphericity criterion in classical and complex normal populations having unknown covariance matrices. *Ann. Math. Statist.* *42*(1971), 764-767 (with B. N. Nagarsenker).
53. On the exact distribution of Hotelling's generalized T_0^2. *J. Multivariate Analysis* *1*(1971), 90-107 (with D. L. Young).
54. An approximation to the distribution of the largest root of a matrix in complex multivariate analysis. *Ann. Inst. Statist. Math.* *23*(1971), 89-96 (with D. L. Young).
55. Asymptotic formulae for the distribution of some criteria for tests of equality of covariance matrices. *J. Multivariate Analysis* *1*(1971), 215-321 (with A. K. Chattopadhyay).
56. An approximation to the distribution of the largest root of a matrix and percentage points. *Ann. Inst. Statist. Math., Supplement* *7*(1971), 61-70 (with G. M. Jouris).

57. On the distributions of a class of statistics in multivariate analysis. *J. Multivariate Analysis 2* (1972), 96-114 (with B. N. Nagarsenker).
58. The max trace-ratio test of the hypothesis H_0: $\Sigma_1 = \ldots = \Sigma_k = \lambda\Sigma_0$. *Commun. Statist. 1*(1973), 57=80.
59. Asymptotic expansions for the distributions of characteristic roots when the parameter matrix has several multiple roots. *Multivariate Analysis III* (1973) (with A. K. Chattopadhyay).
60. The distribution of the sphericity test criterion. *J. Multivariate Analysis 3*(1973), 226-235 (with B. N. Nagarsenker).
61. Distribution of the likelihood ratio criterion for testing a hypothesis specifying a covariance matrix. *Biometrika 60*(1973), 359-364 (with B. N. Nagarsenker).
62. On the max U-ratio and likelihood ratio tests of equality of several covariance matrices. *Commun. Statist. 3*(1974), 29-53 (with D. L. Young).
63. Distribution of the likelihood ratio criterion for testing $\Sigma = \Sigma_0$, $\mu = \mu_0$. *J. Multivariate Analysis 4* (1974), 114-122 (with B. N. Nagarsenker).
64. On the distribution of Hotelling's trace and power comparisons. *Commun. Statist. 3*(1974), 433-454 (with Sudjana).
65. The distribution of characteristic roots of $S_1S_2^{-1}$ under violations. *Ann. Statist. 3*(1975), 773-779.
66. Exact robustness studies of tests of two multivariate hypotheses based on four criteria and their distribution problems under violations. *Ann. Statist. 3* (1975), 617-636 (with Sudjana).
67. Maximization of an integral of a matrix function and asymptotic expansions of distributions of latent roots of two matrices. *Ann. Statist. 4*(1976), 796-806 (with A. K. Chattopadhyay and H. C. Li).
68. Distributions of characteristic roots, Part I (invited paper). *The Canadian J. Statist. 4*(1976), 157-184.
69. Distributions of characteristic roots, Part II (invited paper). *The Canadian J. Statist. 5*(1977), 1-62.
70. Asymptotic distribution of Hotelling's trace for two unequal covariance matrices and robustness study of test of equality of mean vectors. *J. Statist. Plan. Inference 1*(1977), 109-120 (with N. B. Saweris).

71. Exact robustness studies of the test of independence based on four multivariate criteria and their distribution problems under violations. *Ann. Inst. Statist. Math., Part A 31*(1979), 85-101 (with Y. S. Hsu).

72. Power comparisons of two-sided tests of equality of two covariance matrices based on six criteria. *Ann. Inst. Statist. Math, Part A 31*(1979), 185-205 (with S. Sylvia Chu).

73. Distribution of the likelihood ratio statistic for testing sphericity structure for the normal covariance matrix and its percentage points. *Sankhya 41, Series B* (1979) (with B. N. Nagarsenker).

74. Asymptotic formulae for the percentiles and c.d.f. of Hotelling's trace under violations. *Rev. Tec. Ing., Univ. Zulia 2*(1979), 105-122 (with N. B. Saweris).

75. The distribution of the characteristics roots of $S_1S_2^{-1}$ under violations in the complex case and power comparisons of four tests. *Ann. Inst. Statist. Math., Part A 31*(1979), 445-463 (with Y. S. Hsu).

76. Exact and approximate noncentral distributions of the largest roots of three matrices and the max largest root-ratios. *Proc. of the Arab Science Congress* (1980).

77. Some complex variable transformations and exact power comparisons of two-sided tests of equality of two Hermitian Covariance Matrices. *J. Statist. Plan. Inference 4*(1980), 267-290 (with S. Sylvia Chu).

78. The exact distribution of Wilks' criterion. *Sankhya, Series B 42*(1981), 179-186 (with S. Sylvia Shen).

79. On the exact non-null distribution of Wilks' L_{vc} criterion and power studies. *Ann. Inst. Statist. Math. Part A 33*(1981), 45-55 (with Anita Singh).

80. Further results on the trace of a noncentral Wishart matrix. *Commun. Statist. - Theory and Method II* (1982), 1077-1086 (with A. M. Mathai).

81. On the exact distribution of Wilks' L_{vc} criterion in the complex case. *Sankhya, Series B 45*(1983), 1-8 (with Anita Singh).

82. Hotelling's T^2. *Encyclopedia of Statistical Sciences 3*(1983), 669-673.

83. Hotelling's trace. *Encyclopedia of Statistical Sciences 3*(1983), 673-677.

84. Mahalanobis D^2. *Encyclopedia of Statistical Sciences 5*(1985), 176-181.

85. MANOVA. *Encyclopedia of Statistical Sciences 6*(1985), 20-29.
86. Pillai's trace. *Encyclopedia of Statistical Sciences 6*(1985), 725-729.
87. Percentage points of the largest characteristic root of the multivariate beta matrix. *Comm. Statist. - Theory and Methods 13*(1984), 2199-2237 (with B. N. Flury).

Books

1. *Concise Tables for Statisticians* (1957), The Statistical Center, University of the Philippines, Manila.
2. *Statistical Tables for Tests of Multivariate Hypotheses* (1960), The Statistical Center, University of the Philippines, Manila.

INDEX

The manufacturer's authorised representative in the EU is Springer
Nature Customer Service Centre GmbH, Europaplatz 3, 69115 Heidelberg,
Germany. If you have any concerns regarding our products, please
contact ProductSafety@springernature.com

Printed and bound by CPI Group (UK) Ltd, Croydon, CR0 4YY
25/09/2026
02222839-0005